European Bison

Małgorzata Krasińska · Zbigniew A. Krasiński

European Bison

The Nature Monograph

Second Edition

(with Chapter 4 by Małgorzata Tokarska)

Springer

Małgorzata Krasińska
Mammal Research Institute
Polish Academy of Science
Białowieża
Poland

Zbigniew A. Krasiński
Białowieża
Poland

ISBN 978-3-642-36554-6 ISBN 978-3-642-36555-3 (eBook)
DOI 10.1007/978-3-642-36555-3
Springer Heidelberg New York Dordrecht London

Library of Congress Control Number: 2013933294

First edition ISBN 978-83-907521-8-1: © Mammal Research Institute, Polish Academy of Sciences, Białowieża, 2007

Translation from Polish (first edition) and translation of updated text second edition: James Richards

English text first edition revised by Matthew W. Hayward

Translation of F. Kobryńczuk's poems of & excerpt of M. Hussowski's Song on Bison: friends of MRI PAS

Figures: Karol Zub
except: Fig. 3.11 & 3.12 – K. Bartoń, Fig. 11.1 & 12.6 – M. Szlachciuk, Fig. 17.4 – R. Kowalczyk, Fig.16.13 – K. Stachura, Fig. 14.11 & 23.4 – T. Samojlik

Drawings (below the title of chapters 10, 13 and 14): B. Bańka

Printed on acid-free paper

Springer is part of Springer Science+Business Media (www.springer.com)

This book is dedicated to all who have contributed to the preservation and restitution of that master of our wild forest, the European bison. Today, as it has through the Ages, the bison of Europe make their way through the forest. But if they are to persist in the face of changing natural conditions, their species will always need to remain the subject of special care and attention

Preface

We have linked up our whole adult lives with Poland's Białowieża Forest and its bison king. Following our graduation from the Faculty of Veterinary Science in Warsaw, 1961 saw us starting work at Białowieża. Małgorzata was at the Mammal Research Institute, Polish Academy of Sciences, while Zbigniew was with the Białowieża National Park. And that is how things have remained to this day.

The animal to which we have devoted this sizeable chunk of our lives is a truly fascinating one: huge, mysterious, unbowed and saved from extinction by some miracle. For many years, we have sought to gain a better understanding of its habits, biology and links with the environment. And the tracing of this primordial beast's efforts to survive in and (re)adapt to environmental conditions has been a source of enormous satisfaction for us.

The monograph's authors—Małgorzata Krasińska and Zbigniew Krasiński—in the Hybrids Reserve in 1975. *Photo* from the authors' archive.

While the European bison is emerging victorious from all this, for now at least, there are dangers on all sides. So one of the leading goals of our activity has been to keep ourselves fully acquainted with the threats in question, and to make sure that action continues to be taken to protect the species. Our work centres on the Białowieża Forest, taking in the whole area cut across by the state border between Poland and Belarus. However, we have also studied the other populations of European bison present in the north–east of Poland. The result of all our work comprises some 200 publications on the subject of European bison.

We should also mention studies devoted to the crossing/hybridisation of the species with domestic cattle. The animal created by way of these experiments was first known loosely as a hybrid, but later came to termed the "*żubroń*". It was primarily to this issue that Małgorzata devoted herself over her first 15 years at Białowieża.

Our work has had its moments of danger, there being times when bison have decided to treat us to a swift dose of their horns or hoofs—but they always did this in defence of their independence.

Our greatest wish will have been fulfilled if the publication we offer here helps ensure that the relict animal it deals with will still be there in future for our descendents to see and enjoy, in the tiny scrap of wild nature we have left for it to inhabit.

Acknowledgments

from Wróblewski 1927

We would like to extend our sincerest thanks to all those who supplied advice or assistance at each stage of our research and work on the preparation and assembly of materials. The late and much missed Professor Zdzisław Pucek was the initiator of research on the ecology of the European bison, and it was he who supported us with his invaluable advice through the whole time we were carrying out studies. We also have the most grateful memories of Prof. Kazimierz Krysiak, Anatomist and Founder of the Centre for Research on Bison Anatomy in Warsaw. Then there is Prof. Stanisław Piwowarczyk, whose practical advice was so very helpful to us as we embarked upon our work, Prof. August Dehnel of the Mammal Research Institute PAS at Białowieża and Jerzy Szymczak, M. Eng., Director of the Białowieża National Park at the time we started. We are in turn grateful to Prof.

Aleksander Demiaszkiewicz, for his remarks regarding bison parasites; and to Dr. Joanna Izdebska for help in preparing the table concerning the external parasites of bison. We also have much to thank staff of the Białowieża National Park for, as well as technicians and scientific personnel from the Mammal Research Institute PAS at Białowieża—especially for the assistance they gave and the possibilities for joint work they helped open up. We are also grateful to a host of other colleagues too numerous to mention, who have been carrying out scientific work at various institutions in Poland and abroad, and have thus made it possible for us to engage in joint studies. For many years, our colleagues and friends from the Belarussian part of the Białowieża Forest joined us in implementing research tasks that sought to better understand and protect bison across the entire Białowieża Forest area. More specifically, we would also like to thank Tomasz Kamiński and Tomasz Samojlik, employees of the Mammal Research Institute PAS at Białowieża, for the help they gave us in resolving complicated matters associated with the computer graphics underpinning this book.

Likewise, we are indebted to Karol Zub, who went to great lengths to produce and update the figures for our work, and are eternally grateful to our reviewers—Profs. Bogumiła Jędrzejewska and the aforementioned Zdzisław Pucek, whose suggestions and remarks have had such a major impact on the final shape the Monograph has assumed. We would also like to address our sincere thanks to Dr. James Richards, Translator of the Polish version of the first edition of this book into English and updater of the text for this second edition; as well as Dr. Matt Hayward, for the valuable remarks he made on the first-edition English text.

As we have noted, there is no possible way in which we can list all those with whom we have worked, from whom we have gratefully accepted help, and on whom we have always been able to count, down the years. The best we can do is to offer here our most sincere words of thanks.

Contents

Chapter 1
Introduction

The greater part of this monograph is devoted to the European bison of Poland's Białowieża Forest, being based primarily on the results of our own long-term research carried out there, if also taken account of information relating to other free-ranging herds. In its first part, this study offers concise information on the systematics of the subject species, as well as its contemporary history, genetics and morphology. Particular attention is paid to the formation of the Polish population and changes in its abundance over 60 years, as well as to population structure and mortality. The development of the bison populations inhabiting the Białowieża Forest on the Polish and Belarussian parts are also compared. Further subjects dealt with are reproductive biology, the rhythm characterising activity in the species, habits, diet and social organisation. The issue of the strategy by which the species utilises its environment in turn takes in the population's spatial structure, a characterisation of home ranges of individuals and groups, migration, and biotope selectivity. The final part of the book goes over the principles by which the population has been managed and the threats it faces, obviously with emphasis on the process by which the free-ranging herds in different parts of the world have been reinstated, along with a brief characterisation of the populations of this kind currently in existence. This book is augmented by a timeline of key events and dates associated with the story of the restitution of *Bison bonasus*, Appendices section *inter alia* documenting the reinstatement of bison on the Reserves at Białowieża in the years 1929–1952, as well as supplying statistical data pointing to the development, reproduction and structure of the population of European bison inhabiting the Polish and Belarussian parts of the Białowieża Forest. An issue not dealt with at all here is the origin of the European bison, since the phylogenesis of the species has already been presented: in the monograph entitled *Bison*, published in Russian, in the chapter therein elaborated by Flerov (1979); in the work by Pucek (1986), in the chapter "*Bison bonasus* (Linnaeus 1758)—Wisent"; and in the monograph *Mammals of Europe, Part II (even-toed ungulates)*, published in German. Two other areas not discussed as they have been dealt with in detail elsewhere are the anatomy and physiology of the European bison (see Chap. 18). A bibliography for the genus *Bison* taking in the publications on the subject up to the year 1950 was compiled by Jaczewski and Korona (1994).

M. Krasińska and Z. A. Krasiński, *European Bison*,
DOI: 10.1007/978-3-642-36555-3_1, © Springer-Verlag Berlin Heidelberg 2013

DR. KONRAD WRÓBLEWSKI

ZUBR
PUSZCZY
BIAŁOWIESKIEJ

MONOGRAFJA

1 9 2 7

POZNAN

WYDAWNICTWO POLSKIE

Fig. 1.1 The first book entirely devoted to the European bison living in the Białowieża Forest in the late 19th and early 20th centuries

The world's first reinstated population of Lowland European bison—in the Białowieża Forest—is the subject of unique documentation as to the process by which it formed and developed, and the changes its social and spatial structure has

JAN SZTOLCMAN

ZUBR

JEGO HISTORJA, OBYCZAJE. i PRZYSZLOSC

WARSZAWA ═══ ROK 1926
NAKŁADEM CENTRALNEGO ZWIĄZKU
POLSKICH STOWARZYSZEŃ ŁOWIECKICH

Fig. 1.2 In his book, Sztolcman presented the process by which the bison retreated as civilisation advanced. This ended with the extermination of the last population of lowland European bison in the Forest

undergone over the 60 years of its existence (1952–2012). Over the 40 years of his work at Białowieża NP, one of the authors of the monograph (ZAK), personally engaged in the monitoring of changes in the abundance, age and social structure of the bison population, as well as working on the breeding record, mortality and state of health of the animals, thereby assembling the rich documentation existing as regards this population. In turn, from the mid-1970s onwards, alongside the co-author of the monograph (MK), and a team of workers from the Mammal

Research Institute PAS in Białowieża, he participated in parallel in long-term studies of bison ecology. This work allowed for the compilation of sufficient knowledge to enable the present nature monograph devoted to the European bison to be written.

The life of Lowland bison in their natural environment in the Białowieża Forest was documented as regards the late 19th and early 20th centuries by Karcov (1903) and Wróblewski (1927). These authors' observations offer a source of unique data on the last natural population of bison in the Forest (see colour insert and Fig. 1.1).

In the 1926 book, *Żubr: jego historia, obyczaje and przyszłość* (The European bison: its history, habits and future) (Fig. 1.2), Jan Sztolcman was so concerned for the future existence on Earth of *Bison bonasus* that he recognised that saving of the species might only be achieved through the combined efforts of states, plus the individuals in possession of the handful of bison that survived extermination during the First World War. It was for this reason that he used the occasion of the International Congress of Nature Protection convened in Paris in June 1923 to announce his intention to establish an international organisation that would devote itself to saving the European bison.

It is our hope that this book, going so far in documenting the contemporary 60-year history of the reintroduced bison of the Białowieża Forest, will itself become a recognised source of knowledge on this species successfully reinstated in nature.

Chapter 2
Systematics, Registration and Nomenclature

English: European bison or wisent, Polish: żubr, French: bison d'Europe, German: Wisent, Russian: Zubr,

Order: Cetartiodactyla—even-toed ungulates
Sub-Order: Ruminantia—ruminants
Family: Bovidae—hollow-horned animals
Subfamily: Bovinae
Genus: *Bison* (Hamilton Smith 1827)
Species: *Bison bonasus* (Linnaeus 1758)

European bison are artiodactyls (even-toed ungulates), the order Artiodactyla extending to include several hundred different mammal species. The "even-toed" nature reflects the fact that the last digits of the limbs are divided, ending in a double hoof.

Bison are in the sub-order of ruminants, whose representatives have a four-chambered stomach comprising the three pre-stomach compartments of the rumen, reticulum and omasum, as well as the abomasum serving as the true stomach. It is in the pre-stomach, especially the massive rumen (whose volume in adult bison may exceed 100 L), that plant food is gathered for periodic regurgitation as "the cud is chewed".

Horn structure qualifies European bison as a member of the cow family in which horns are hollow inside. This is in contrast with less closely-related animals like red and roe deer or moose, whose bony antlers have no empty space inside. The horns of European bison are composed of a bony core (cornual process) and hard sheath of horny material.

Ruminants have 32 teeth, lacking as they do the upper incisors and canines, while their molars with their broad folded crowns are adapted for chewing plant material.

The genus *Bison* includes large and massive herbivorous mammals once present across the two continents of Europe and North America. The genus first appeared several million years ago in the Pliocene (at the end of the Tertiary) in South and East Asia. In the Pleistocene Ice Ages of the Quaternary, the genus

M. Krasińska and Z. A. Krasiński, *European Bison*,
DOI: 10.1007/978-3-642-36555-3_2, © Springer-Verlag Berlin Heidelberg 2013

extended its range into other parts of Asia and Europe (Flerov 1979). Shapiro et al. (2004) estimate that *Bison* first entered eastern Beringia from Asia during the middle Pleistocene between 300,000 and 130,000 years ago, subsequently spreading southward into central North America between 130,000 and 75,000 years ago. The final separation of *Bison* living in North America and Asia occurred when the ice sheets retreated a little over 10,000 years ago, the land linking Asia and North America previously giving way to what is today the Bering Strait.

The genus is today represented by two species—the European bison or *Bison bonasus* (L.) and the (North) American bison (or popularly "buffalo") *Bison bison*. Further differentiation took hold across the huge areas inhabited by the latter species, the conditions of the treeless prairie favouring the emergence of the prairie bison *Bison bison bison* (Linnaeus), as distinct from the forest bison *Bison bison athabascae* Rhoads 1897, which evolved in forested areas of the northern part of what is today Canada.

A similar situation occurred in the European bison; the Lowland or Białowieża bison (*Bison bonasus bonasus* L.) being one subspecies, while the Caucasus boasted another, the mountain or Caucasian bison, *Bison bonasus caucasicus* Turkin et Satunin 1904 (Fig. 2.1).

When Europeans discovered America in the late 15th centuries, the prairies still teemed with bison. It is estimated that there were some 50 million individuals there. At the same time, the species' European relative was already so rare that action was being taken to ensure its protection through limits on hunting. Alas, this activity did nothing more than delay the process of retreat before an advancing human civilisation. What is more, the process of extermination on both sides

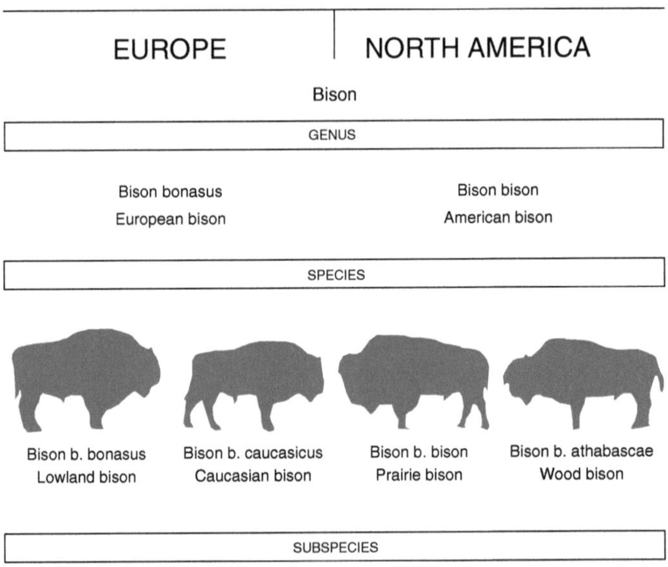

Fig. 2.1 The systematics of the genus *Bison*. *Source* Krasiński (1999a)

of the Atlantic ensured that, by the beginning of the 20th centuries, both species were literally on the verge of extinction. How lucky it is, then, that we do not have to imagine what these animals looked like, on the basis of drawings and descriptions—as we do with that other extinct bovine, the aurochs. Yet even here, our optimistic outlook needs to be tempered by harsh reality. The Caucasian bison subspecies did not make it, and neither—in the true sense—did the forest bison of North America. Ultimately, the former subspecies was brought down to just a single individual, while the vast territories once occupied by the latter were subject to the import of more than 6,000 prairie bison in the years 1925–1928, ensuring that the genetic makeup of the two American forms was subject to irrevocable mixing.

2.1 The Species and Its Subspecies

The Lowland (Białowieża) bison, ***Bison bonasus bonasus*** (Linnaeus 1758) (Fig. 2.2). In historical times, this still occupied forested areas of western, central and even south-eastern Europe, as far east as the River Don (Heptner et al. 1966).

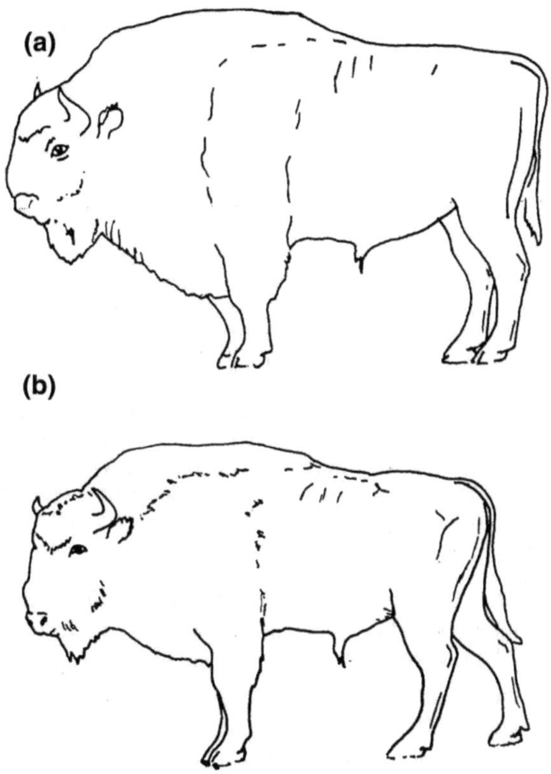

Fig. 2.2 **a** Bull 45 PLEBEJER. A sketch from a photograph taken at Pszczyna and included in *Das Zuchtbuch* of 1932. *Source* Krasiński (1994a). **b** Bull 100 KAUKASUS. A sketch from a photograph taken at Stellingen near Hamburg and included in *Das Zuchtbuch* of 1932. *Source* Krasiński (1994a)

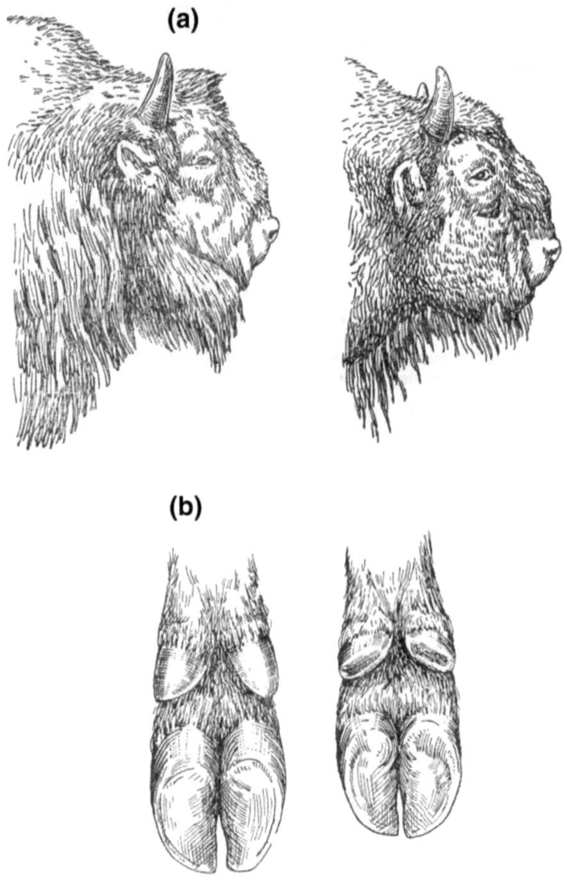

Fig. 2.3 Differences in the structure of the heads and hoofs of the lowland European bison *Bison bonasus bonasus* (**a**) and the Caucasian European bison *Bison bonasus caucasicus* (**b**). *Source* Flerov (1932). **a** Białowieża bison, the largest living representative of genus *Bison*, average height at the withers of a bull c. 1.85 m, larger than in the Caucasian bison. Hoofs elongated, hair across the body almost *straight*, hair at the front of the body relatively long, colouration paler than in the Caucasian bison, grey–brown with a hint of *ochre–brown*. Tail covered in long hairs. *Source* Flerov (1932) **b** Caucasian bison, height at withers about 1.6 m, smaller than the Białowieża bison. Hooves short, high and rounded. Hair over whole body frizzy, hair at front of body considerably shorter than in Białowieża bison, colouration darker, *brown–grey* with a touch of *chocolate* colour. Tail covered with short hair and with a bunch at the end. *Source* Flerov (1932)

The Caucasian (mountain) bison, *Bison bonasus caucasicus* Turkin et Satunin 1904 (Fig. 2.2) was a resident of the forest zone of the northern arc to the main ridge of the Caucasus Mountain massif plus foothills. On the southern arc of the ridge, bison were only present in the western area, up to the border with Abkhazia. There are thought to have been some 2,000 bison of this subspecies in the 19th century. However, war in the Caucasus plus the colonisation of foothill areas by the Cossacks combined with the development of cattle rearing, the felling of forest and hunting, were to force bison back into the area between the Belaya and Laba rivers (Heptner et al. 1966). By the 1890s, there were just 442 European bison in the Caucasus Mountains (Kulagin 1919). However, game protection extended to the species by the Tsar at that time succeeded in bringing the decline in numbers to a halt. Unfortunately, a further slide towards extinction progressed rapidly at the beginning of the 20th century, the situation worsening still further in 1919, thanks to an epizooty carried by domestic cattle. By 1920, there were just 50 animals, and even the creation of the Caucasus Reserve in 1924 could do nothing to save this form of the European bison. The last Caucasian bison was killed here in 1927 (Bashkirov 1939; Nemtsev et al. 2003). Differences between the Białowieża and Caucasian bison are presented in Fig. 2.3.

Certain authors, like Flerov (1979) and Pucek (1986), recognise a third subspecies known as the Carpathian bison, ***Bison bonasus hungarorum*** Kretzoi 1946. This form was described by the Hungarian researcher Kretzoi on the basis of a small piece of neurocranium from a single adult male in the collection of the National Museum in Budapest. The collection has not survived, falling victim to destruction in the course of the 1956 Hungarian Uprising. The bison in question lived in the Carpathians and Transylvania (a historical region of central Romania). Szalay (1913) wrote that this subspecies became extinct in 1790, although some sources say 1762 (Nahlik 1991). Since there are no reliable data as to when the Transylvanian bison might finally have disappeared, a general reference to the second half of the 18th century is more appropriate. There is also a lack of distinguishing features sufficient to justify the separating off of the bison living in Transylvania as a separate subspecies. A likeness of this bison included in Jickeli (1927) hardly serves to resolve this matter (Fig. 2.4).

2.2 Breeding Lines

2.2.1 The Lowland or Białowieża Line

The descendants of the still-extant Lowland bison derive from the last natural population present in the Białowieża Forest in the 19th century. Slatis (1960) determined that there had been seven founders of the population of Lowland-line bison: four males: M 15 BEGRÜNDER, M 45 PLEBEJER, M 87 BILL and M 147 BISMARCK, and three females: F 16 PLAVIA, F 42 PLANTA and F 89 BILMA (Table 2.1). That said, it should be noted that 3 of the founders, i.e. F 16 PLAVIA, M 15 BEGRÜNDER and M 147 BISMARCK left only one descendent—the female F 524 BESTE. This is why Belousova (1993) considers today's European bison to have descended from just five founders, i.e. F 42 PLANTA, M 45 PLEBEJER, M 87 BILL, F 89 BILMA and F 524 BESTE. This last female was described by Belousova (1993) as a "pseudo-founder".

Fig. 2.4 A Carpathian bison. *Source* Jickeli (1927)

In Poland, Lowland-line bison receive names beginning with the letters PO (PORANEK, POCIESZNA, etc.).

To be distinguished within the Lowland line is the Pszczyna line, to which pure Lowland bison are assigned. These derive from 4 (1, 3)[1] bison brought to the Pszczyna forests in 1865 from the Białowieża Forest. They were obtained by the owner of an estate in the Upper Silesia region of Poland from Tsar of Russia Alexander II—in exchange for 20 red deer supplied. These bison were held in a large enclosure close to the locality of Pszczyna. The bison kept here obtained names beginning with the letters PL—from the German name Pless for Pszczyna. The bull PLISCH, originating from Pszczyna and brought into Białowieża in 1936, is the founder of almost the entire population of bison now resident in the Białowieża Forest.

2.2.2 The Lowland-Caucasian Line

Not a single pure-blood representative of the Caucasian European bison made it through to modern times, although one specimen of this subspecies caused a

[1] Throughout this book, the names of bison are preceded by a large letter F in the case of females or an M in the case of males, plus a pedigree number.

Table 2.1 Founders of the contemporary world population of the European bison

No.	Sex M, F	Pedigree No	Name	Year of birth–death	Place of birth	Main place of breeding
1	F	42	**PLANTA**	Ca 1904–1931	Pszczyna	Pszczyna (von Pless)
2	M	45	**PLEBEJER**	1917–1937	Pszczyna	Pszczyna (von Pless)
3	M	87	**BILL**	1913–1929	ZOO Budapest	From: 1916 ZOO Stockholm
4	F	89	**BILMA**	1913–1939	Białowieża	From: 1916 ZOO Stockholm → From: 1935 Białowieża
5	M	15	**BEGRÜNDER**	1903–1919	ZOO Berlin	ZOO Berlin
6	F	16	**PLAVIA**	1906–1932	Pszczyna	From: 1919 ZOO Berlin
7	M	147	**BISMARCK**	1925–1934	ZOO Schönbrunn (Austria)	From: 1927 ZOO Berlin From: 1928 ZOO München
8	M	100	KAUKASUS	1907–1925	Caucasus	From: 1908 Stellingen near/ Hamburg From: 1922 Boitzenburg (von Arnim)
9	F	96	**GATSCHINA** (GATCZYNA)	1911–1932	Białowieża	From: 1916 Scharbow (von Beyme) From: 1924 ZOO Poznań From: 1931 Białowieża
10	F	95	**GARDE**	1907–1922	Białowieża	From: 1909 Stellingen near/ Hamburg From: 1922 Boitzenburg (von Arnim)
11	F	35	**PLEWNA**	1912–1922	ZOO Frankfurt	ZOO Frankfurt
12	F	46	**PLACIDA**	1918–1926	Pszczyna	From: 1921 Scharbow (von Beyme)

Data from: Groeben (1932), Slatis (1960), Krasiński (1994a) and Olech (1989). The Lowland line comes from: 7 founders (Nos 1–7), the Lowland-Caucasian line from: 12 founders. M male. F female. *Source* Krasiński and Krasińska (2007)

certain amount of consternation at various European breeding centres. A male from the free-ranging Caucasian population called KAUKASUS (pedigree no. 100) was brought to Germany in 1908, where it mated with captive females of the Lowland subspecies. Bison descending from KAUKASUS are thus kept separate and assigned to the so-called Lowland-Caucasian line. The animals of this line are descended from all 12 founders of identities determined by Slatis (1960) (Table 2.1). The bison of this kind bred in Poland received names starting with the letters **PU** (PUMA, PUSZCZAN, etc.). Today, the central part of the Caucasus is inhabited by a population of European/American bison hybrids deriving from 5 hybrid European bison brought there from Askania Nova in 1940 (Heptner et al. 1966). It is not permissible for these to be included among pure-bred European bison.

The convention in Poland is to draw a distinction between the European bison of the Lowland and Lowland-Caucasian lines, in accordance with the wish that pure-blood Lowland-line bison be retained. Any crossing will result in the loss of the genes of the five founders characteristic for the Lowland line (Olech 2006). In the European Bison Pedigree Book, bison of the Białowieża line are listed in bold print.

2.3 The European Bison Pedigree Book

During the International Congress of Nature Protection held in Paris on June 2, 1923, Polish delegate Jan Sztolcman (1854–1928)—an explorer, academic and Deputy Director of the State Zoological Museum in Warsaw—announced his "Appeal concerning the necessity of saving the Wisent". In response, the Congress passed a resolution expressing the wish to establish a society forthwith, between those countries on whose territories European bison were still present. The inaugural sitting of the International Society for the Protection of the European Bison took place on August 25–26, 1923 in Berlin (Figs. 2.5, 2.6). Its President was Kurt Priemel, Director of the zoo at Frankfurt-am-Main, while the Society's Board included Władysław Janta-Połczyński of Poznań. Private individuals and institutions also joined up. Collective membership was taken out by the US-based American Bison Society, upon which the European Bison Society was in fact modelled. From the Polish side, those signing up included the Polish Hunting Society in Warsaw, and Poznań Zoo, as well as Sztolcman himself, Urbański and Wróblewski from Bydgoszcz. It was in fact Poznań Zoo plus Janta-Połczyński and Urbański (Head of the bank in Poznań) who would initiate the purchase—for the not-insignificant sum of 11,000 Deutschmarks—of the first two bison known as F 96 GATCZYNA[2] and M 101 HAGEN (see Footnote 2), these being transferred to Poznań in 1924. The first goal of the Society was to compile a register of all surviving European bison. This census (Das Zuchtbuch) was published in 1932, under the editorship of Gerd von der Groeben, in the journal *Berichte der*

[2] Throughout this book, numbers given in brackets describe the numbers of bison by gender, with males first, followed after a comma by females.

KOMUNIKAT

o założeniu

Międzynarodowego Towarzystwa
Ochrony Żubra

w dniach 25 i 26 sierpnia 1923
w Berlińskim Ogrodzie Zoologicznym

Złożony na zlecenie zarządu przez
Dr Kurta Priemela, Frankfurt nad Menem

COMMUNIQUE

concerning the establishment of

The International Society for the Protection
of the European Bison

on 25th and 26th August 1923
at Berlin Zoological Gardens

Founded at the instigation of the Board by
Dr Kurt Priemel, Frankfurt am Main

Fig. 2.5 The communiqué on the founding of the International Society for the Protection of the European Bison

Internationalen Gesellschaft zur Erhaltung des Wisent, in five volumes (Fig. 2.7). This first census gave a total for the end of 1924 of 66 (33, 33) European bison. Careful checking led to the striking off the list of several more animals whose origins were uncertain, leaving a final total of just 54 animals (29, 25) (Olech 2009). The Pedigree Book distinguished the Białowieża line and within it the Pszczyna line and Białowieża-Caucasian line. Also designated were zoo-reared bison belonging to the two lines. The first pedigree number was assigned to PLANET, who had been born in 1881 in Pszczyna, at the menagerie of the Duke von Pless;

Statut
Międzynarodowego Towarzystwa Ochrony Żubra

§ 1.

Nazwa i siedziba

Nazwa Towarzystwa brzmi:

„Międzynarodowe Towarzystwo Ochrony Żubra"

Siedzibą Towarzystwa jest miejsce zamieszkania urzędującego prezydenta.

§ 2.

Cele Towarzystwa

Głównym celem Towarzystwa jest zachowanie żubra poprzez planowe rozmnażanie i hodowlę. Później, po udanym rozmnożeniu stada, planuje się ponowne wprowadzenie żubrów do odpowiadających im dużych kompleksów leśnych.

* * *

Statute of
The International Society for the Protection of the European Bison

§ 1.

Name and seat

The name of the Society shall be:

„The International Society for the Protection of the European Bison" and the Seat of the Society shall be the place of residence of the incumbent President thereof.

§ 2.

Objectives of the Society

The main objective of the Society shall be the preservation of the European bison through planned breeding and rearing. Following successful rebuilding of the herd, it is planned to reintroduce European bison to large forest complexes suitable therefor.

* * *

Fig. 2.6 The first two paragraphs of the Statute of the International Society for the Protection of the European Bison

while the second went to BELLONA, born in 1883 at Berlin Zoo. The first volume also included all of the European bison that had played a part in the reintroduction of the species (Table 2.1; Slatis 1960; Krasiński 1994a; Olech 1998). The two next volumes of the Pedigree Book—for the years 1933 and 1937—were compiled and published in the same edition of the aforementioned journal by Erna Mohr of

Berichte
der
Internationalen Gesellschaft
zur Erhaltung des Wisents

Herausgegeben im Auftrage der Gesellschaft
von Prof. Dr. HERMANN POHLE, Berlin

Zuchtbuchband

Band 5

110 und IV Seiten Text und 20 Tafeln
42 Abbildungen

Berlin 1932/37
In Kommission bei Dr. W. Stichel, Berlin-Frohnau

Fig. 2.7 The first editions of the European Bison Pedigree Book 1932/1937 (das Zuchbuch). Editors G. von Groeben and E. Mohr

Berlin, who was to become so renowned for her services to the species' reintroduction (Fig. 2.8).

An enormous amount of work on the post-war publication of the Pedigree Book was done by Jan Żabiński (1897–1974), Director of Warsaw Zoo,

Fig. 2.8 The Bison reserve at Białowieża in September 1937. Dr Erna Mohr of Berlin and Bull M 163 BORUSSE. *Photo* from the authors' archives

a renowned expert and a well-known populariser of nature (Fig. 2.9). Jan Żabiński worked together with Erna Mohr in establishing and verifying the pedigrees of the European bison that came through World War II, publishing the first post-War edition of the Book. The last book under Żabiński's editorship was in turn the 1973 one.

The early post-War years saw data on bison from Western Europe being accumulated by Erna Mohr, and then they were submitted to Editor Żabiński (Pucek 1984). Likewise, the pedigree data for the European bison on what was then Soviet territory were collected together by Mikhail Zablocky (1956), a scientific worker at the Main European Bison Breeding Centre (Priokskijj Zapovednik) (Fig. 2.10). He passed this on to the editorial office for the European Bison Pedigree Book through to the 1960s (Zablocky 1965; Pucek 1984). The years 1974–1986 denoted regular changes of address for the editorial office, the actual Editors of the Pedigree Book in this period being: Krysiak, Woliński, Pilarski and Giergiel. Such regular changes made contacts with breeders more difficult, and volumes began to come out less regularly. However, the locating of the editorial office in Białowieża National Park and the conferment of the editorial task

Fig. 2.9 Dr. Jan Żabiński (1897–1974) takes the floor at the conference devoted to the reinstatement of the European bison. It was he who promoted the idea of bison being released into the wild. To be seen in the foreground is engineer Andrzej Jaroński, who led the breeding work under the auspices of the Ministry of Forestry and the Wood Industry in the 1960s. *Photo* from the authors' archives

Fig. 2.10 Michail A. Zablocky (1912–1996) was the pioneer of work to reinstate European bison in the former Soviet Union. It was from the Prioksko-Terrasnyj Central Bison Reserve, that he directed, where most of the bison for the newly created free-ranging populations in this area derived. *Photo* from the authors' archives

Fig. 2.11 The *European Bison Pedigree Book* is the world's oldest publication documenting the population of an endangered species. A European bison entered on the register receives a pedigree number that is a kind of certificate confirming the place of origin and pedigree data

upon Jan Raczyński provided for better organisation and the regular issue of volumes from 1987 onwards. In the Books, all bison whose existence is reported by breeders are given pedigree numbers. Naming is in turn a matter for the breeders, although in line with a key that is agreed with the Editorial Office. The Pedigree Books also note (Fig. 2.11) the total numbers of bison in the free-ranging populations. The fact that a bison is noted in the Book offers a guarantee of purity of blood, and also supplies important information needed by breeders.

2.4 Nomenclature

An adult male bison is a **bull** (*byk* in Polish, never *buhaj*). The adult female is in turn a **cow** (or *żubrzyca* in Polish), while the young are—predictably enough—**calf**. The highest, arch-shaped elevation along the back of a bison is known as the **hump**. Bison live in **herds** or **groups**. Different Polish words are applied to groups of red deer and wild boar as opposed to bison. A male on its own is described as a **lone** bull, the old word *odyniec* having been applied in the literature long ago (as in Karcov 1903). Special Polish words also exist for the places where bison rest, as well as the places in which they take sand baths.

Polish hunters make widespread use of their own vocabulary (Krawczyński 1947; Hoppe 1981). In his “Dictionary of Hunting Language”, Hoppe (1981) supplies words referring to bison. While only a few of these are actually used in practice, those that are better known include *cielę*, for a bison up to one year old. The word *żubrza* given by Hoppe is not in fact in use. Also, not making it through into even slightly more widespread vocabulary is Hoppe’s “typical” *kądziel*—a well-maned and hairy nape and neck, *tałok*—for the dangling skin on the neck and *pyza*, as a term for the muzzle.

Chapter 3
Morphology and Longevity

3.1 Features of Build and Senses

The European bison is Europe's largest extant mammal. The strong and compact outline gives the impression of the animal being thick-set, but there is a sense of majesty emanating from the beast, as well as of the force of primaeval nature. The powerful body is covered in a reddish-brown coat that blends remarkably well with the animal's surroundings. Even experienced bison-watchers can mistake a bison for the fallen trunk of a tree.

Early literature tended to emphasise the role of the European bison as a dangerous beast. This was how Konrad Celtis presented the animal in his poem *Quattour libri amorum* from 1502:

> "…Abrupt in its manner the animal terrifies
> Shoots, past curved horns, fire from its eyes,
> Head and body in shag, black is its coat,
> A woolly beard to cover its swelling throat…"

In turn, in his 1,523 poem entitled "*Pieśń o żubrze*" ("Song of the Bison"), Mikołaj Hussowski wrote that three men may sit between the horns of a bison. He had clearly been carried away by fantasy. On 15 and 16 February 1830, Professor Feliks Jarocki of Warsaw University was present in the Białowieża Forest, having received permission from Tsar Nicholas I to kill two bison. These animals were to become exhibits in the University's Zoology Section. Following his appearance in the Białowieża Forest, a lone bull was brought to bay and shot by local riflemen Tomasz Szpakowicz. Jarocki, the academic, succumbed to the same temptations as Hussowski centuries before him, again writing of: "this huge beast, between whose horns two medium-sized people might sit" (Jarocki 1830). Faced with a further description of the same kind, one might be moved to wonder if such huge specimens really did exist in earlier times. However, the specimens from the expedition in question persisted into our era and were on show at the Museum of Nature and the Forest in Białowieża. They were in fact of relatively limited dimensions, such that there is no possibility of even "medium-sized" people being able to perch between those horns.

M. Krasińska and Z. A. Krasiński, *European Bison*,
DOI: 10.1007/978-3-642-36555-3_3,

The difficulties encountered more generally with correct presentation of the species' appearance were outlined humorously by Professor of Anatomy-cum-poet Kobryńczuk (1999):

"Dear oh dearie me, drawing bison's quite a battle
Despite your pains they always resemble cattle
How can you obtain a bison from a cow
Add a thick brown furry coat that's how
And a hump's a feature of your average bull
Now sketch a straggly beard and the picture's full!"

In fact, bison display marked **sexual dimorphism** in their build. Bulls are markedly larger and more heavily built than cows, with a greater disproportion between the hind and forequarters of the body (Fig. 3.1). The bull's particularly tall spinal processes of the thoracic vertebrae are well-muscled (Krysiak and Świeżynski 1967), the effect being to form the hump conferring that very impressive appearance upon males. The humps of cows are much smaller (Fig. 3.2). The powerful head is in turn held up by a sinewy *ligamentum nuchae.*

The **trunk** of a bison is relatively short and narrow, a potentially adaptive feature since it may facilitate movement through the forest. The **head** is large and low-set, while the profile is somewhat convex. The **ears** of males are relatively broad and rather hard to make out amidst the animal's hair. The **hairy coat** of an adult bison is of a fawny-brown colour, though the sides of the head and lower parts of the legs are darker. The head, neck and front of the body are covered in long hair, this forming the so-called "beard" located along the lower part of the throat and the upper part of the chest. The rear part of the body is covered with short hair only, although the end of the tail again supports long hairs extending down to the level of the hock.

Fig. 3.1 A typical Białowieża bull. *Photo* ZA Krasiński

Fig. 3.2 A typical Białowieża cow. *Photo* from the authors' archives

Adults are brown in colour, although individual variability in shade is present among adults (with some more fawn-coloured or darker brown), as Wróblewski (1927) noted. New-born calves are red-brown, although the first moult taking place 3–4 months after birth leaves them with a more typical coat of grey-brown. However, calves born late (in late autumn or early winter) immediately have coat colour as in spring-born calves after their first moult.

A bison's coat has been shown to comprise three kinds of guard hair (outer), bristles and downy (wool hair or underhair). The guard and bristle hairs determine the density, length and colour of the coat as a whole. They are of various lengths on different parts of the body. The numbers of bristles, and in particular downy, hairs change from season to season, increasing in winter to provide warmth. Downy hairs have no core, while guard and bristles hairs do (Pilarski 1962; Sokolov 1979a). The longest hairs of all (reaching lengths of 50 cm) form the tuft at the end of the tail (Wróblewski 1927; Pilarski 1962).

Bison begin **to moult** in late winter and early spring, most often at the beginning of March. In adult males, the process begins around the head and neck. Depending on the condition of a given animal, the moult may be a brief affair, or may become protracted, maximally lasting for as much as a year. Kiseleva (1974) notes that the bison in the Okskii Reserve begin moulting in mid-February. In males, the moult continues for 138 days on average, in females for 183 days. Adolescents in turn take 157 days, and calves 163. As they moult, bison encourage the loss of their winter coat by way of intensive rubbing against windthrown trees, rough tree trunks and stumps, which have been dubbed "bison combs" by Wróblewski (1927). In spring, stumps become "polished" by all the rubbing, and are covered with hanging wisps of shabby coat (Fig. 3.3).

Fig. 3.3 A "bison comb". Rubbing against various natural objects helps bison to dispense with their winter coats. *Photo* ZA Krasiński

The **skin** of the European bison is at once thick and flexible. It is thickest (14 mm) in the middle neck area, although it also reaches 9.4 mm near the *manubrium sterni* and 8 mm on the sides of the neck (Pilarski 1962). The thinnest (1.8 mm thick) skin is in turn that of the paramedian surface of the carpus. Wróblewski (1927) held that the skin on the head reached thicknesses of 15 mm. A bison's skin is exceptionally elastic, and hence extremely damage-resistant. While blows from horns often leave their mark in a bison's coat, it is rare for the skin to be pierced. There are even cases in which the rumen is found to have ruptured by an aggressor's horns while the skin above it remaines intact. Likewise, a bison hit by a train still had an undamaged pelt, notwithstanding extensive damage done to the internal organs (Fig. 3.4).

The **muzzle** is black in colour, surmounted by a silvery-white ring of hair. Likewise, the **cloven hooves** of the hind and forelegs are black, although they can sometimes be of a lighter colour resembling ivory. The female's udder is small and set high up the body, while the male's prepuce ends in a tuft of hair, and his scrotum is small and set close to the underbelly.

European bison of both sexes have **horns** which emerge close to the posterior edge of the head. Horn buds are present in calves immediately after birth, although

Fig. 3.4 A bison went beneath the wheels of a train on the Hajnówka-Białowieża line on 7 November 1979. The result was the derailment of two of the locomotive's wheels and several carriages. While the internal damage visible in the deceased animal was inevitably severe, the pelt was undamaged, attesting to its enormous elasticity. *Photo* ZA Krasiński

it is only in two-year-olds that the horns begin to turn inwards. The curvature increases with age, in females in particular. Thus, while the horns of bulls feature only a gentle inward curvature, those of cows turn inwards to a far greater extent. Horns are black in colour, although certain specimens of the free-ranging population of the Białowieża Forest have ivory-coloured horn-tips. In turn, the tips of the horns of old bulls are often wiped or damaged on account of the number of times they have come into contact with various kinds of obstruction like downed tree-trunks and windthrown trees, or else been deliberately employed in the uprooting of young spruce trees as an element of rutting behaviour. Bulls' horns are set more widely apart, and are also thicker and longer, than those of females. The sickle shape of horns ensures that the many brushes with rivals that animals get themselves into rarely have more serious consequences for the animals' health.

The **eyes** are relatively small, brown in colour and located deep in telescopic orbits. The protruding orbits protect a bison's eyes from being damaged by branches as the animal moves about through the forest. A person is looked at with a lack of confidence, and even hostility. The appearance of threatening flash of gold there deserves respect on the part of the observer, and should be taken to be discouraging of excessive familiarity. Grazing bison look much less severe, as their meaty grey

Fig. 3.5 A single bison's tracks in the snow. *Photo* ZA Krasiński

tongues draw clumps of grass into the shapely mouth. Bison do not have particularly good **eyesight**, and among trees in particular it is possible for them to fail to note an observer, especially if the latter is appropriately dressed and standing still.

Bison wandering around a forest leave distinct **trails** (Figs. 3.5, 3.6). The prints left by the front limbs are larger than those made by the rear ones. The prints left by males are apparently larger than those of females. Bison of both sexes and all ages release a **musky smell** that remains in places they have frequented long after they have departed. Since the whole pelt of a bison is impregnated with this musky

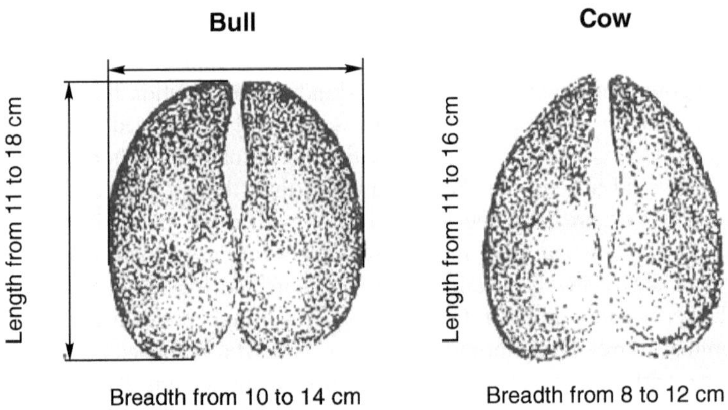

Fig. 3.6 A sketch of the tracks of adult bison. *Source* Krasiński 1993, modified

smell, it is believed that this is released by sebaceous glands in the skin, whose ducts discharge into the hair follicle area.

The **call** of a bison is a grunt. The sound is most frequently heard as cows communicate with their calves, while the male's bellow is mainly associated with the rut. In large winter aggregations, cows still recognise the cries of their own offspring as opposed to others flawlessly, while the calves find their mothers as they call them in to suckle. A bull's voice is lower and hoarser than of a cow. Bison never roar. In his 16th century "Song of the Bison", the poet Mikołaj Hussowski writes that bison do indeed fill the air with a penetrating roar, but this was either a flight of fancy or a misidentification with the roaring a red deer stag makes in the rutting season.

A bison's best-developed sense is its **sense of smell.** Rutting males are adept at tracking down herds, sniffing the scent of cows that was left behind on the forest floor. The same behaviour applies where an individual becomes separated from its group for a period of time.

European bison are fast runners, although they are more in the nature of sprinters than long-distance operators. They usually come to a panting stop having covered just a couple of hundred metres. It was Wróblewski (1927) who drew attention to certain characteristic features of the way a bison moves. As it steps, it extends a straightened foreleg before it, before transferring to it its body's centre of gravity. And when that leg stands firmly upon the ground, the weight of the body is transferred onto it, while the other leg is put in motion. The shoulder muscles are in action in the process (Fig. 3.7). A bison on the march leaves footprints 84 cm

Fig. 3.7 The characteristic way in which an adult bull moves about. *Photo* ZA Krasiński

Fig. 3.8 Bison at full gallop. *Photo* from the authors' archives

apart (Krasińska 1993), while running increases this distance to 1.5 m and galloping to 2.0 (Wróblewski 1927). Bison gallop with tails raised. Analysing such a gallop on film, Gambaryan and Sukhanov (1979) concluded that it was diagonal in nature, which is to say that it takes place with a diagonal body. In the course of the gallop cycle, the rear and fore rebound is followed by crossed-over support of the two hind legs, jump, the supporting of three limbs and finally propping on a single extremity (Fig. 3.8). In jumping, a bison mainly employs the rear of its body.

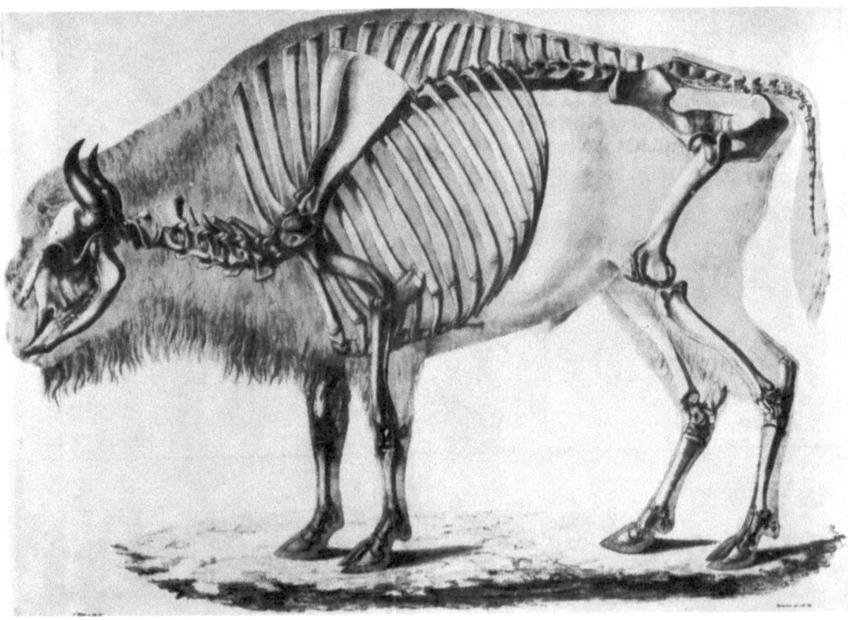

Fig. 3.9 The silhouette of a bison with the skeleton visible. This fine drawing was an 1827 work by Bojanus entitled *De uro nostrate eiusque sceleto commentatio*

While European bison have 14 pairs of ribs, domestic cattle have 13 pairs (Fig. 3.9). This fact was first drawn attention to in 1827, by expert Vilnius-based zoologist and anatomist Ludwik Henryk Bojanus.

Bison have a well-developed system of fasciae (sheets of fibrous tissue), ensuring that even heavily-pregnant females are not characterised by an excessively pendulous abdomen.

The bison's large head is held up by a sinewy nuchal ligament of considerable durability. This ligament and the *ligamentum flavum* between the spinal processes of the thoracic vertebrae allow for the holding in place and raising of the head as food is obtained from the forest floor (Krysiak 1951a; Kobryńczuk 2000). The powerful muscles of the front part of the body in turn interact with the aforementioned long spinal processes of the thoracic vertebrae to form the hump so characteristic for a bison's outline. It is in this place that the line of the back reaches its highest point (of up to 188 cm) above the ground.

3.2 Body Mass

Body masses of bison have been determined by reference to data on 152 (86, 66) specimens originating from the Białowieża Reserves, as well as 492 (248, 244) from the free-ranging herd in the Białowieża Forest (Krasińska and Krasiński 2002). At birth, calves are quite small, weighing between 15 and 35 kg. This in part accounts for the few complications during birthing that female bison face. The birth weight of a male calf—at 28 kg on average (range 16–35)—is greater than that of a female—24 kg on average (15–33) (Table 3.1). The Belarussian population is characterised by even lower birth weights in the case of calves of both sexes, the average being 20.4 kg (range 17.5–24) (Krasochko et al. 2004; Bunevich 2006; Kozlo and Bunevich 2009). The first doubling of this weight is achieved by the time a calf is aged 3 months. In turn, adolescents up to two-years old have similar body masses irrespective of gender. Only among 3-year-olds (of both the free-ranging and captive populations) do males begin to attain significantly greater body weights than females. Subsequent years of postnatal development maintain this difference in body weight associated with gender. Ultimately, bull bison aged 6 years and over attain far greater weights than their female peers (Table 3.1), while those from enclosures are considerably larger than those roaming free. This difference begins to make itself felt as early as in the sixth month of life. Despite what popular opinion may say, bison of the free-ranging population do weigh less than animals regularly in receipt of supplementary feeding and deriving from breeding enclosures. In the free-ranging population of the Białowieża Forest, body masses among adults (aged 6 and over) are of 634 kg on average in the cases of males (range 436–840), and of 424 among females (range 340–540). Body masses among Reserve-bred bison are of 747 kg on average among males (range 580–920), and 460 kg among females (range 320–640) (Table 3.1). The maximum weight noted for a Reserve-bred bull was 920 kg

Table 3.1 Comparison of body masses (in kg) of European bison living in free-ranging populations and at captive breeding centres

Age (months, years)	Free-ranging population			Captive breeding		
	Mean	Min	Max	Mean	Min	Max
Males						
0	–	–	–	28	16	35
1	–	–	–	52	32	83
2	41	35	48	–	–	–
3	53	32	70	–	–	–
4	76	46	96	83	82	84
5	82	48	108	–	–	–
6	99	90	160	143	131	160
7	101	78	140	150	–	–
8	103	70	180	170	130	180
9	109	80	140	–	–	–
10	–	–	–	–	–	–
11	–	–	–	187	155	220
12	124	80	170	175	130	220
2 yrs	220	130	300	331	220	453
3 yrs	303	240	390	387	310	502
4 yrs	366	260	430	502	380	609
5 yrs	467	345	660	591	572	610
≥6 yrs	634	436	840	747	580	920
Females						
0	–	–	–	24	15	33
1	–	–	–	30	–	–
2	–	–	–	–	–	–
3	47	32	70	–	–	–
4	66	43	81	66	65	67
5	72	48	100	–	–	–
6	91	52	120	111	105	118
7	97	62	148	129	120	148
8	111	70	190	159	148	170
9	124	96	146	–	–	–
10	–	–	–	–	–	–
11	–	–	–	200	–	–
12	119	96	140	190	170	220
2 yrs	213	170	280	228	170	306
3 yrs	252	150	310	227	220	240
4 yrs	341	260	458	400	378	432
5 yrs	407	340	460	439	420	458
≥6 yrs	424	340	540	460	320	640

Data from: Krasińska and Krasiński (2002). *Explanations*—lack of data, yrs–years.
Source Krasińska and Krasiński (2007)

Fig. 3.10 The fine bull PORANEK (pedigree number 980) was one of the most impressive specimens known. His body weight reached 920 kg—a record never beaten at Białowieża. *Photo* ZA Krasiński

(Fig. 3.10) and the largest cow weighed 640 kg. This compares with maxima for the free-ranging herd of 860 and 540 kg, respectively (Table 3.1). Similar body masses characterise adult males and females in the Belarussian population of the Białowieża Forest, as well as the Białowieża-Caucasian bison inhabiting the Oka Reserve (Russia) (Sipko et al. 2002; Krasochko et al. 2004; Bunevich 2006; Kozlo and Bunevich 2009). Body mass in males increases in line with age until 7 years have passed; thus, sexual dimorphism in body mass is something that progresses slowly in postnatal development, to the extent that it is not marked at all until the age of 3, and thereafter becomes marked up until the end of the animals' lives.

Very few references concerning the weights of the bison living in the 18th and 19th centuries are available to us. The obelisk erected at Białowieża to commemorate King August III's hunt here in 1,752 notes that the biggest of the 42 bison killed then weighed 14 quintals 50 pounds, or around 720 kg. Bojanus (1827) in turn notes that a specimen killed in 1,612 weighed more than 1,770 pounds, or a little over 700 kg. According to the data from Wróblewski (1927), an author who had the opportunity to weigh the bison shot within the Białowieża Forest in the years 1906–1909, the body masses of adult males reach 519 kg on average, the range being between 368 and 640 kg. Among cows, in turn, an average of 247 kg conceals rather the actual range of weights noted—between 123 and 560 kg. The low weight of these females is noteworthy, since the figures seem clearly below the averages noted for adult bison being culled currently in the Białowieża Forest. This is a further indication that the true condition of the bison living in the Forest in the early 20th century was poor. The occasional remarks on beasts of huge weights and dimensions would therefore seem to have been founded on nothing more than subjective assessments.

3.3 Body Size

At birth, a calf's trunk is short and tall. It is only in the 8–10 months of life that those proportions undergo change, such that the thorax becomes as long as it is high. Throughout development, this part of the body is shorter in females than in males (approximated to the height at the hump).

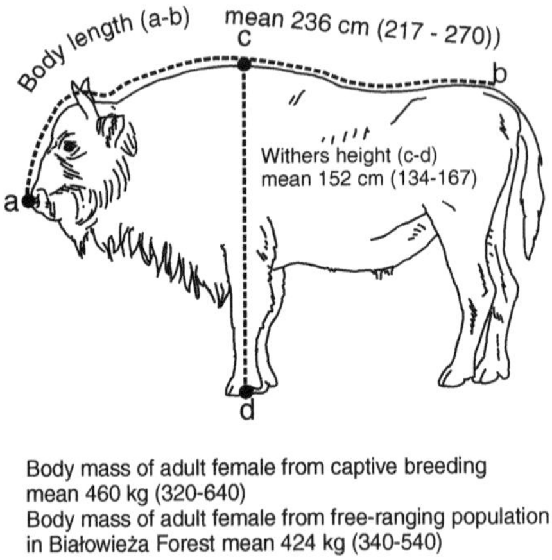

Fig. 3.11 The silhouette of an adult female (a cow). Illustration by Bartoń

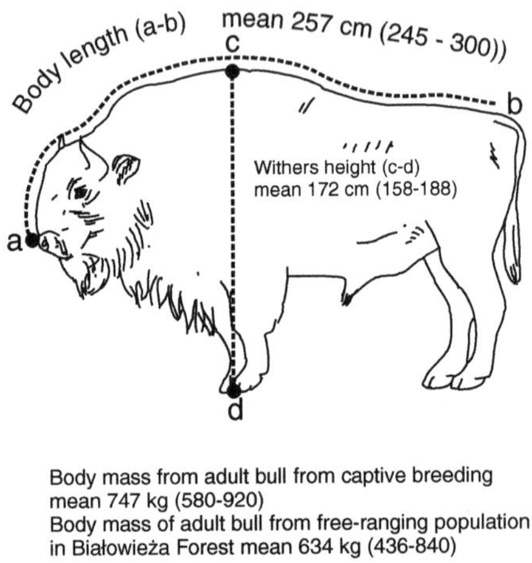

Fig. 3.12 The silhouette of an adult male bison (a bull). Figure by Bartoń

At birth, bull-calves are somewhat larger than heifers. Sexual dimorphism becomes apparent from age 3 onwards, and is retained to the end of life. Maximum body dimensions in adult bulls aged 6 and over are: 188 cm in height at the hump, 300 cm in length of the trunk and a 280 cm circumference of the thoracic cavity. Among cows, these respective dimensions are of 167, 270 and 246 cm (Krasińska and Krasiński 2002). The two most important dimensions of the bodies of adult bison are presented schematically in Figs. 3.11 and 3.12. Similar dimensions characterise the adults present in the Belarussian part of the Białowieża Forest (Sipko et al. 2002; Kozlo and Bunevich 2009).

Consideration of the age-related changes in the masses and dimensions of the body leads to the conclusion that a bison's physical development comes to an end at the age of 5 among females and 7 among males. Likewise, anatomical studies show that the development of bison ends at 5 years of age (Empel and Roskosz 1963; Krysiak and Świeżyński 1967). Jarocki (1830) also noted that the development of bison lasts for up to 6 years. Brincken (1826) in turn wrote that a bison continues to grow until it is 6 or 7 years old.

Thus, in comparing body masses and sizes among the European bison alive today or present in former times, we may conclude that the high level of inbreeding among Lowland bison (Olech 2003) has not actually reflected unfavourably upon the animals' development, build or size. At the time they reach physical maturity, the bison alive today are characterised by proper development, and have achieved a mass and size no smaller than the bison living in the wild state prior to the World War I (Krasińska and Krasiński 2002).

3.4 Longevity

The authors of the past regularly exaggerated the bison's length of life, having no documented data (only subjective assessments) to base themselves on. Brincken (1826) resembled many other authors in informing his readers that bison in general could live for 30–40 years, while bulls sometimes made it through to 50. In the assessment of Wróblewski (1927), 35 was the top age for a bison. He viewed any figure of 50 as a definite exaggeration. Bojanus (1827) in turn went for a maximum longevity figure of just 20 years.

Precise determination of length of life in bison can be achieved by reference to data contained in the European Bison Pedigree Book. A record age of 30 was reached in 2010 by the French-based cow F 4661 DIPPE. Other old females were F 5111 SPINNE (28) and F 5352 SPINNERIN (27), with both originating from the German Reserves. Among the group of founders, a record age of 27 years was achieved by the female F 42 PLANTA (1904–1931), although the date of birth in this case was obtained by approximation. Three other cows—F 16 PLAVIA (1906–1932), F 106 FRIGGA (1921–1947) and F 89 BILMA (1913–1939) made it to 26 years of age. Female F 524 BESTE (1939–1963) died at the Białowieża Breeding Reserves aged 24, having broken the head of her femur. The cow F 1131

KANIA (1958–1981) in turn passed away aged 23, having produced calves in the 14 consecutive years 1961–1974.

Males are shorter-lived in terms of both maxima and averages. The Białowieża-Caucasian line bull M 101 HAGEN (1911–1933) was 22 when he died at Białowieża, while M 45 PLEBEJER (1917–1937) made it to 20. Also at Białowieża was the fine bull M 980 PORANEK, who left a great many offspring but still only lived to be 20. Among the well-known bison, M 229 PLISCH survived for the 18 years 1933–1951, while the only Caucasian bison M 100 KAUKASUS (1907–1925) also lived to be 18.

Analysis of longevity among bison kept in captivity in the years 1891–1946 revealed that females had lifespans of up to 27 years, males—even exceptionally—to just 20 (Koch 1956). Jaczewski (1958) notes that, in a sample of 67 bulls, only 3 lived beyond 20, the oldest being 23 at the time of death. In contrast, 11 out of 66 cows passed the 20-year mark, 4 lived beyond 25 years and the four oldest all made it to the age of 26.

In the free-ranging population, the oldest marked female lived to be 24—far beyond the maximum of 18 noted for marked bulls. Overall, then, females can be said to pass the 25-year mark only rarely, while bulls rarely exceed 20 years of age. Bison in the wild have shorter lives than those in Reserves, although the determination of age using teeth and horns is obviously a less exact science than reference to known dates of birth.

Chapter 4
Aspects of the Genetics of European Bison

By Małgorzata Tokarska

4.1 Chromosome Number

Bison have 60 chromosomes, of which 58 are acrocentric autosomes and two are the sex chromosomes (a large sub-metacentric X and a small sub-metacentric Y) (Melander 1959; Koulischer et al. 1967; Fedyk and Sysa 1971).

M. Krasińska and Z. A. Krasiński, *European Bison*, DOI: 10.1007/978-3-642-36555-3_4, © Springer-Verlag Berlin Heidelberg 2013

4.2 Genetic Background to the Recovery of the Species

Today's reinstated population of the European bison (*Bison bonasus*) derives from individuals living in captivity in the 1920s. The registration list of all the European bison alive on 31st December, 1923 contained 54 individuals considered to be Lowland bison (*Bison bonasus bonasus*), as well as one bull of the Caucasian subspecies (*Bison bonasus caucasicus*). Only a part of these animals, descendants of just 12 individuals, including Caucasian bull KAUKASUS, were successfully used in the recovery effort and left descendants. It was on account of the presence of KAUKASUS that newly-bred animals were divided into two genetic lines: the pure Lowland line, and the Lowland-Caucasian line which also obviously contains Caucasian bison genes. The two lines have been kept in isolation from one another from the outset.

Unfortunately, while the Lowland-Caucasian line draws on all 12 of the aforementioned founders, the Lowland line derives from a subset of just 7. Furthermore, the share of genes of individual founders is extremely unequal in the two lines. The genes of M 45 PLEBEJER and F 42 PLANTA are nevertheless predominant in both lines, constituting over 80 % of those in the Lowland line, and half in the Lowland-Caucasian line (Olech 2003).

4.3 The Genetic Status of the Contemporary Population

The small number of founders ensured a high degree of relatedness among members of the first generations, and hence extremely far-reaching inbreeding of their progeny. The degree of inbreeding in the Lowland line is at 44 % (which is to say that 44 % of the genes of each individual are identical in terms of origin), while the comparable figure for the Lowland-Caucasian line is 26 % (Olech 2003; Pucek et al. 2004).

While a high level of inbreeding is often associated with inbreeding depression symptoms that are potentially very serious and even lethal, this seems not to have given rise to any life-or-death problems in the case of the extremely inbred Lowland line bison. Quite unexpectedly, in the Lowland line, there is a strong positive effect of the inbreeding coefficient of a cow on the survival of her calves during the first 30 days of life (Olech 2003). In comparison, negative effects of inbreeding (reduced fertility, survival, milk production and calving performance) may be evident in cattle when the inbreeding coefficient exceeds just 12.5 % (Sewalem et al. 2006). Nevertheless, it is not possible to preclude the idea that the lack of signs of inbreeding depression in the Lowland line is a reflection of the difficulties with detecting inbreeding depression effects under natural conditions (Reed et al. 2003). Signs of inbreeding depression (neurocranium shortening and elongation and narrowing of the splanchnocranium) were found in the lowland-Caucasian line, despite a lower value for the average inbreeding coefficient than in the Lowland line (Kobryńczuk 1985). The lack of visible signs of inbreeding depression in the Lowland European bison may be due to rapid demographic recovery of the species after a bottleneck, which minimizes the negative impact on genetic variability by purging the genetic load. It has been shown that this process may even increase species fitness (Nei et al. 1975). In contrast, slow recovery may result in decreased viability of a species due to more marked reductions in heterozygosity (Bouzat et al. 1998; Westemeier et al. 1998). Other authors claim that the effectiveness of the purging effect is more dependent upon the genetic basis of the genetic load (inbreeding depression) (Hedrick 1994), and that rapid inbreeding in very small populations cannot be relied upon to eliminate the deleterious effects (Frankham et al. 2001). Another explanation of the lack of visible signs of inbreeding depression could be that genetic depletion took place before the recovery event, or maybe it was simply that the genomes of founders of the Lowland line were by chance free from significant genetic load (Tokarska et al. 2011).

It should be added here that any estimations made on the basis of the European Bison Pedigree Book (EBPB) proceed on an assumption that the founder individuals were unrelated, simply because there is no further information about their ancestors. Obviously, this assumption is not necessarily true. For example, the predominant founders of both genetic lines F 42 PLANTA and M 45 PLEBEJER come from the same Pszczyna line, which was based on just 12 individuals (Krasiński 1996a). So the actual inbreeding coefficients known for the European bison, high though they already are, are most probably still underestimations.

An additional problem in the Lowland line is the existence of just a single male line in the population. This means that all living bulls have the Y chromosome of M 45 PLEBEJER. The Lowland-Caucasian population in turn has three male lines, founded by M 45 PLEBEJER, M 100 KAUKASUS and M 15 BERGRÜNDER (Pucek et al. 2004; Tokarska et al. 2011). A disease affecting 5–10 % of bulls of different ages every year (Gill 2002; Lehnen et al. 2006) and often instinctively connected with Y-chromosome monomorphy is a necrotic inflammation of the penis and prepuce known as *posthitis*. The disease seems to be conditioned by either a high level of inbreeding or by Y-chromosome monomorphy, but so far there is no scientific data in support of these opinions. On the contrary, studies show that the occurrence of *posthitis* is not correlated with the level of inbreeding (Matuszewska et al. 2004). There is also no significant association between DRB3 alleles or genotypes and susceptibility to *posthitis* (Radwan et al. 2007). There have been variable hypotheses as to the aetiology of this disease. The latest suggests that environmental and endemic factors are responsible for its occurrence (Krasińska and Krasiński 2010), since *posthitis* is only observed in animals living in, or originating from, the Białowieża Forest.

Low genetic variability (a low level of expected heterozygosity) in the European bison has been confirmed by numerous authors using different genetic markers such as blood serum proteins (Gębczyński and Tomaszewska-Guszkiewicz 1987; Hartl and Pucek 1994; Sipko et al. 1997), blood-group systems (Sipko et al. 1995), milk protein genes (Sipko et al. 1994; Burzyńska and Topczewski 1995; Kamiński and Zabolewicz 1997), mitochondrial DNA (Tiedemann et al. 1998; Burzyńska et al. 1999; Wójcik et al. 2009), MHC genes (Udina et al. 1994; Udina and Shakhaev 1998; Radwan et al. 2007; Babik et al. 2012), microsatellites (Gralak et al. 2004; Luenser et al. 2005; Tokarska et al. 2009a, b) and SNPs (Pertoldi et al. 2009a, 2010; Kamiński et al. 2012).

Analyses of the complete, 1,429 bp sequence of the control region (D-loop) performed on nearly 200 individuals shows that there are just three haplotypes corresponding to the three female founders (Wojcik et al. 2009). Although it is reassuring that all the female founders' haplotypes are still hopefully present in the Lowland bison population, their frequency is very uneven: the most common haplotype has been found in 93 % of the Lowland individuals studied, the next commonest in 6 % and the least common in just 1 % (Wojcik et al. 2009). The threat of the rarest being lost is thus a major one.

Estimated values for the expected heterozygosity of the Lowland line (describing the genetic variability as the mean percentage of heterozygous loci in the population) vary from 23 to 39 % (Gralak et al. 2004; Luenser et al. 2005; Tokarska et al. 2009a, b); while the mean number of alleles per *locus* (another basic parameter of genetic variability) based on the analysis of 17 microsatellite markers is estimated at $A = 2.3$. Both statistics reflect the dramatic demography of the species (Tokarska et al. 2009a). The above values are approximately half as high as those estimated for the American bison *Bison bison* (62–67 %; Schnabel et al. 2000; Tokarska 2010), which also experienced a bottleneck. Studies on genetic variability among the Lowland bison born in the Białowieża Forest over 50 years after reintroduction reveal that—due to the considerable size of the population now—the bison no longer mate with their closest relatives. Nevertheless, the number of founders was so small that the actual inbreeding coefficient increases with each generation.

Studies on a 260 bp-plus sequence of exon 2 of the MHC II DRB3 locus, as one of the most variable genes of Major Histocompatibility Complex, revealed the presence of only four alleles in Lowland-line European bison, compared with at least 15 found in the American bison (Traul et al. 2005; Radwan et al. 2007).

The basic parameter of genetic variability—mean expected heterozygosity, calculated on the basis of microsatellite data, has been estimated as 0.29 (29 %) for the Lowland line and 0.35 (35 %) for the Lowland-Caucasian line (Tokarska et al. unpublished). The impoverishment of genetic variability in the Lowland line of the European bison is so great that microsatellite markers simply do not work as a reliable tool in identity analyses or any related issues (assessment of reproductive success, pedigree control). Out of 276 individuals with 17 polymorphic microsatellite markers studied, some 43 (16 %) lack unique multi-locus genotypes (Tokarska et al. 2009b).

As the European bison is closely related to domestic cattle, a powerful, cattle-designed, SNP analysis system (BovineSNP50 BeadChip Illumina®) has been applied successfully in genetic analyses of the species. Of the more than 52,000 cattle SNP loci in the BovineSNP50 BeadChip amplified in European bison, only about 1,000 SNPs were found to be polymorphic. In comparison, over 1,500 were polymorphic in the American wood bison (*Bison bison athabascae)* and over 1,000 in the American plains bison (*Bison bison bison*) (Pertoldi et al. 2010).

Contrasting with the unsuccessful application of microsatellites to identity analysis in Lowland bison is the work by Tokarska et al. (2009b) showing that a selected set of 50–60 polymorphic and informative SNP loci provides sufficient power for identity and paternity in European bison to be identified. SNP genotypes also allow for the reconstruction of evolutionary relationships between the different bison subspecies (European bison, wood bison and plains bison), and cattle (Pertoldi et al. 2010).

The effective population size[1] (Ne) is a parameter, function of the harmonic mean, influenced markedly by the minimum population size reached by the population (Lynch and Walsh 1998; Pertoldi et al. 2007). This explains well why the Ne value for the European bison from the Polish part of the Białowieża Forest is very low, at 23.5, 95 % confidence interval (CI) = 23.1–23.9; and only slightly increased beyond the level noted in the 1950s (Tokarska et al. 2009).

Free-living European bison from the Białowieża Forest on Polish and Belarussian parts of the Forest respectively, are isolated from one another by a border fence built 30 years ago. Both populations have been regarded as Lowland line (Raczyński 2010), so there are suggestions (Daleszczyk and Bunevich 2009) that these should be allowed to merge in the interests of the species.

In this regard, it is interesting to note how comparative genetic analyses of microsatellite data for the European bison in the two parts were able to detect substantial differences between them. The effective population size (Ne) of the Belarussian population of European bison is considerably larger than the Polish population (Ne = 32 and Ne = 23.5, respectively), and there are several genetic variants found in the European bison of Belarus that have never been found among their Polish counterparts ($N = 275$), even though only a relatively small number of Belarussian bison have been analysed ($N = 50$).

As the alleles in question have been found in the Lowland-Caucasian line, a hybrid origin for the European bison across Belarus is indicated, since at least eight out of nine of that country's populations originate from the Belarussian part of Białowieża Forest (Tokarska 2010).

Thus, Poland's European bison from the Białowieża Forest are seen to constitute the only 'demographically safe' (with over 100 individuals) world population of the Lowland European bison in the world. The idea of removing the border fence and merging both populations from Białowieża Forest becomes therefore a problematic one that needs to be thoroughly thought through and discussed before the idea is propagated further.

---

[1] Effective population size refers to the number of individuals that theoretically take part in reproduction. The Ne value is usually much lower than the actual population size, and is very much dependent on the level of genetic variation within the population.

4.3.1 Is There Anything We Can Do to Improve Genetics of the European Bison?

As has been noted, genetic variability in this species is at a very low level, since all existing animals derive from just 7 (Lowland line) or 12 (Lowland-Caucasian line) founders. There is no “fresh blood” available, because there are no European bison alive today that are descended from alternative founders. In these circumstances, the only remaining room for manoeuvre relates to the following applications of science as regards species management.

4.3.2 Save the Rare Maternal (mtDNA) Lines

The fact that three mtDNA haplotypes are still present in the Lowland line may be explained by the relatively long generation times in the species, but may not be sustainable without intervention. Targeted restocking of Lowland European bison herds to form groups carrying the rare haplotypes should therefore be undertaken, to reduce the high probability of one or more being lost (Wójcik et al. 2009).

4.3.3 Improve Ne and Genetic Variability

Application of sophisticated technologies (SNP chips, RAD-tag sequencing) may allow for accurate reconstruction of the pedigree of the European bison, and could also then be used in estimating the actual relatedness between individuals kept in captivity and recorded in the Pedigree Book. Breeding pairs are usually selected to minimize inbreeding by reference to pedigree information, which describes the expected relationship among individuals, while saying nothing about actual genetic relationships at the individual level. Information obtained from the above technologies could offer a truer indication of this, and then be applied in mating plans to improve both effective population size and genetic variability (Pertoldi et al. 2009a).

4.3.4 Stop Culling to Sustain Genetic Variability

From a genetic point of view, the culling of European bison is most often a careless action. It may of course be understood in cases of certain diseases like *posthitis*, a broken leg or blindness, when it is then a mercy gesture towards a dying and suffering animal. But the culling of any other animals (even if in poor condition or with body defects) represents the potentially irreparable removal of rare alleles or haplotypes whose retention in the population will always be a priority.

Chapter 5
The Original Range of the Species *Bison bonasus* (L.)

Mammals of the genus *Bison* Hamilton Smith, 1827, colonised huge areas of two continents, Europe and North America (Heptner et al. 1966). The original range of *Bison bonasus* in fact took in the greater part of the European continent (Fig. 5.1).

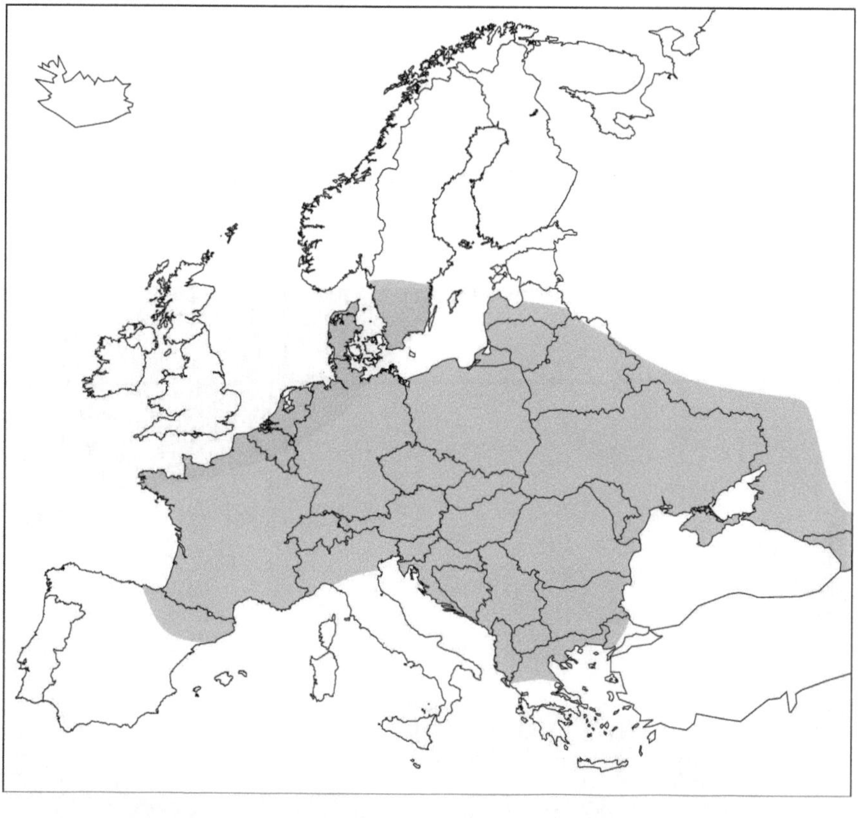

Fig. 5.1 The primary range of the European bison *Bison bonasus* (L.) in early historical times. Limits are set using data from Heptner et al. (1966), Pucek (1991). *Source* Pucek (2004), modified

DOI: 10.1007/978-3-642-36555-3_5,

The Białowieża Forest is actually at the centre of this area, whose limits were set by the coast of the Atlantic Ocean, and the North and Baltic Seas, although the southernmost part of Scandinavia was included. In the east, the range limit ran from the Gulf of Riga in a south-easterly direction but not going as far as either today's Moscow or Ryazan areas. In the south, the range extended as far as the Sea of Azov and Black Sea, thereby encompassing the massif of the Caucasus Mountains. The southern range limits were further set by the Black and Mediterranean Sea coasts, with the Italian and Iberian Peninsulas not occupied (Heptner et al. 1966). Bison were present in the Pyrenees, however, as the cave paintings in that area attest—notably at the famous Altamira in Spain (see colour insert). Southern France also boasts a number of caves in which bison and other animals have been drawn (Gstalter and Lazier 1996). Especially, many likenesses of animals have been found in the Dordogne area of south-western France. The Font-de Gaume cave includes a rock painting of bison (see colour insert). It was in this area in 1910 that D. Peyrony found a fragment of reindeer horn into which the outline of a lying bison had been carved (Fig. 5.2). However, Benecke (2005) made an analysis of the locations of archaeological sites making reference to bison in one way or another, drawing the conclusion that the range of the species in Western Europe in the Middle and Late Holocene may not have extended as far west as had been thought previously.

As the second millennium got underway, European bison were still present in the extensive forests covering Europe. However, as the area under forest declined

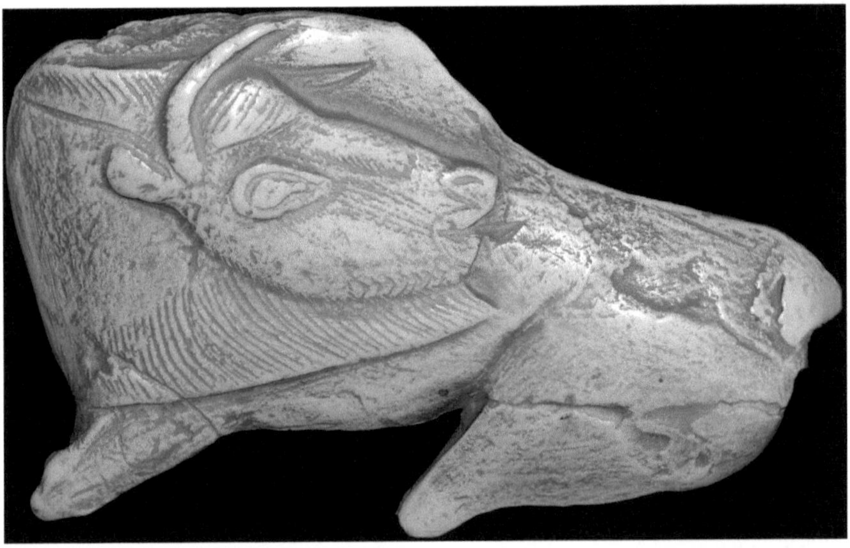

Fig. 5.2 A copy of a 15 cm-long carving presenting a bison with a turned head, licking its flank. Done on a reindeer antler, it was found in the Madeleine Cave by the bank of the River Weser. It dates back some 11,000 years. This figure presents a copy of it. *Source* Krasińska and Krasiński (2007)

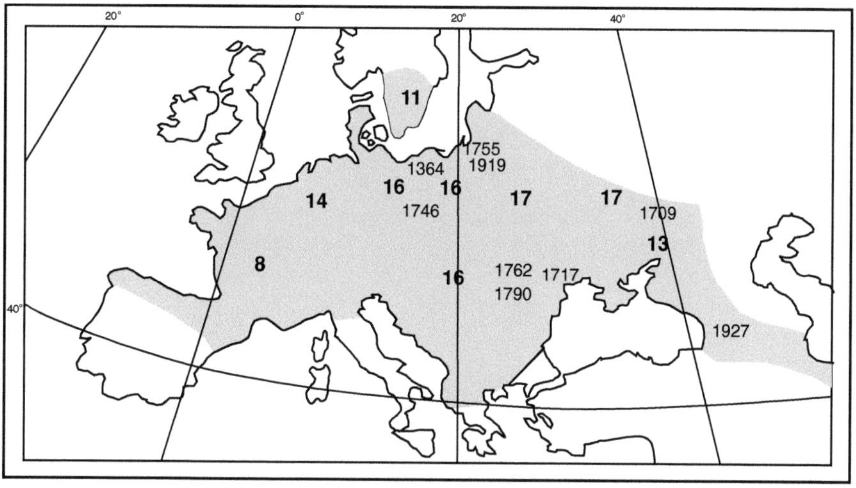

Fig. 5.3 Process of extinction of the European bison *Bison bonasus* in the Holocene and in the early part of the historical period. The Roman and Arabic numbers respectively, denote the century or year in which the last bison are known to have been present in the given area. *Source* Pucek (1991, 2004)

and the cleared land was taken over for field cultivation, the areas occupied began to shrink. The process of gradual decline towards extinction had begun as early as in the 11th century, and was maintained without reverse up to the 19th century (Fig. 5.3).

Chapter 6
The European Bison and the Aurochs on Polish Territory

In the glacial epoch (Pleistocene), it was possible to distinguish between long-horned and short-horned representatives of the genus *Bison*. The long-horned bison, *Bison priscus* (Bojanus 1827), survived into the Younger Pleistocene, ultimately becoming extinct in the Dryas of the Late Pleistocene, around 11,000 years ago (Niezabitowski-Lubicz 1937; Kowalski 1967). Remains of bison standing out in terms of their powerful cornual processes have been found in eastern and central Poland, in the Małopolska, Wielkopolska and Pomeranian regions (Niezabitowski-Lubicz 1937; Krysiak 1951b). The smaller form *Bison schoetensaki* Feudenberg 1914 is known by some authors as the progenitor of today's European bison (Kowalski 1967), though the alternative candidate species in the minds of other experts is *Bison priscus* (Flerov 1979; McDonalds 1981).

In historical times, the modern species of European bison was found across the whole of Poland. However, the species had become quite rare and confined to the larger forest complexes as early as the 11th to 13th centuries where it was reported from the Niepołomice, Sandomierz, Rawa and Mazovian Forests. A description of a bison hunt in a 12 century chronicle otherwise devoted to King Bolesław the Wrymouth's capture of the town of Szczecin allows us to conclude that the species was still present in what is today north-westernmost Poland at the beginning of the 12th century (see the Legend of Sieciech) (Guillaume 2003). The species was also present in the forests of Greater Poland (*Polonia Major* or Wielkopolska) in this period, as well as in Prussia.

Remains of bison living in the Middle Ages have been found in archaeological excavations on what is today Belarussian territory. Bison accounted for some 16–22 % of what the game hunters were found to have taken, on the basis of digs carried out in and around the castles of contemporary Belarus (Aleksandrowicz 1999). Peak numbers of bison bones uncovered in Grodno coincide with the 15th to 16th centuries, while digs in Nowogródek have yielded bison bones dating back to the 17th century. In general, the results of the archaeological work show that the remains of bison are present most frequently in layers dated back to the 16th century (Aleksandrowicz 1999). According to Długosz, the 15th century range of the

DOI: 10.1007/978-3-642-36555-3_6, © Springer-Verlag Berlin Heidelberg 2013

bison extended beyond the Białowieża Forest into Wołyń, along the River Prypeć, in the forests between the Rivers Wisła and San (today's Poland) (Sztolcman 1926). In turn, Jędrzej Święcicki's 1634 description of Mazowsze (historically Mazovia) only makes reference to bison in the context of the River Szkwa (Skwa), which is to say the area of today's *Puszcza Kurpiowska* forest complex. The westernmost limit of occurrence is given there as the River Omulew—which empties into the Narew near Ostrołęka (Sztolcman 1926). Today's Lithuania certainly retained bison to the end of the 18th century, the area in question being part of Prussia at the time. The last bison in East Prussia was killed in 1775. In turn, in the 19th century, the Lowland bison only persisted in Poland, in the Białowieża Forest (Sztolcman 1926).

Living in the Polish lands alongside the European bison was that second huge ruminant, known as the aurochs: *Bos primigenius* Bojanus, 1827. In his 1827 description, Bojanus drew a clear distinction between the bone structure of the aurochs, as opposed to domestic cattle and the European bison. Nevertheless, naturalists would long argue about the very existence of the aurochs. Many learned figures maintained into the 19th century that the European bison and aurochs were one and the same animal, simply going under alternative names (Jarocki 1830).

Likenesses of both aurochs and bison were included in Herbstein's work *Rerum moscoviticarum commentari*, which was published in Basle in 1556 (Figs. 6.1, 6.2). These outline sketches present two markedly different animals. The back of the bison has a "carplike" shape, while that of the aurochs is flat. The forequarters of the bison are covered with long hair, and markedly long hair in the lower part of the neck, creating the so-called beard. In the aurochs, only the forehead bore a tuft of red hair known in old Polish as the "aurochs-lick" (*turzy wicher*). The tail of the aurochs ends in a distinct "brush". The horns of the bison and aurochs in the two diagrams are very similar. However, the likeness of the "Augsburg aurochs" originating in the same period shows much longer, forward-pointing horns that turn up slightly towards the ends. Łukaszewicz (1952) speaks of the considerable variability to the size and shape of the horns of specimens of the aurochs that had been dug up. Thus, various sources and documents allow it to be concluded that coat colour also varied markedly in the aurochs, from black, brown or red through to light-coated forms. Łukaszewicz (1952) estimated that a bull stood 170–185 cm tall at the hump, a cow 165 cm tall. Bulls weighed in at 600–800 kg. Many authors have written of the huge strength and volatile nature of this animal. The saying "strong as an aurochs"—not dissimilar to the English "strong as an ox"—remains in use in Polish, as does the phrase *chłop jak tur* ("a lad like an aurochs").

Łukaszewicz (1952) cites three important historical references concerning the aurochs. In 1288, Duke Bolesław of Mazovia gave permission for various animals to be hunted, but not the bull known as the aurochs. A similar reservation on what could be hunted was applied in 1359 by one of Bolesław's successors to the title of Duke of Mazovia, Ziemowit. In turn, in his famous "Chronicle" of Poland dated 1422, Jan Długosz noted that King Władysław Jagiełło "…travelled to to Wiskitki in Mazovia, to give himself over to hunting for the forest ox, which in Polish is called the *tur*." It follows from these entries that—by the 14th

Fig. 6.1 The likeness of a European bison from the 1556 work by Herberstein. The Latin inscription reads "*BISONS SUM, POLONIS SUBER, GERMANIS BISONT; IGNARI URI NOMEN DEDERANT*". (I am the bison, in Polish *suber*, in German *Bisont*—the ignorant call me *Urus*). *Source* Floeriche (1930)

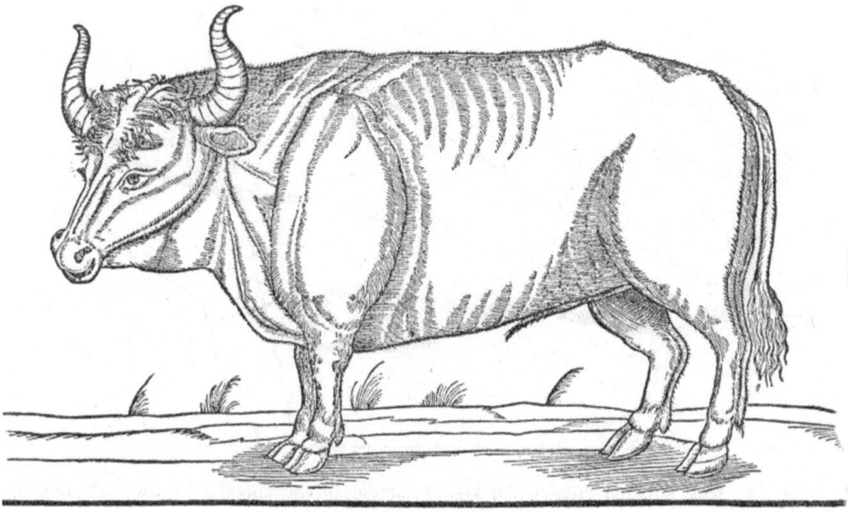

Fig. 6.2 The likeness of an aurochs from the 1556 work by Herberstein. The Latin inscription reads "URUS SUM, POLONIS TUR, GERMANIS AUROX; IGNARI BISONTIS NOMEN DEDERANT" (I am the urus, in Polish *tur*, in German *Aurox*—the ignorant call me *Bisont*). *Source* Floeriche (1930)

and 15th centuries—the aurochs was already confined to the forests of Mazovia and was even then rare and consequently subject to any protection that the then rulers were capable of extending (Fig. 6.3). The tragic situation that the species was facing by the 16th century is made clear by documents still be to consulted in the Central Archives of Historical Records in relation to the inventorying of the area administered by the *starosta* of Sochaczew. We learn from a survey carried out in 1520 that the village of Kozłowce on the edge of the then Jaktorowska Forest was home to "hunters that guarded forest and aurochs alike…", who also "ensured that… hay was brought into the aurochs" (Łukaszewicz 1952). Subsequent inventorying of 1526 and 1564 mentioned that there were 38 head of aurochs. The herd comprised 22 cows, 3 younger animals, 5 calves and 8 bulls

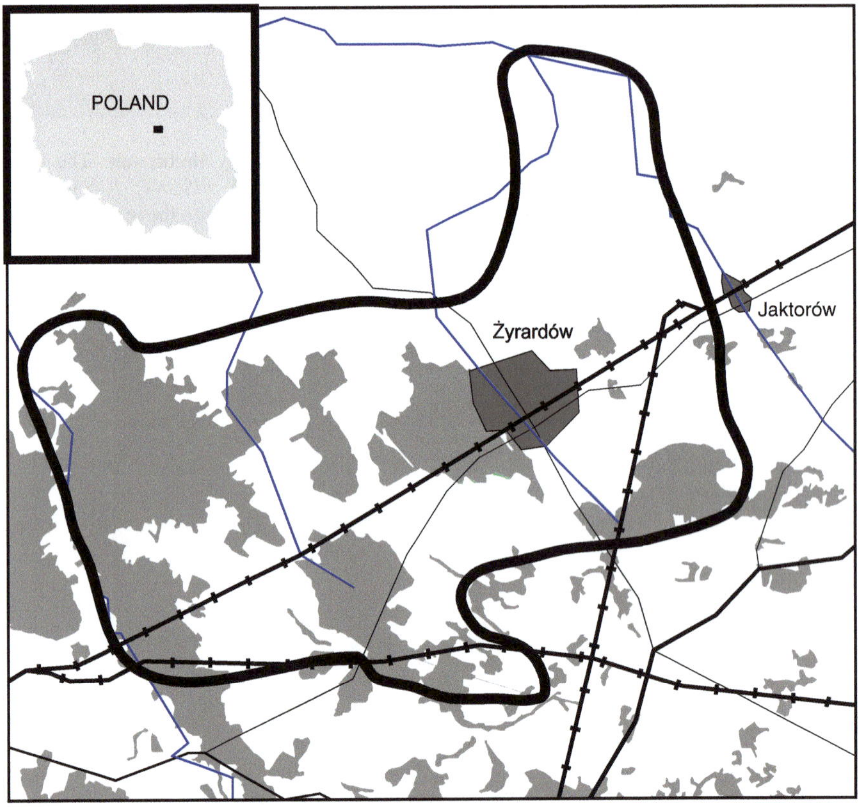

Fig. 6.3 The reconstructed area of the last specimens of the aurochs in Mazovia. The area encompassed the then Jaktorowska, Wiskicka and Bolimowska Forests. Today it retains only much-transformed patches of forest west of Żyrardów, as well as the Bolimowska Forest and Mariańska Forest (*marked gray*). The area around Jaktorów is now highly urbanised and includes the town of Żyrardów. There was no possibility of the aurochs living on here beyond the beginning of the 17th century. Data from Łukaszewicz (1952), van Vuure (2005). *Source* Krasińska and Krasiński (2007)

each walking alone (Łukaszewicz 1952). This description confirms the idea that the animals went around in herds, while males were loners—except presumably at rutting time. The small numbers of adolescents and calves seem noteworthy, and was in part accounted for in the survey, wherein we read that… "there are many herds of horses and cattle also wandering hither and yon at the same time as the aurochs, who thus lack good rearing…". The unfavourable situation the aurochs found itself in gained further confirmation in surveys conducted in 1565 and 1566, 1570 and 1577. By 1599, only 24 aurochs were left alive. The *coup de grace* came with a disease passed on by cattle, such that by 1601 there were just four aurochs left—three males and only one single cow. By that time the fate of the species was sealed, yet the last cow survived in the Jaktorowska Forest until as late as 1627. Two precious souvenirs of the great beast have survived into our times: the horn of an aurochs set in silver from 1532, which was once presented to King Zygmunt the Old by the miners at the famed Wieliczka Salt Mine (and is now in the collections of the Kraków branch of the National Museum), as well as another—again with silver—which has an outside curve length of 47 cm and was used as a hunting horn. The Polish-language inscription around the mouthpiece reads: "Horn of the last aurochs from the Sochaczewska Forest from Voivod of Rawa Stanisław Radziejowski, then Staroste of Sochaczew. AD 1620". This was the horn of a male (Collections of the Royal Armoury in Stockholm).

Ironically enough, the same fate almost met the European bison, as some three hundred years later the species was only saved at the eleventh hour by a near-miracle. And the two great ruminants both survived longest on Polish territory, other countries having exterminated them long before. This attests to the care extended to such "relict" animals in Poland above all.

Although today the epitome of a wilderness refuge, the Białowieża Forest has not thus far been associated with any document making mention of the aurochs, so we may only conclude that the species was not present there, at least not from any time after the beginning of the 15th century.

Chapter 7
The Natural Population of Lowland European Bison in the Białowieża Forest

7.1 In Historical Times

We have no way of knowing how the populations of completely wild bison of old would have functioned. One thing is for sure—the numbers of European bison never approached those of the species' American cousin, otherwise unhelpfully termed the buffalo. It seems likely that numbers of the species present in Europe were always held back by severe winters and the considerable losses they could incur among animals of the youngest and oldest age classes.

A certain selective pressure was also exerted by two carnivores, the wolf and brown bear (Figs. 7.1 and 7.2). Human impacts on the European bison grew

Fig. 7.1 Bison encircled by a pack of wolves. Illustration by A. S. Hronov. *Source* Karcov (1903)

M. Krasińska and Z. A. Krasiński, *European Bison*,

DOI: 10.1007/978-3-642-36555-3_7, © Springer-Verlag Berlin Heidelberg 2013

Fig. 7.2 A bear on the back of a bison. Illustration by N. S. Samokish. *Source* Karcov (1903)

steadily as the centuries passed. What was originally a hunt with bow and arrows to stave off starvation (Fig. 7.3) became a far more effective and indiscriminate culling once firearms came along.

The Białowieża Forest is widely known for its status as the last place in the world in which a wild herd of Lowland (otherwise Białowieża) European bison was able to persist. Yet this was not solely decided by the suitability of the area's natural features, being as much a reflection of the care and protection afforded to the bison by both the Kings of Poland and the Tsars of Russia. The Białowieża Forest was being relatively well-administered and protected by a forest service of the Royal Estates by the 16th century. From the 18th century Białowieża Forest was divided up into 13 sections, each being taken care of by a forest guard, who could count on the assistance of between 5 and 15 riflemen. The Forest as a whole was thus a place of work for some 100–123 riflemen (Jarocki 1830; Hedemann 1939). They kept guard over the Forest and the game within it. Belonging to these sections were the villages located along the forest edge. Peasants originating from these villages were known as beaters (there being some 217–277 of them). They participated in royal hunts as well as performing a variety of different work in the forest, such as scything the meadows and stacking the hay so that it might serve as winter fodder for bison. Both the riflemen

Fig. 7.3 A man with a spear by a dead bison. Illustration by N. S. Samokish. *Source* Karcov (1903)

and the beaters took their wardening of the Forest seriously, not least because the work was associated with privileges and was handed down from father to son (Jarocki 1830; Hedemann 1939). The forest service organised in this way took care, not only of the forest, but also of the game within it, the bison included. The hay designated for feeding the bison was augmented by hay laid out on meadows by the local people and earmarked for their own livestock, in fenced-off stacks or sheds built of wooden boards. As Jarocki (1830) noted: "since bison are perfectly well able to sniff out good hay, they use their horns to break down any obstacles to their getting it".

7.2 Royal Hunts

The Białowieża Forest was the favourite hunting ground for many of Poland's Kings. This status was justified by the sheer abundance of game, as well as by the relative proximity of centres of administration in Vilnius, Kraków and Warsaw. Records of Polish monarchs' hunting expeditions to the bison's "lair" of Białowieża Forest in search of big game survive to this day. King of Poland and Lithuania, Władysław Jagiełło (reigning 1386–1434), treated the Forest as a kind of larder full of all kinds of game. So it was here that he came in 1409—hunting to gather

Fig. 7.4 King of Poland Władysław Jagiełło hunting in the Białowieża Forest (15th c.). Illustration by N. S. Samokish. *Source* Karcov (1903)

supplies for his army in advance of the Battle of Grunwald (Fig. 7.4). The King went into isolation to avoid the plague here in 1426, indulging his passion for hunting at the same time. Around 70 years old by then, the King broke his leg while on a hunt. His grandson, Aleksander Jagiellończyk (1492–1506)—who married Tsar Ivan III's daughter Helena—would ride out from Vilnius or Kraków to hunt in the Forest. In his poem "Song of the Bison", Hussowski (1523) mentioned how a 1504 hunt for bison resulted in damage to a bower in which Helena was resting.

"... A tribune on four great posts was raised"
To please spectators keen to see the hunt.
Among the dignitaries assembled was the queen
Along with her retinue of courtiers and ladies in waiting..."

and further...

"... Spooked by the din of their shouts, the bison
Forced the hunters back with all their might,
But one of the beasts, already wounded by a bullet,
Stood firm as his instinct guided him right,
Shaking his head, he glared with eyes that bulged...
... Battered one of the posts with his massive brow,
Uprooting that timber supporting the dais.
The weight of the scaffolding hung in the balance..."
"... the panic this caused was beyond description!..."

By the times of King Zygmunt I "The Old" (1506–1548) the first Forest Protection Act (of 27th February 1538) was passed. In turn, the first fenced-off park known as the "Wielka Kletnia" was probably established in the days of King Stefan Batory (1576–1586) (Fig. 7.5) (Karcov 1903; Zaręba 1962). It was probably from this enclosure that Warsaw and Kraków each received one bison that had

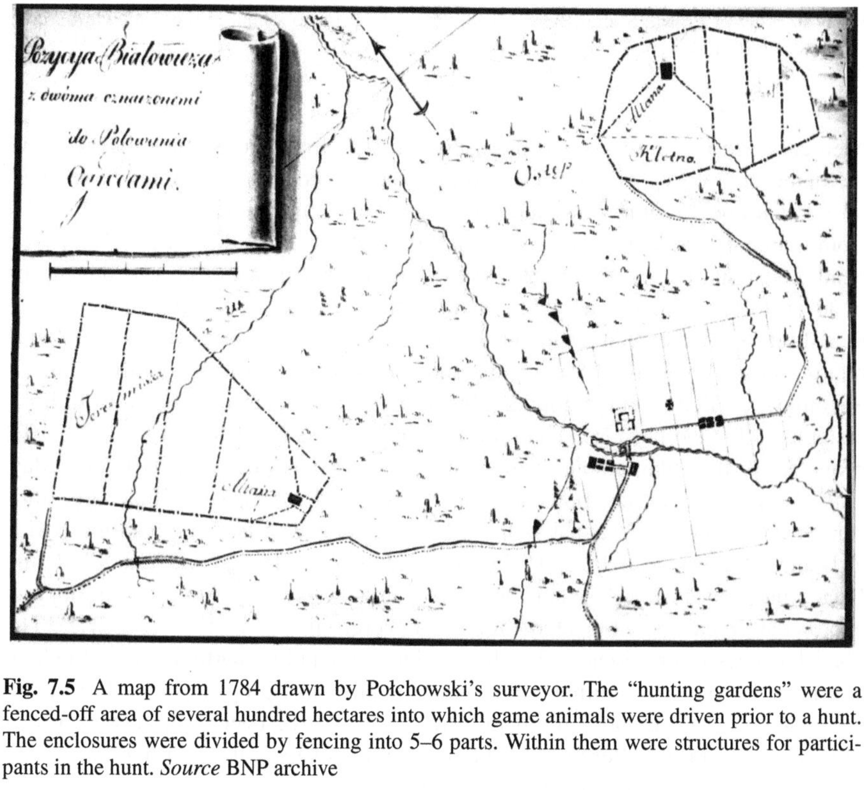

Fig. 7.5 A map from 1784 drawn by Połchowski's surveyor. The "hunting gardens" were a fenced-off area of several hundred hectares into which game animals were driven prior to a hunt. The enclosures were divided by fencing into 5–6 parts. Within them were structures for participants in the hunt. *Source* BNP archive

Fig. 7.6 King of Poland Stefan Batory at the gathering prior to a hunt in the Białowieża Forest (16th c.). Illustration by VN Navozov. *Source* Karcov (1903)

been captured in the forest and then sent off. It is from those times that we have the name of Batory's Hill to describe the place where the King must have taken up his hunting position. The rather low elevation in question is in today's Forest Compartment 449, at the place in which the road from Białowieża to Hajnówka bends most sharply. Stefan Batory went hunting twice in the Forest, during the build up to his military campaign against Pskov (today in NW Russia) in 1581, then three years later (Samojlik 2006) (Fig. 7.6). Jan Kazimierz (1648–1668) went hunting in the Forest in 1650, while a change of fashion ensured that the Kings of Saxony who ruled in Poland opted to shoot at specially corralled game from bowers erected for the purpose. The hunt of King August III of Saxony (1733–1763) provided for the enlargement of the game park existing since Stefan Batory's day, with the aforementioned bowers being constructed there. The whole area thus came to be known as "August's Orchard" (*Augustowski Sad*). The Hunting Lodge at Białowieża was destroyed in 1832, its place being taken by the Tsar's Hunting Palace built on an elevation in the years 1889–1894. King August III's hunting exploits are recalled in the sandstone obelisk erected here in 1752 (Fig. 7.7). The obelisk carries inscriptions in Polish and German detailing participants in the hunt and numbers of game animals taken. Specifically, 43 European bison fell victim back then, 13 moose and 2 roe deer. There were no red deer in the Forest at that time. Happily, in the aftermath of the hunt, much of the game that had been shot was given away to the residents of Białowieża village (Brincken 1826).

Fig. 7.7 The obelisk standing on the embankment between the Białowieża ponds recalls the hunting by King of Poland August III in 1752. In the background is the oldest building in Białowieża—the residence of the Governor of Grodno (Hrodna) dating back to 1845. It was in this building that a hunting pavilion was erected when Tsar Alexander II came here in 1860. The inscription in Polish and German tells how August III killed 42 European bison (among other species), in the course of a day's hunting on September 27, 1752. *Photo* ZA Krasiński

Inscription on the obelisk

«Dnia 27 Septembr. 1752 Nayiaśnieysze Państwo August III, Król Polski, Elektor Saski, z Królową Jeymością i Królewiczem Ichmość Xawerym i Karolem tu mieli polowanie Żubrów i zabili: 42 Żubrów; to jest: 11 wielkich, z których nayważnieyszy ważył 14 centnarów 50 funtów, 7 mnieyszych; 18 Żubryców; 6 młodych; 13 Łosiów; to jest 6 z których nayważnieyszy ważył 9 centnarów 75 funtów; 5 samic; 2 młodych; 2 sarny. Summa 57 sztuk. Byli przytomni: JW. JPan Branicki, Hetman Wielki Koronny, JW. Hrabia de Brühl pierwszy Minister J. K.M., JW. JPan Wielopolski, Cześnik Koronny. J W. Hrabia de Brühl Koniuszy naywyższy Jego Królwskiej Mości, JW. Marszałek de Biberstein, Jeneral poczty Koronney. JW. de Szönberg Marszałek Nadworny J. K. M., JW. i Wiel. JPanowie Poniatowski, Wilczewski, Węgierski, Starzewski, Pułkownicy; i Sapieha Oberszter Lieutenant.

Polowaniem dyrygował JW. JPan de Wolfersdorff Łowczy naywyższy J. K M. Assystowali przy polowaniu: W. JPan de Gablenz Podłowczy J. K. M., W. JPan de Arnim Szambelan. W. JPanowie de Geusau i de Leipziger Paziowie od polowania. Strzelcy nadworni JPan Pflug, Stokmann, Schreyer i Pezold. Nadleśniczowie: Breyter, Bormann, Eckhard. Leśniczowie: Szubert, Angermann, Stockmar, Richter. Eichler, Homburg, 3. Kozłowscy, Wap, Rode, Prokopowicz, Urepkiewicz»

«Den 27 Septemberr 1752 haben Ihro Majestaet August III Koenig in Pohlen und Churfürst zu Sachsen nebst Dero Koeniglichen Gemahlin Majestaet ingleichen Ihrer Königliechen Hoheiten Prinz Xaver und Prinz Karl allhier ein Auerjagen auf dem Lauf gehalten, und geschossen: 42 Auer, naemlich: 11 Haupauer, wovon der schwerste gewogen 14 Centner 50 Pfund; 7 Mittelauer; 18 Auerthiere; 6 Auerkälber. Ferner 13 Elend, naemlich:6 Elendhirsche, wovon der schwerste gewogen 9 Centner 75 Pfund; 5 Elendhiere; 2 Elendkälber. Ferner 2 Rehe. Summa 57 Stück.

Hierbey waren zugegen: Jhro Excellenz Herr Kron-Gross Feldherr Branicki. J. E. Herr Premier Minister Graf Brühl. J. E. Herr Kron-Mundschenk Graf Wielopolski, I. E. Herr Oberstallmeister Graff Brühl. J. E. Herr General Kron- Postmeister Marschall von Biberstein. J. E. Herr Hofmarschall von Schoenberg. Die Herren Obristen Poniatowski, Wilczewski, Wegierski und Starzewski. Der Herr Obrist Lieutenant Sapieha.

Das Jagen haben dirigirt: Ihro Exc. Herr Oberhofjägermeister Graf Wolfersdorff. Beym Jagen haben Dienste geleistet: Herr Kammerjunker und Obersforstmeister von Gablenz. Herr Kammer und Jagdjunker von Arnim, Herr Jagdpage v. Geusau und v. Leipzyger. Jagd Volontair Pflug, Wildmeister Stockmann, Hofjäger Schreier, Leibschütz Pezold. Die Oberfoerster: Breyther, Bermann, Eckhard. Die Foerster: Schubert, Angermann, Stockmar, Richter, Eichler, Homburg; 3 Kozlowski, Wap, Rode, Prokopowicz, Urepkiewicz ».

The hunting session engaged in 1784 by the last King of Poland, Stanisław August Poniatowski (1764–1795), closed the chapter in the Forest's history involving visits by Polish monarchs (Fig. 7.8). In fact, King Stanisław August had stopped off at the Białowieża Forest while en route for the parliamentary session in Grodno. The collections of the University Library in Warsaw include an illustration of a hunting bower on which there is a handwritten note to the effect that it had been built for: "His Majesty the King, Stanisław August…" (Samojlik and Jędrzejewska 2002). We know that there were two "hunting gardens" in the Forest back then: the *Wielka Kletnia* at the mouth of the River Orłówka (today's area of strict protection within the National Park), and the *Teremiska* located between the village of Budy and what is today the main road leading to

Fig. 7.8 The bower built for the visit of King Stanisław August Poniatowski of Poland to the Białowieża Forest on 31 August 1784. *Source* Samojlik and Jędrzejewska (2002)

Hajnówka (Fig. 7.5) (Zaręba 1962). Samojlik and Jędrzejewska (2002) presume that the bower of King Stanisław August presented on the illustration was located in the Teremiska enclosure.

7.3 The Hunts of the Tsars

from Karcov 1903

In the year 1795, in the aftermath of Poland's partitioning from existence by Russia, Prussia and Austria, the Białowieża Forest found itself within the Russian area. Chaos reigned initially where the Forest's administration and management were concerned. Empress Catherine the Great and then her son Pavel opted to give large parts of the Forest over to whichever favourites they might be cultivating at the time. Only later was a stop put to such unplanned activity. In 1802, Tsar Alexander I issued a decree in law requiring the protection of both Forest and bison. Anyone caught poaching a bison would face a fine of 2000 roubles, assuming they could afford to pay. Those who could not had Siberia to look forward to (Jarocki 1830). What remains unsure (since the chronicles of the day are silent) is whether these penalties were ever handed down in practice. What we do know is that Tsar Alexander saw to it that the bison herd was recounted each year (from 1809 on), with signs of growth in numbers being looked for, and numbers and causes of any deaths noted (Karcov 1903). A surprisingly long 65-year period passed with no hunting by monarchs anywhere in the Forest. This meant that Alexander II's announcement of his intention to go hunting at Białowieża necessitated a large amount of preparatory work. The house originally designed for the Governor of Grodno was made over to accommodate the Tsar. Furthermore, a huge 600 ha enclosure was delineated, this later being enlarged to 800 ha (see Fig. 14.11). Members of the forest service reinforced by no fewer than 2000 peasants then proceeded to herd game animals into the enclosure over a period of three weeks prior to the imperial arrival. They managed to squeeze in some 117 bison, 3 moose, 14 fallow deer, 25 wild boar, 36 roe deer, 17 wolves and a number of smaller game animals (like foxes, hares and badgers). Ultimately, the tally when the two days of hunting were over was 28 bison. Two years later, the part of the Białowieża Palace Park now known as the *Osada Zwierzyniec* became home to a cast-iron bison of natural size made in St. Petersburg

Fig. 7.9 The life-size cast-iron bison erected to recall Tsar Alexander II's hunting at Białowieża on October 6 and 7, 1860. Among the animals taken were 28 bison. The statue was placed in a small park close by the hunting settlement, at a place called Osada Zwierzyniec. It was taken to Moscow at some point during World War I, only to return to Poland (Spała, rather than its original location) between the Wars. *Photo* ZA Krasiński

(Fig. 7.9). The accompanying inscription in Russian on the high plinth states that it stands: "In memory of Tsar Alexander II's hunt of 6th and 7th October 1860." Russia's Grand Dukes would in turn come here to hunt in the years 1865, 1875, 1880, 1881 and 1885 (Fig. 7.10).

In 1888, the Forest passed out of state management and into the hands of the Tsar's own estates under the auspices of what was known as the *Udelov* department. The administrators were in the so-called *Udel'noe Vedomstvo*. In this period, the rule of the Tsars over the Białowieża Forest was characterised by a focus on hunting, the main policy aim being to boost numbers of game animals. To this end, the period 1881–1907 brought the transfer into the area of red deer, fallow deer and roe deer from Russian lands and other countries (Karcov 1903). Such activity had to have negative impacts on both game and Forest, and indeed the combination of game management of this kind plus grazing and browsing by several thousand head of cattle brought near-catastrophe for the Forest. The process of natural regeneration was arrested (Wróblewski 1927; Hedemann 1939; Faliński 1968; Jędrzejewska and Jędrzejewski 1998). Yet even supplementary feeding could not stop the local game from going hungry, their debilitation only encouraging the spread of disease, to the point where a degeneration of both bison and red deer was referred to (Büchner 1900).

Tsar Alexander III came here to hunt in 1894, the occasion being the grand opening of the newly-built Palace at Białowieża. However, no fewer than 12 days of hunting yielded "only" 7 bison. Embarrassingly, none of these were actually bagged by the Tsar himself. A more fruitful affair was the hunt of

Fig. 7.10 Russia's Grand Duke Mikhail Nikolaevich obtains the spruce twig as a symbol of a successful bison hunt in the Białowieża Forest (1886). *Source* Karcov (1903)

1897—coinciding with the extension of the railway from Bielsk Podlaski to Białowieża. The 10 days of hunting engaged in on that occasion produced a tally of 37 bison, Tsar Nicholas II accounting for no fewer than 7 of these. We may learn a little about the then distribution of bison in what we today call the Białowieża Forest from the fact that Hajnówka Forest District yielded 18 of the 37, Browsk 11, Królowy Most 5 and Starzyna 3. Tsar Nicholas would pay return visits to the Forest in 1900 and 1903, this time to mark the construction of the strategically important Purana—Białowieża road. The Tsar's last hunt of all came in 1912, while the last crowned monarch of any kind to hunt here did so just a year later—the distinguished guest in question being the then Prince of Monaco, who shot 2 bison. So it was that another epoch of hunting for bison at Białowieża came to an end. Yet the era it gave way to (of the World War I) would in no way leave the area better favoured. Rather, a period of chaos, lawlessness and unrestrained poaching had been ushered in.

7.4 Transfers of Bison

In the 19th century, a habit cultivated by Tsars was to send bison off around the globe as imperial gifts. It seems certain that the bison in question came from the game park area established within the Forest from 1860 onwards. And what is for sure is that it is only thanks to these transferred bison and their descendents that the saving of the species in the first half of the 20th century would prove possible at all. The process began in 1848, when a pair of bison ware sent off to London Zoo, and another to Tsarskoe Selo. An animal collection in Vilnius then received a pair in 1860, while 1861 saw transfers of three animals to a landed gentleman called Leopold Walicki (who was seeking to hybridise them with cattle), as well as of two to an animal collection in Gatchina (Russia). Two more specimens went off the Amsterdam Zoo in 1862, with no fewer than four being sent to the Schönbrunn Zoo in Vienna, as well as 2 to the Paris *Jardin des Plantes*. Three bison were despatched to Moscow in 1864, and four to Pszczyna the following year. Gatchina obtained 7 more specimens in 1867, while 1873 saw 2 individuals each making their way to zoos in Constantinople and Berlin. After a gap until the beginning of the 20th century, a pair was sent to Askania Nova in 1902, 3 animals went to augment the Potockis' menagerie at Pilawin in 1905 and 4 bison were transferred to the Crimea in 1913 (Karcov 1903). European bison sent from Białowieża also gave a start to the Duke of Bedford's breeding centre at Woburn Abbey. Also of particular subsequent importance was the decision to transfer 4 (one male and 3 female) bison to Pszczyna (Pless) in 1865. The bison raised there in isolation over several decades gave rise to the so-called Pszczyna line of bison.

7.5 The 19th Century Abundance of the European Bison

Statistics concerning the dynamics of the Białowieża population in the 19th century and the first ten years of the 20th were included in the 1903 monograph by Karcov, and are based on the reporting engaged in by the then administration in line with devised methods still applied in the period of the First Republic (Hedemann 1939). When snow cover was renewed, forest guards counted the tracks animals left within their own particular forestry sub-districts. The subordinates of each forest guard would then report there to, comparing their observations and passing on information as to the real numbers of bison present. Data coming from each section of Forest were submitted in a report to the person in charge of the forest as whole, whose task it was to present a summary report to the Governor's Office (Jarocki 1830).

The abundance of European bison in the Białowieża Forest can be determined from the administrative data concerning the years 1809–1902 included in Karcov's monograph, as well as data concerning the years 1903–1914 supplied by Wróblewski (1927) (Table 7.1). There are authors who question the reliability of data from any period in which there appear to have been more than 1,500 animals in total. Indeed, the year 1857 brought a peak apparent population of as many as 1,898 animals. However, it needs to be recalled that, in the period between the reigns of the Kings of Saxony in Poland (i.e. the mid to late 18th century) and 1865, there were no red deer in the Forest at all—a situation of eased competition in which a large(r) population of bison could presumably have been present.

Table 7.1 Numbers of European bison in the Białowieża Forest, 1802–1919. Data from: Karcov (1903). Wróblewski (1927)

Number of individuals							
Years	N	Years	N	Years	N	Years	N
1802	200	1838	906	1863	874	1888	?
1809	350	1839	932	1864	?	1889	380
1813	363	1840	817	1865	724	1890	403
1816	483	1841	946	1866	?	1891	479
1817	497	1842	?	1867	?	1892	491
1818	546	1843	984	1868	559	1893	494
1819	504	1844	993	1869	541	1894	496
1820	528	1845	1025	1870	542	1895	561
1821	508	1846	1095	1871	528	1896	561
1822	477	1847	?	1872	?	1897	637
1823	515	1848	1264	1873	527	1898	662
1824	572	1849	1354	1874	536	1899	661
1825	?	1850	1560	1875	558	1900	727
1826	653	1851	1642	1876	561	1901	747
1827	690	1852	1748	1877	559	1902	665
1828	?	1853	1802	1878	565	1903	703
1829	711	1854	1824	1879	574	1904	708
1830	772*	1855	?	1880	579	1905	651
1831	657	1856	1771	1881	574	1906	663
1832	770	1857	1898	1882	600	1907	741
1833	768	1858	1434	1883	592	1908	696
1834	810	1859	?	1884	384	1914	727
1835	845	1860	1575	1885	433	1919	0
1836	858	1861	1447	1886	427		
1837	860	1862	1251	1887	438		

Explanations: ?—lack of data. *—Data from: Eichwald (1830). *Source* Krasińska and Krasiński (2007)

It is significant that the period 1850–1860 also brought reports of more calves born than at any time before—in excess of 150 a year (and no fewer than 170 in 1850).

Once the game park had been built in 1860, both bison and other species were kept there. In relation to the needs of the given moment, there were between 6 and 43 bison present (Büchner 1900; Karcov 1903). This is important to know, since the counts of bison in the Białowieża Forest included those kept for a period in the game park (Table 7.1). The coefficient of births in this period was of 10.3 %—in fact lower than today (seeChap. 10). The official statistics also recorded kills among bison by the Tsar's high-profile guests. In every case, just single individuals here and there were involved. However, on each of the occasions on which the Tsar himself went hunting (in 1860, 1897 and 1900), he and his entourage accounted for between 33 and 45 animals, these being included in a special count category following the hunt (see Sect. 7.3). Bison deaths due to old age or disease were also registered—the numbers involved in the year 1840–1891 being between three and nine annually. Indeed, there were some years in which no casualties at all could apparently be noted (attesting to some imprecision with the recording or compiling of the statistics). Only from the late 19th century onwards did the deaths of larger numbers of bison (up to 50) begin to be recorded, but this undoubtedly coincided with such real events as outbreaks of *Pasteurellosis bovum* disease. In the period in question, the selection pressure exerted on the bison population by carnivores was very much limited. Eight cases of wolves mauling bison were noted in 1833 and 1847, other years bringing just 1–3 such cases annually. This state of affairs was associated with a resolute extermination campaign waged against wolves throughout the period. In the years 1835–1873, twelve cases of bear predation on bison were stated. Several cases of bear predation on bison were noted in last decades in Caucasus, and also four cases in free-ranging population BUKOVYNSKA in east Carpathians (Ukraine), and two other in Bieszczady mountains (Poland) (Paszkiewicz 2008). Moreover, at the beginning of February 2012, a bison calf was killed by a small pack of wolves (M. Januszczak pers. comm), the same fate meeting a 3-year-old individual of the Belarussian BEREZINA population in 1999 (Kashtalian 2009).

7.6 The Extermination of the Lowland Bison

The data from K. Wróblewski (1927) show that there were 727 European bison inhabiting the Białowieża Forest on the eve of the outbreak of World War I. Thereafter, poaching flourished out of control during the war years which, combined with the passage through the area of different armies, ensured a steady reduction in the number of all game animals. Being large and rather easy to track, bison proved a particularly easy target.

The last European bison in the Białowieża Forest was killed in April 1919. A great deal of misunderstanding has arisen in regard to the exact date on which this last bison fell, so renewed reference to the facts needs to be made if the matter is to be accounted for. It all started with the incorrect noting of the date by Wróblewski in his 1927 monograph, which holds that: "*in 1919, there were still 9 animals alive*", while... "*in 1920, just one single female bison was left, she being killed on 9 February 1921 by a former forest gamekeeper, a peasant from Stoczek in Białowieża, one Bartłomiej Szpakowicz*". However, Wróblewski was not present in the Forest area at that time, his statements relying solely on reports sent into him. Nevertheless, in his 1932 study, Wróblewski came back once more to the issue of the date of death of the last bison. He this time cites eyewitness Herman Knothe, with whom he himself had conducted an interview on the subject of the death of the last bison. Knothe confirmed what had been established previously—that no one had seen either bison or traces thereof at any time after April 1919. So it is simply impossible that bison could have lived on unnoticed in the Forest for the two more years to 1921. Thus the legend of Bartłomiej Szpakowicz's role just like a lot of other was created without any reference to the true facts. There is just no evidence whatever to confirm it. The true items of source material in this respect are the reporting of Herman Knothe, who came to the Forest himself in 1919, with a view to acquainting himself with the bison's situation, as well as the administrative report cited by Sztolcman (1924, 1926). Knothe expressed things in the following way: "*in April of that year (1919)—which is to say in the month after my first arrival, I came again to Białowieża... I learnt that several more bison had fallen to poachers...The people were going hungry...so it was nothing surprising that a bison capable of giving such a mass of meat was something very desirable. The rest of the animals died then*". **The last tracks of 4 bison, and then signs of a poached individual, were found on April 4th 1919** (Sztolcman 1926), and these were the last traces of the species to be seen in the Białowieża Forest.

Further items of source material are the reports filed by Prof. Władysław Szafer on his travels to the Białowieża Forest just after wartime activity ceased in 1919. The result of these visits was the initiative to found a nature park (National Park) within the Forest. As a botanist, Szafer was more interested in the overall state of the area's natural environment, although he did also pay some attention to the bison, noting that: "...it started with concern for the fate of the European bison and the desire to determine on the spot at Białowieża whether this largest of Europe's mammals had come through the war". To this end, an expedition comprising Professors Szafer and Eugeniusz Kiernik, as well as Engineer Jan Kloska, had made its way to Białowieża. The impressions from their stay in the Białowieża Forest in **April 1919** were published by Prof. Szafer in *Sylwan* (1919) and in the monthly *Orli Lot* (1920). Writing in Kraków in June 1919, Szafer put it as follows: "*To be found at the site were extensive areas of felling, with bison bones scattered across the Forest. There were no living animals in the Forest... "In the second half of February and March of this year the Forest was subject to such wild, incessant and noisy shooting, that it was a danger to make one's way into the forest, it being so easy to take one or more of the bullets straying from all over the place. That royal beast, half tame as it was, had no means of escape to more far-flung areas in order to save its own life, so it fell to the deadly bullets as it wandered around helplessly in circles, in the areas known to it and to which it had become accustomed... In those first days of April, I walked the Forest, seeing no poachers, but also coming across no living game. During my three-day tramp around the fasts of the Forest, I saw three red deer on one occasion, 1 hare and a crane! The Forest was as good as dead. Apparently the whole Białowieża Forest is still inhabited by several tens of red deer, fallow deer and roe deer, with quite a few boar still left... And so, amidst the turmoil of the greatest war that ever shook the world, amidst the clashing encounters between alien weapons, amidst the blows of German axes bringing down giant trees nurtured by whole generations,* ***there was lost from the fasts and backwoods of Białowieża, the European bison, that king of the Polish forest***".

Nevertheless, Wróblewski's error of 1927 concerning the date on which the Forest's last bison was killed was repeated by Żabiński (1960) in his materials for the Congress of the International Union for the Conservation of Nature and

Natural Resources held at Białowieża in 1960. These were in English and thus widely disseminated, with the result that the same error found its way into the press, articles, and latterly also into the work by Kossak (2000). An attempt to explain the whole affair was made by Okołów (1966).

Finally, it needs to be noted that—in truth—it is now no longer possible to determine the exact date in 1919 on which the last bison died, or the person responsible for the barbarous act involved. Let the name of whoever it was remain in inglorious oblivion.

Chapter 8
The Restitution of the European Bison in Enclosures Within the Białowieża Forest

The years 1920–1928 were those of the European bison's absence from the Białowieża Forest. Precise data on the species' restitution (restoration of the species) at Białowieża in the subsequent period 1929–1952 have been presented in Krasiński (1994a), as well as in Appendix 1 to this monograph. These sources detail all the bison raised in the years in the European bison breeding centre at Białowieża, as well as the state of the population at the end of each year. Also presented are pedigree data on the Lowland bison present at Białowieża. It is possible to divide the bison participating in the restitution of their species at Białowieża into 3 groups in relation to their place of origin (or birth).
Born in:

Pszczyna		Berlin Zoo	Stockholm Zoo	
M 229	PLISCH	F 524 BESTE	F 93	BISERTA
M 546	PLUVIUS II		F 161	BISERTA
F 547	PLESSE			
M 575	PLATO			
F 631	PLARKA			
M 789	PLAMIEC			

In 1929, bison restitution began in Poland when K. Hagenbeck purchased three (1, 2) European bison, as well as four (2, 2) animals that would later be found to be European/American bison crosses. After spending a short time at Warsaw Zoo, the animals were placed in a specially-built enclosure at Białowieża. On September 19, 1929 two bulls arrived at Białowieża—a pure-blood bull called M 163 BORUSSE who had been born in 1927 at Boitzenburg (Germany) to a mother of Białowieża origin and a father characterised by a mixture of Caucasian blood; and a male named Kobold who was a European/American bison hybrid (Fig. 8.1). The next batch of animals were brought to Białowieża on October 10, 1929 and were the pure-blood F 93 BISERTA (born 1924) and the female European/American bison hybrid Faworyta. Both of these animals had been raised at Stockholm Zoo. At the end of August 1930, the herd was joined by pure-blood female F 161 BISCAYA (Fig. 8.2), as well as the female

M. Krasińska and Z. A. Krasiński, *European Bison*,
DOI: 10.1007/978-3-642-36555-3_8, © Springer-Verlag Berlin Heidelberg 2013

Fig. 8.1 Bull M 163 BORUSSE having just left his travelling cage on September 19, 1929. This was the first bison to return to the Białowieża Forest following a 10-year absence of the species. *Photo* J. J. Karpiński. *Source* the BNP archive

Fig. 8.2 Bison in the reserve at Białowieża in 1932: M 163 BORUSSE was brought in on September 19, 1929 and the cow F161 BISCAYA on August 29, 1930. *Photo* J. J. Karpiński. *Source* the BPN archive

European/American bison hybrid Stolce (Groeben 1932; Krasiński 1994). The last of the purchased group of bison—Kobalt—remained at Warsaw Zoo. In 1930, the first bison was born at the Białowieża Reserve—the male M 189 PUHACZ by BORUSSE out of BISERTA. The 2 females of pure Lowland blood (BISCAYA and BISERTA) were joined in 1935 by further arrivals from Stockholm Zoo, i.e. the 6-year-old bull M 186 BJÖRNSON and the 22-year-old cow F 89 BILMA. Alas, these bison in no way fulfilled the hopes invested in them, BJÖRNSON dying on July 18, 1935 in the course of a famous fight with BORUSSE, and BILMA dying at age 26 leaving no offspring.

A reserve for American bison offered by the Polish community in Canada was established in 1934 at Smardzewice near Spała. This reserve offered a suitable home for Białowieża's European/American bison hybrids, whose presence posed a constant threat to the maintenance of a population of pure-blood European bison. All the hybrids were thus transferred out of Białowieża in the years 1935/1936. The last such animal, Huron, was taken to Warsaw Zoo in March 1936 (Krasiński 1968a, 1994a).

The three European bison transferred to Białowieża earliest—i.e. BISERTA, BISCAYA and BORUSSE—were joined by the male M 229 PLISCH brought in from Pszczyna on 17 April 1936, in playing the key role in the restitution of bison to the Białowieża Forest, and to the wilds of Poland in general. The bull PLISCH fathered 45 calves at Białowieża, including 14 (5, 9) that were Lowland bison taking part in the further reintroduction effort (Krasiński 1994a). In contrast, from among the offspring obtained from the two females BISERTA and BISCAYA, only the two females born in 1939 (i.e. F 519 POGANKA and F 520 POLANA) played any part in the further reintroduction work. PLISCH was the father of both of these females, their mothers being BISERTA and BISCAYA, as noted (Krasiński 1994a). The first calf by PLISCH was born out of BISCAYA in 1937 (Fig. 8.3).

Fig. 8.3 Bull M 229 PLISCH and the first calf, F 479 POLKA of the Lowland line, born at the Białowieża reserves in 1937. *Photo* E. Mohr. *Source* the BNP archives

Long-term work then commenced at the Białowieża Reserves with the objective of the Lowland or Białowieża bison being reinstated. As of 1939, there were 16 animals at the site, of which 7 (1, 6) were pure Lowland-line bison and 9 (3, 6) animals of the Lowland-Caucasian line. No fewer than 26 calves were born at the Białowieża Reserve during World War II (years 1940–1944 inclusive). As the German front came ever closer, all the bison were let out of the Reserve (Fig. 8.4). Once the front line had passed through, the recapture effort made it clear that 11 animals had been lost, including those specimens of such value to the restitution effort, BORUSSE and BISCAYA. In the years 1939–1944, in total 25 bison died in Białowieża centre. Thus, by 1944, the Białowieża population was back down to 17 (8, 9), of which 4 (2, 2) were pure Białowieża (Lowland) bison, and 13 (6, 7) animals of the Lowland-Caucasian line (Appendix 1).

The breeding work at the Białowieża centre sought, not merely to build up the population of European bison, but also to remove animals carrying an admixture of Caucasian bison blood. Thus, the last animal of the latter line, M 694 PUSTELNIK, was transferred off to Warsaw Zoo on September 25, 1950. Overall, the post-War period saw some 30–40 animals kept on the Białowieża Breeding Reserves.

Fig. 8.4 As German forces withdrew from Białowieża in July 1944, they opened the gates of the bison reserves and animals emerged from their enclosures. For bison remaining in their enclosures might have been an easy target for marauders or poachers. The red army entered Białowieża on July 17, 1944. The former guards were quickly able to herd the surviving bison back into the reserve. *Photo* Ebeling, Białowieża, July 1944. *Source* Mohr (1951)

8.1 The Białowieża European Bison Breeding Centre

The Białowieża Centre comprises two Breeding Reserves and the Show Reserve. As of 31 December 2010, these Reserves had 43 (14, 29) head of European bison.

8.1.1 The Breeding Reserves

Built in 1929, along the road linking Hajnówka and Białowieża, the enclosed area covered 22 ha and was divided into three enclosures, of which two were small and one large. The whole area was fenced off to a height of 2 m, using a closed fence of 3-inch planks. The few visitors coming in at that time could view the animals from a special platform located in one of the small enclosures. The Reserve was enlarged in 1932, when a new 37-ha enclosure was built on to the north-eastern corner (Fig. 8.5). After World War II, the desire to create still better conditions for the developing breeding centre led to the establishment of another Reserve covering as much as 123 ha, this bordering directly on to the old Reserve and giving a total enclosed area of 203.23 ha. In line with their tasks, both Reserves (nos. 1 and 2) were termed Breeding Reserves (Krasiński 1968b; Krasiński and Krasińska 2001). They represented an enclave among the forests of Białowieża Forest District, comprising parts of Forest Compartments 420BC and 421AB therein.

The complex of Breeding Reserves was divided into 9 enclosures overall, these being separated by corridors 3 m wide, so that any contact between animals from adjacent enclosures might be avoided (Fig. 8.6). The areas of the different enclosures

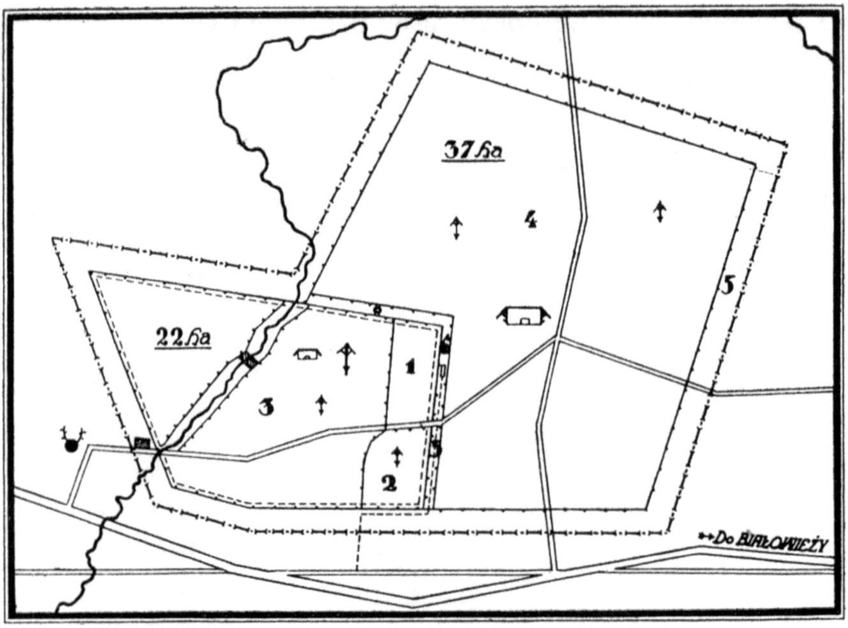

Według Zakł. Dośw. Lasów Państw.

Fig. 8.5 Plan of the 22 ha pens for bison opened in 1929 and enlarged by 37 ha in 1931. *Source* Grochmalnicki (1933)

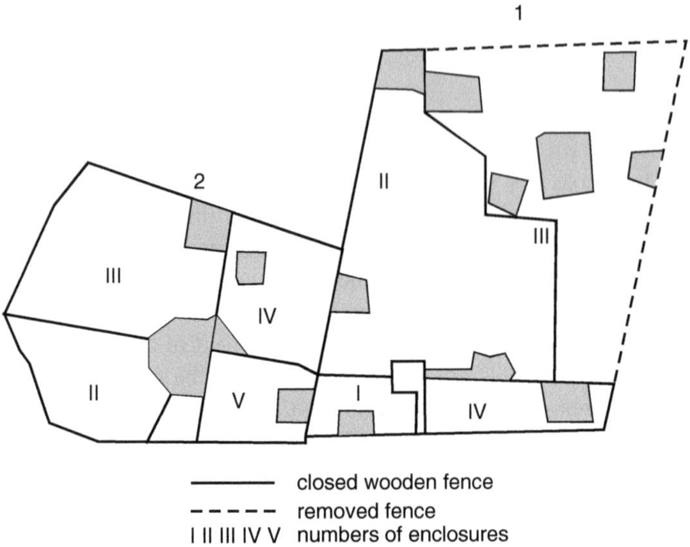

Fig. 8.6 The plan of breeding reserves Nos. 1 and 2 in the Białowieża Forest (as in 2002). Forest clearings are marked in *grey*. *Source* Krasińska and Krasiński (2007)

Insert I. 1, 2 The monograph on the Białowieża Forest published in 1903 in Saint Petersburg. Author: Grigorii P. Karcov

Insert I. 3, 4 Rivals squaring up to each other before fight (*above*), and making first contact with heads well down (*below*). *Photo* ZA Krasiński

Insert I. 5, 6 The aim of each competing bull is to throw the rival over or to achieve contact with the horns. *Photo* ZA Krasiński

Insert I. 7, 8 A drawing of a European bison in the Altamira Cave (Spanish Pyrenees). The length of the original was 205 cm. It dates back around 14,000 years. *Photo*: A Leroi-Gourdan. The likeness of a bison from the Font-de Gaume Cave (Dordogne, SW France). Painted in a dark brown colour and engraved, it is 110 cm long and dates back some 14,000–14,500 years. *Photo*: P. Paillet

ranged from 3.5 do 80 ha. In 1988, a 42 ha area was separated off from Reserve no. 1, this area being designated for the supplementary feeding of the free-ranging bison herd.

Within the Breeding Reserve area, mixed and fresh deciduous forest accounts for some 39.5 % of the total, while wet deciduous forest and alderwood habitats account for 56.5 %. The three dominant trees are: alder 42.6 %, spruce 34.2 % and oak 16.6 %. The large share taken by alder reflects the flow of forest streams. The forest here renews itself naturally, the only forestry measures pursued involving sanitation felling. The area of clearings was once as much as 41.62 ha, but this has declined to just 10 ha, since some of the original open area was left beyond the fenced-off parts of the Reserves. Each enclosure has sites for the provision of food as corn, as well as hay racks and a water trough.

8.1.2 The Show Reserve

After it was built, this Reserve gained the name of the Tarpan Reserve on account of the work then being done here to recreate the race of wild horse known as the tarpan. The area was then renamed Reserve no. 3, before becoming known as the Show Reserve, since the main aim here has consistently been the presentation of animals associated with the Białowieża Forest. At times the name European Bison Show Reserve has also been used, on account of the fact that the most important species on show is—not unnaturally—the European bison. On February 9, 1936, an initiative launched by Prof. T. Vetulani bore fruit in the arrival at Białowieża of 6 primitive horses purchased in eastern Poland and placed in a temporary enclosure within the then forest sub-district of Gródki. With a view to the Polish horse (konik) being bred, a breeding station was constructed within the

Fig. 8.7 The bull M 816 POMRUK was one of the first two bison to be released back into the wild in 1952, having been born on May 22, 1951 in the BNP Breeding Reserves. His parents were M 229 PLISCH and the cow F 520 POLANA. The release date was September 13, 1952, but recapture for the Show Reserve—where he would spend 10 years—took place on August 15, 1955. *Photo* ZA Krasiński

Białowieża Reserve, horses being introduced into it in July 1937. Under the direction of Prof. T. Vetulani, it was planned to recreate the original forest tarpan in the natural conditions of the Białowieża Primeval Forest, using appropriate selective breeding measures.

The Show Reserve is located in the Forest District of Białowieża, within the sub-district of Krzyże. It is 5 km from Białowieża itself, and 750 m off the road leading from Białowieża to Hajnówka. The Reserve covers 27.76 ha and is located in Forest Compartment 425C, as surrounded by the forests of Białowieża Forest District. Bordering on to it from the east is Reserve No. 4, also known as the experimental reserve. In 1955, an enclosure was delineated within the Show Reserve, the aim being to make this accessible to tourists wishing to see bison. The bison were at that time located close to the Reserve entrance, and they could be viewed from the platform built for the purpose. It was this enclosure that for 10 years held the pair M 816 POMRUK (Fig. 8.7) and F 909 POCIECHA. The Reserve was remodelled in the years 1979–1988, over which period it came to assume its present shape (Fig. 8.8). Four enclosures are now given over to bison, these being located centrally and covering a total area of 6.24 ha. Other enclosures now house wolves, red deer, roe deer, wild boar and the bison–cattle hybrid.

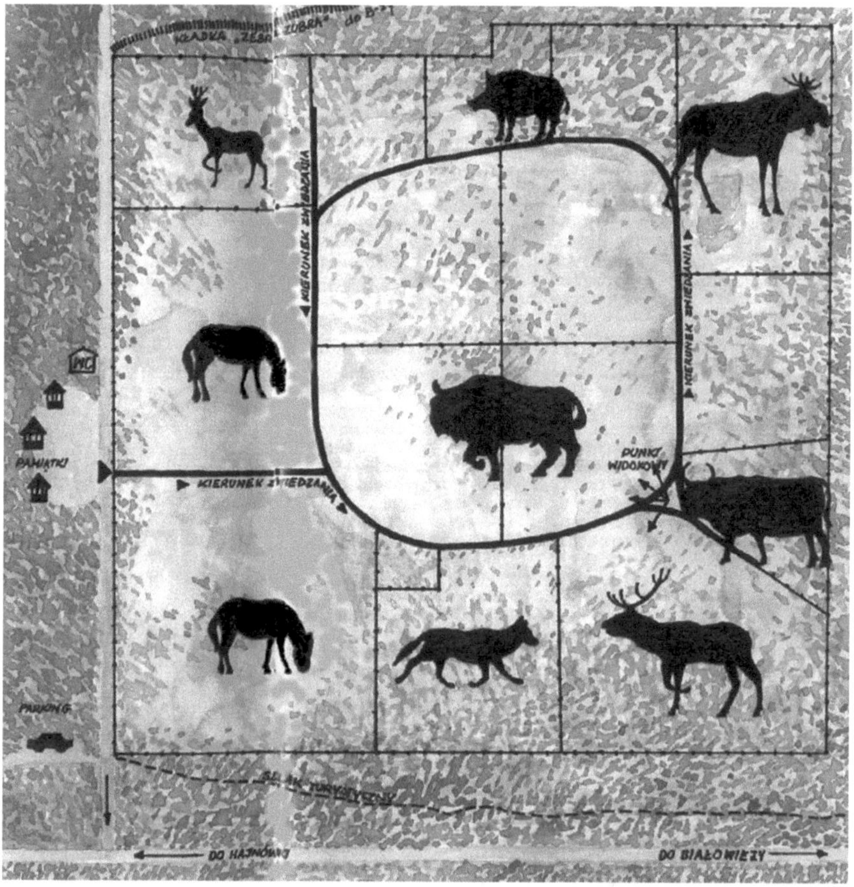

Fig. 8.8 Plan of the show reserve at Białowieża. Illustration by L. Nowacki. *Source* Krasiński (1999b)

The European bison is nevertheless the principle source of interest for tourists visiting the Show Reserve, and it has been necessary to reconcile the expectations of these visitors with the Reserve's possibilities. A valuable feature of the Show Reserve is its presentation of bison in their natural Białowieża Forest environment. For many years the habit at the Show Reserve has been to display a group of bison comprising one large male, plus 4–6 females with calves. While this does not fully meet the expectations of tourists, a larger group of bison would rapidly lead to further degradation of an already impoverished vegetation. The Show Reserve as it exists currently should not hold a group of more than 10 bison (Krasiński 1999b).

Since the Białowieża Forest area boasts both captive and free-ranging herds, the possibility remains open for individuals to be exchanged between the two forms of breeding, in justified cases.

The captive breeding facilities at Białowieża are sufficient to allow four breeding groups to be maintained: three in the Breeding Reserves and one in the Show Reserve. A breeding group comprises individuals of reproductive age plus calves. It should be made up of one male, between 2 and 7 cows and their young. There should be no more than 35 bison in the captive-breeding population overall, and the experience to date suggests that the optimal number is 25–30.

Chapter 9
European Bison in the Białowieża Forest: A Characterisation of the Environment

The fates of the bison and the Białowieża Forest in which it lives have been linked inseparably throughout their existence. The fact that it was here, rather than anywhere else, that European bison clung on longest in the wild is something we owe, not only to the exceptionally rich and diverse forest, but also to the tender loving care extended to both by the Kings of Poland. Yet, the Forest was not spared a number of different catastrophes and wars, and the state it finds itself in today is also a reflection of the improper game management pursued here in the 19th century, overexploitation of the forest in the 1920s and in the first decades after World War II, and the grazing of cattle that continued here into the 1970s. The changes that the Forest has undergone have been reflected unfavourably, not only in the tree stands themselves, but also among the animals (most especially the ungulates) those stands sustain. The Forest has nevertheless retained its original character at least in part, making it possible for a wealth of plants and animals to survive here up to the present day. Eliminated from the area at the beginning of the 20th century, the European bison was saved from total extinction by some minor miracle, and has proved capable of re-occupying the fasts of its Primeval Forest home. And there is indeed no way of describing the life of the bison without discussing—even if briefly—the environment in which it exists, of the Białowieża Forest.

M. Krasińska and Z. A. Krasiński, *European Bison*,
DOI: 10.1007/978-3-642-36555-3_9, © Springer-Verlag Berlin Heidelberg 2013

9.1 Location Size and Climate

The Białowieża Forest is a contiguous forest complex covering 1,500 km^2 (Fig. 9.1), at least one that is contiguous except for its being bisected by the Poland-Belarus state border. The eastern (Belarussian) part is larger, covering 875 km^2, while the western part located in Poland covers 650 km^2. The Forest extends for 55 km in the East–West direction and 51 km from north to south. The area is delineated by the coordinates 23° 31′–24° 21′ E and 52° 29′–52° 37′ N (Sokołowski 2004). It is within the lowlands belt of Eastern Europe. The terrain is almost entirely flat and the average altitude is 170 m a.s.l. The larger northern part of the Forest comes within the drainage basin of the Narew River, while the southern part is drained by the River Bug. The main rivers in the area itself—the Narewka, Narew and the Leśna, are augmented by a number of smaller rivers and streams, as well as many periodically active watercourses taking away the spring meltwaters (Sokołowski 1998).

The Białowieża Forest is characterised by the presence of a lowland type of forest in which a major role is played by deciduous trees (i.e. those shedding their

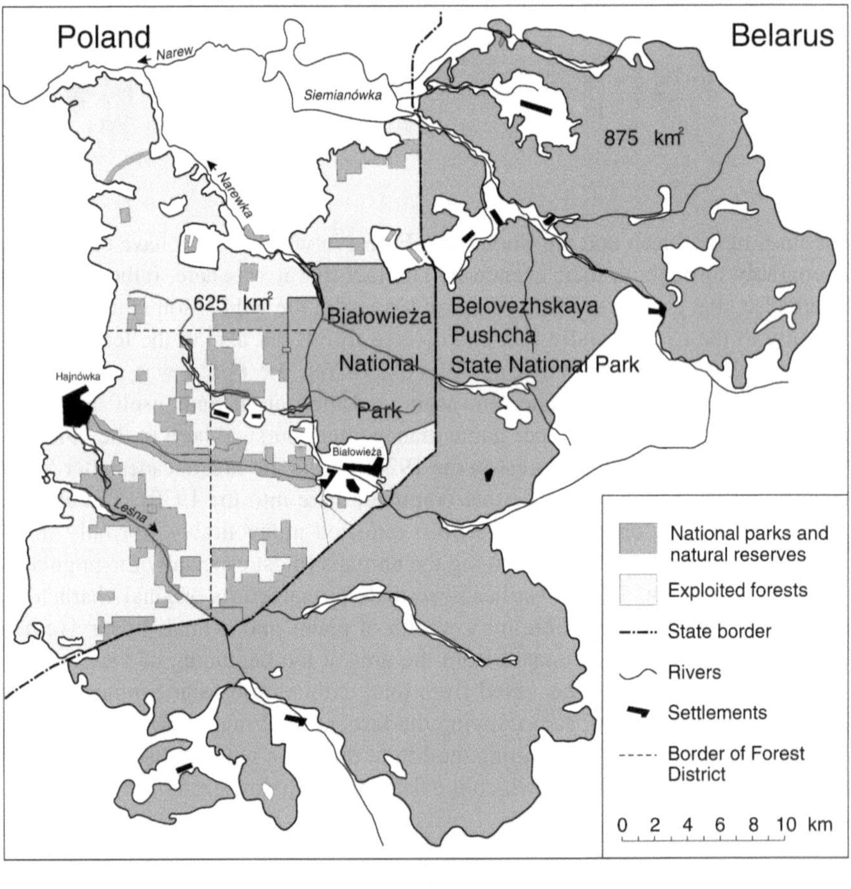

Fig. 9.1 Map of the Białowieża Forest. *Source* Krasińska and Krasiński (2007)

leaves in winter), alongside evergreen coniferous species. The most important kind of forest is the oak, lime and hornbeam associations, while oak, spruce and Scots pine associations are also common. There are also some pure pine forests. Although accounting for a small share of the total, the spruce forests here are also of significance. Compared with forests elsewhere in Poland, the Białowieża Forest stands out for its large share of deciduous and mixed stands, including some of primary origin that are characterised by a diverse structure in terms of age, layering and the species present (Faliński 2001). Some 44 % of the total forest area in the Białowieża Forest comprises old stands of natural origin (i.e. not planted). The greater part of these stands are in the central part of the Forest area, which now enjoys strict protection within the Białowieża National Park, as well as in the adjacent and analogous area on the Belarussian side of the border (Fig. 9.1). Other remnant stands of a natural character persist in the north-eastern and south-western parts of the Belarussian part. There are far fewer such stands in the Polish part of the Forest. Outside the area under strict protection within the National Park, natural stands account for less than 30 % of the forest area, and are dispersed (Jędrzejewska and Jędrzejewski 1998).

The Białowieża Forest is within the zone of temperate climate. Most winds here blow from the SE (i.e. are south-easterlies) or the SW (south-westerlies). Analysis of the Forest climate over the period 1780–2000 reveals a mean annual air temperature for Białowieża of 7 °C. The coldest month is January, in which the mean daily temperature is −4.8 °C (range −16.8 to −1.8 °C in different years), while the warmest is July—with an average of 18.4 °C (15.5–22.5 °C). Mean annual rainfall in the years 1948–2000 was 631 mm, which ranged from 425 to 940 mm (Jędrzejewska and Jędrzejewski 1998). The period of the last 200 years has not presented any discernible directional trend in mean annual temperatures, although there has been a reduction in the amplitude of mean July and January temperatures.

Snow cover in the area persists for 90 days on average, although the duration varies markedly from year to year. On average, there is a permanent covering of snow from 17th December onwards which melts by 12th March. In the years 1960–1991, the number of days of snow cover ranged from 38 (in winter 1974/1975) to 130 (in winter 1969/1970). The average thickness of the layer of winter snow is 17 cm, as compared with a maximum of 95 cm in winter 1969/1970. The years 1971/1972–1985/1986 were characterised by a downward trend for the mean number of days with snow cover, when set against the figures for the period 1955/1956–1969/1970 (Olszewski and Żarnowiecki 1992). While the long-term climatic cycle for the Białowieża Forest shows no directional changes, the differences from year to year are of significance where the behaviour of ungulates mammals is concerned, in that they influence the food base, the spatial structure of the population and indices for mortality.

Over the 50-year period of development of the population in question (1952–2002), neither the mean annual temperature (of 7.2 °C) nor the mean annual precipitation total (620 mm) differed from the long-term average figures. Major departures in winter (December–March) conditions were apparent, however. While the winters in the period in which the bison population was just taking shape (the 1960s and 1970s) were only rarely warmer or characterised by smaller amounts of snow, those in the most recent 10-year period of 1992/1993–2001/2002 included just one single more severe example (winter 1995/1996, during which the mean daily temperature was of just −6 °C, while snow lay for 122 days at an average thickness of 25 cm). The remaining 9 winters in that period had average daily temperatures in the range +0.25 to −2.36 °C, while snow lay for just 24–65 days, and was of limited thickness (only 0.6–9 cm on average).

It is obviously also a matter of importance for the vegetation when the last ground frosts occur in the Forest in spring. Different forest communities in BNP have their last frosts at different times throughout May or even into very early June. In turn, the first frosts of autumn have already arisen by 15th September. The period without frosts thus varies in length from community to community, across the range 114–158 days (Faliński 2001).

In line with climatic criteria (the number of days of temperature >5 °C), the growing season in the Białowieża area lasts 206 days on average. Phytophenological criteria (relating to the development of seasonal vegetation) yield a much shorter period of 185 days, however (Faliński 2001). Trees and shrubs begin to green up in March, but leaves are only fully developed in May and June. By September, the leaves are in turn taking on their autumn colours, only to have fallen completely by the end of October–early November. It is generally in autumn that the seeds of deciduous trees mature (Jędrzejewska and Jędrzejewski 1998). Those of the oak (i.e. acorns) are of special significance to ungulates, but they are not generated in the same quantities in each successive year, being rather subject to "masting" every 6–9 years. They are taken very readily by bison when present in abundance.

The vegetation of the forest floor begins to appear first in the oak-lime-hornbeam forest (especially the fresh and humid forest types)—mostly in the 2nd or 3rd ten-day periods of April (Falińska 1973). This coincides with the final disappearance of any more persistent snow cover (Olszewski 1986). However, the latest possible time for the emergence of forest-floor vegetation may fall in May. Typically, though, the first flowers are present amidst the ground cover in April, flowering continuing for different periods of time depending on the forest community involved. It is in the bison's favourite habitat of deciduous forest that flowering of successive forest-floor herbs and shrubs goes on longest—even for a period of 5–6 months. Flowering is followed by fruiting, which reaches a peak in late June and July. Seeds are dispersed in the first and second 10-day periods of August, in September in broadleaved forest only. By late autumn and obviously in winter, there are very few forest-floor species still presenting green leaves in either deciduous or mixed forests (Falińska 1973). Indeed, the first falls of snow bring an abrupt end to the growing season for the forest-floor vegetation as a whole.

9.2 The Western (Polish) Part

The Polish part of the Forest takes in the Białowieża NP—currently of 105 km^2, plus managed forests under the administration of the 3 Forest Districts of Białowieża, Browsk and Hajnówka and privately owned forests.

The total area is of some 650 km^2 (Gątkiewicz and Zbrożek 2003; Sokołowski 2004). The Białowieża NP includes an area under strict protection extending across some 47.5 km^2 of the central part of the Forest. Since its establishment in 1921, this area has not been subject to any forest management or hunting activity at all. In turn, the part of the Forest under State Forests management includes no fewer than 21 Nature Reserves covering a total of 120.1 km^2 (Sokołowski 2004).

Forests cover over 96 % of the Polish part of Białowieża Forest with the remainder covered by meadows (0.1 %), fields, wastelands and roads and forest clearings (0.2 %). Since the end of the 1970s, the Forest meadows have not been utilised, and so are gradually becoming subject to the processes of secondary forest succession.

Even the forests utilised economically are dominated by habitats in which broadleaved species dominate, being either mixed deciduous forest or fresh and moist deciduous forests. Together these account for 61 % of the forest area, with alderwoods taking 13 %. The coniferous forest habitats are in turn dominated by fresh mixed coniferous stands (over 17 % of the area) (Gątkiewicz and Zbrożek 2003). The composition of different tree species within the Polish part of Białowieża Forest is Norway spruce 25.3 %, alder 19.6 %, Scots pine 27 %, oak 11.9 %, birch 8 %, hornbeam 3.5 %, ash 2.8 %, lime 0.7 %, aspen 0.8 % and maple 0.4 % (Sokołowski 2004). The part of the Forest included within the Białowieża National Park has a slightly higher share of deciduous (mixed deciduous, fresh and moist deciduous) forests (65 % of the forest area), while alderwoods comprises another 15 %. Among the coniferous forests, the prevalent fresh mixed coniferous forest type occupies an area that is proportionally smaller (at 8 %) than is the case for the forests utilised economically. The composition of different tree species within the National Park is Scots Pine 21 %, Norway spruce 18 %, alder 17 % and oaks 16 %, as well as hornbeam 10 %, birches 6 %, ash 6 %, and aspen, limes and maple at about 2 % each. While the high proportion of Scots Pines seems noteworthy, it primarily reflects the 1996 incorporation into the Park of the Hwoźna Protective Forest, in which stands are mainly of pine. Overall, the Polish part of the Białowieża Forest is 25 % coniferous, with 61 % under deciduous broadleaved forests and 14 % in alderwoods (Sokołowski 2004). The Polish part is traversed by two larger rivers in the shape of the Narewka and Leśna, as well as a host of smaller watercourses. Artificial watering places dug out in the 19th century also occur.

A major influence on the plant cover in the Forest is of course exerted by the animals living in it. Until the early 1990s, the Polish part of the Forest sustained a high-density population of ungulates. The hunting off-take was doubled in the mid 1990s, however, and populations declined as a result. As of the beginning of 2012, the population of ungulates in the Polish part was estimated (by means of tracking and driving) at: c. 3,672 red deer, 1,120 roe deer, 80 moose and 4,280 wild boar (information from the Regional Directorate of the State Forests in Białystok). This compared with a figure of 481 for the number of bison in the Forest at the end of 2011 (Table 9.1).

Table 9.1 Numbers of ungulates in the Białowieża Forest at the beginning of 2012, only for strictly protected area of BNP from 2009/2010

Species	Number of individuals			Density (individuals/km^2)	
	Polish part of the Forest	Belarussian part of the Forest	Whole Białowieża Forest	Polish part of the Forest	Belarussian part of the Forest
European bison	481	428	909	0.7	0.5
Red deer	3,675	2,139	5,814	5.7	2,5
Roe deer	1,120	660	1,780	1.7	0.8
Moose	80	196	276	0.1	0.2
Wild boar	4,280	1,553	5,833	6.6	1.8

Data from: Regional Directorate of the State Forests in Białystok, Białowieża National Park, Belovezhskaya Pushcha State National Park in Belarus. Source: Krasiński and Krasińska (2007), updated

9.3 The Eastern (Belarussian) Part

In 1991, the whole Belarussian part of the Forest was granted the status of National Park under the name "State National Park Belovezhskaya Pushcha". This covers 850 km^2, while its zone of strict protection in which no activity other than scientific research can be engaged in has an area of 157 km^2 (Sokołowski 2004). In the Belarussian part, forest covers 89 %, while the non-forest land share is 11 %. The most common tree stands are those with a predominance of coniferous species, which account for 69 % of the forest area. Deciduous forests occupy only 12 % of the forest area and alderwoods 19 % (Kozul'ko and Zhukov 1999). The northern and southern areas have fresh coniferous forest stands, while the central area is covered in mixed coniferous and deciduous forests. Open land accounts for just 2 % of the area (comprising fields, forest glades and meadows). The latter nevertheless offer an important supplement to the feeding base for ungulates (Kozul'ko and Zhukov 1999; Bunevich 2003c).

The waterways in the Belarussian part of the Forest have been changed by drainage, with the central and southern parts having dried out most. Drained marshes have tended to become meadowlands. The drainage channels that have brought this about extend for 140 km in total, while the Forest here has also gained artificial reservoirs, of which "Lyackie" in Forest Compartments 647–649—which was created in 1965—covers no less than 253 ha. Another artificial body of water located in Compartments 677 and 679 covers 63 ha, while a third—the "Pererov" Reservoir in Compartment 589—extends over an area of 25 ha. All other man-made bodies of water have areas ranging from 2 to 10 ha.

The largest unregulated river is the Narew, a 12 km section of which flows through the Forest. The Forest is also crossed by a regulated 12 km stretch of the River Narewka, as well as by 18 km of the Leśna and 11 km of the Biała. There are also several smaller rivers (Jurkevich and Geltman 1965).

As of the beginning of 2012, the Belarussian part of the Forest supported 428 bison, 2,139 red deer, 660 roe deer, 196 moose and 1,553 wild boar (Table 9.1) (unpublished data of the Belovezhskaya Pushcha State National Park). In comparison with the situation as between 1960 and 1980, this represents fewer red and roe deer.

Chapter 10
The Restitution of a Free-Ranging Bison Population in the Białowieża Forest

B. Bańka

In the aftermath of the World War II, the Białowieża Forest became bisected by a state border, the western part remaining in Poland, while the eastern part found itself in the USSR (within the Belarussian SSR). The successful development of the reserve breeding of European bison after the War—both in Poland and in other countries—bore fruit in an increase in the world population of the species and allowed for the stepwise shift into the next phase of the bison's restitution—the creation of a free-ranging population. The first experimental reintroduction therefore took place in the Białowieża Forest, as homeland of the Lowland bison, and the place in which it had persisted for longest in the wild state.

Poland and Belarus set to work establishing two separate free-ranging populations in their respective parts of the Białowieża Forest, although they did not entirely let the existence of the border get in the way of a respect for the fact that the natural environment could not be divided. They thus strived to devise joint principles of action indispensable in the restitution and protection of bison in the Forest (see Chap. 21).

Nevertheless, the developing populations of bison effectively made no contact with each other, remaining isolated on account of the species' sedentary lifestyle. Only very occasionally was the migration of single bulls across the state border observed. However, the 1981 construction of a fence along the border put a stop to even this limited natural exchange of specimens between the populations.

M. Krasińska and Z. A. Krasiński, *European Bison*,
DOI: 10.1007/978-3-642-36555-3_10, © Springer-Verlag Berlin Heidelberg 2013

As of the end of 2011, the two parts of the Forest were inhabited by 909 bison (481 in the Polish part and 428 in the Belarussian). This is a larger number than that of the last population to live wild in the Forest, at the beginning of the 20th century. Furthermore, it is a population large enough to guarantee the retention of whatever genetic diversity remains.

10.1 The Population in the Western (Polish) Part of the Forest

10.1.1 Year of Establishment, Origins of Bison and Founder Group

The establishment of a free-ranging population in the Białowieża Forest proceeded under specific conditions. All the animals derived from the Reserves at Białowieża, so there was never any need to translocate any in. Animals earmarked for release were simply let out of the Reserves by opening the gates. The first two bison identified for release were year-old males (M 816 POMRUK and M 817 POPAS; Fig. 10.1) that had been placed in a large (80 ha) enclosure in May 1952, and had become accustomed to finding their own food. They were let out into the forest on September 13, 1952. The years from then onwards until 1957 were in effect a trial period, the objective being to study the capabilities of released animals when it came to surviving under natural conditions.

Fig. 10.1 Bull M 817 POPAS was one of the first European bison to obtain its freedom. While bull M 816 POMRUK went around on his own, POPAS tended to keep together with the other bison. In April 1957, he was captured for the Reserve, before being sent to Poznań Zoo in September 1957. POPAS was not included within the founding group of the Białowieża population, since the first calf thereof was only born in 1957. *Photo* ZA Krasiński

Table 10.1 Size of the free-ranging population of European bison in the Polish part of the Białowieża Forest during the period of reintroduction (1952–1966)

Years	Number of introduced animals			Number of animals born			Number of animals caught or dead[a]			Number on 31 December
	n	Males	Females	n	Males	Females	n	Males	Females	
1952	2	2	–	–	–	–	–	–	–	2 (2, 0)
1953	3	1	2	–	–	–	–	–	–	5 (3, 2)
1954	–	–	–	–	–	–	–	–	–	5 (3, 2)
1955	1	–	1	–	–	–	2	1	1	4 (2, 2)
1956	–	–	–	–	–	–	1	1	–	3 (1, 2)
1957	3	1	2	1	–	1	1	1	–	6 (1, 5)
1958	10	4	6	2	2	–	–	–	–	18 (7, 11)
1959	5	–	5	6	5	1	1[a]	–	1	28 (12,16)
1960	1	–	1	6	3	3	1[a]	1	–	34 (14, 20)
1961	–	–	–	11	5	5	1[a]	1	–	44 (19, 25)
1962	5	4	1	8	4	4	1	1	–	56 (26, 30)
1963	–	–	–	13	7	6	1[a]	–	1	68 (33, 35)
1964	5	3	2	16	10	6	1[a] + 1	2	–	87 (44, 43)
1965	4	–	4	15	9	6	4[a]	1	3	102 (52, 50)
1966	1	–	1	20	6	14	3[a] + 1	3	1	119 (55, 64)
Total	40[b]	15	25	98	51	46	19	12	7	

[a] Number of dead animals
[b]Two animals caught and later re-released. *Source* Krasińska and Krasiński (2007)

August 15th, 1955 saw the recapture of POMRUK, who had proved to be too aggressive; while the young PODAREK was removed from the free-ranging herd after he moved beyond the limits of the Forest. Further bison were released in 1957: the 8-year-old male POLEL, plus two cows (Appendix 2, Tables 10.1 and 10.2). In 1958, the largest group was let out of the enclosures, comprising 10 young individuals aged 1–3 (4 males, 6 females). These were followed by five more in 1959. The first calf was born to the free-ranging herd in 1957, although the mother had in fact been covered while still in the Reserve. The first fatality was in turn noted in 1959 (Table 10.1), but it was also from that year on that it was possible to note dynamic development of the population in the wild state in the Białowieża Forest (Krasiński 1967, 1968a, 1983).

If we exclude the animals released but then recaptured (and hence not participating in breeding), the group founding the Białowieża population can be taken to comprise 28 European bison: 9 males and 19 females (Table 10.2, Appendix 2). The pedigrees of five of the males introduced (POLEL, POLAS, POLONEZ, PORUBANEK and POMNIK) feature the bull M 229 PLISCH in both their paternal and maternal lines. The remaining four (PORANEK, POCZET, POŚCIG and POTOP) have this same ancestor in either the paternal or maternal lines, but not both. In turn, the pedigrees of 10 of the introduced females include the bull PLISCH in both parental lines, while the remaining eight have him in one line or the other. Only the female F 912 KAMIONKA derived from the line of bull M 575 PLATO.

Table 10.2 Group of founders of the free–ranging population in the Polish part of Białowieża Forest (Lowland bison released from: captive breeding centres at Białowieża in the years 1952–1966, which took part in the restitution)

Males				Females			
No.	Pedigree number	Name	Year of birth	No.	Pedigree number	Name	Year of birth
1	762	POLEL	1949	1	735	POJATA	1948
2	980	PORANEK	1955	2	812	PORĘBA	1951
3.	985	POLAS	1955	3	827	POGOŃ	1951
4	1,019	POCZET	1956	4	831	POZŁOTA	1951
5	1,022	POŚCIG	1956	5	910	POGODNA	1953
6	1,202	POTOP	1959	6	912	KAMIONKA	1953
7	1,225	PORUBANEK	1959	7	940	POŁUDNICA	1954
8	1,231	POLONEZ	1959	8	1,025	POWABNA	1956
9.	1,301	POMNIK	1960	9	1,028	POHULANKA	1956
				10	1,032	POGARDA	1956
				11	1,036	PONOWA	1956
				12	1,072	POSAŻNA	1957
				13	1,085	PODKOMORZANKA	1957
				14	1,137	PODOLANKA	1958
				15	1,140	POCIESZNA	1958
				16	1,144	POGRANICZNA	1958
				17	1,145	POMYŁKA	1958
				18	1,146	PODOBNA	1958
				19	1,387	POROHA	1961

Source Krasińska and Krasiński (2007)

What proved to be a rather ineffective move was the introduction of 8–17 year-olds over the period 1964–1968, the older males in particular not always showing a readiness to join herds, but simply going on to adopt a solitary lifestyle (Krasiński 1976, 1983). Rather, it was young bison aged 1–3, released in line with a 1:1 male:female ratio, that emerged as the most beneficial in creating the free-ranging population.

10.1.2 Population Dynamics

Two calves were born in 1958, and subsequent years were characterised by sufficient natural growth of the population to provide for a rapid increase in numbers. Indeed, by the 1970s (when the population exceeded the 200-animal threshold once more), the herd consisted solely of individuals born in natural conditions (Krasiński 1967, 1978a).

The growth of the free-ranging population of bison in the Białowieża Forest from the early 1960s onwards was exponential in nature (Fig. 10.2). By 1965, the population had exceeded 100 individuals, and by 1970—200. There were already 250 bison in 1973 (Appendix 3). A stabilisation of population size began to take

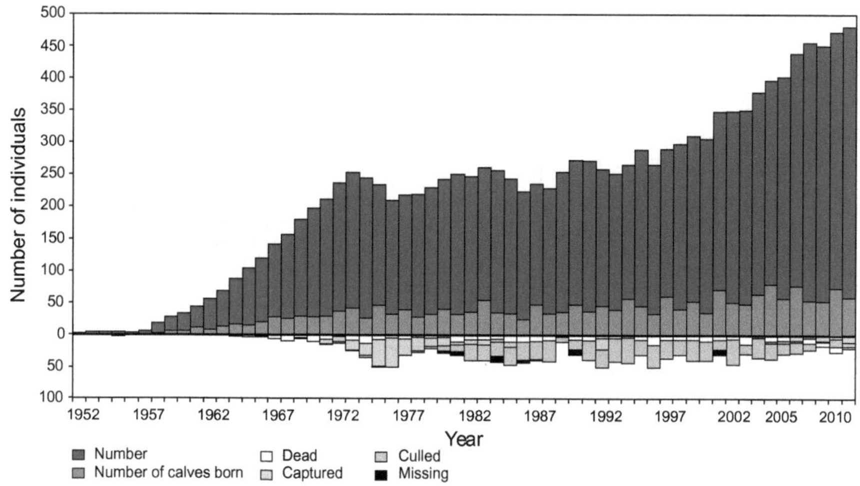

Fig. 10.2 Dynamics of the European bison population in the Polish part of the Forest, 1952–2010. *Source* Krasińska and Krasiński (2007), updated

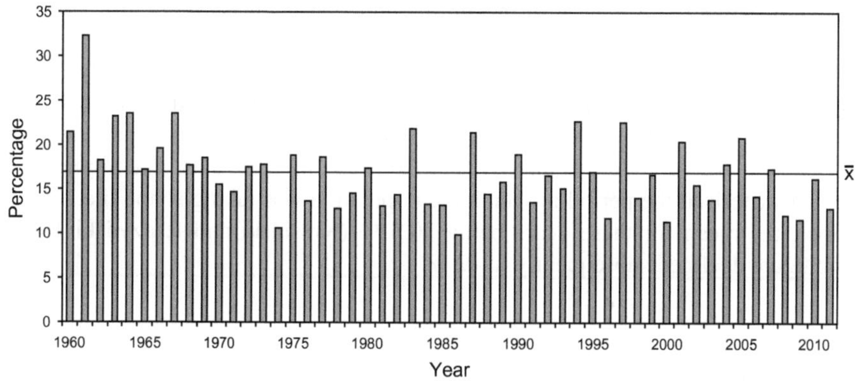

Fig. 10.3 The coefficient of birth, i.e., the number of calves born as a percentage of the overall population of bison in the Polish part of the Białowieża Forest, 1960–2010. *Source* Krasińska and Krasiński (2007), updated

hold from 1971, although initially this simply reflected deliberate captures and transfers to other centres. Culling was also practised later on. In the light of these measures, the population fluctuated between 220 and 481 across the whole period 1973–2011 (Appendix 3, Fig. 10.2). Overall, the years 1957—2011 saw some 2070 (982, 1088) calves born to the Białowieża population (Appendix 4).

In the years 1960–2011 inclusive, the average value for the coefficient of births (number of calves born as a percentage of the overall population size at the beginning of any given year) was 16.9 %. The lowest value was the 9.9 % noted in 1986, the highest the 32.3 % characterising 1961 (Appendix 4, Fig. 10.3.). The numbers of calves born to the Białowieża population are correlated significantly

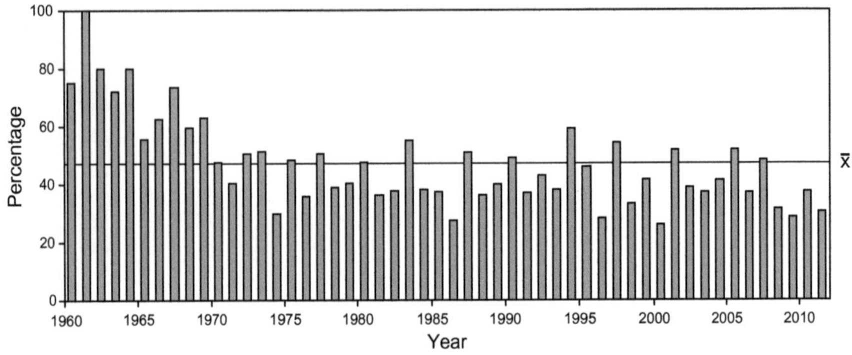

Fig. 10.4 The coefficient of fertility, i.e., the number of calves born as a percentage of the numbers of cows capable of breeding (aged 4 and over) within the bison population of the Polish part of the Białowieża Forest, 1960–2010. *Source* Krasińska and Krasiński (2007), updated

with the numbers of cows aged ≥4 years. In contrast, there is no significant relationship with the number of adult bulls—a consequence of the polygynous mating system existing in the European bison. The numbers of calves born have a significant influence on population size.

Over the period 1960–2011, values for the coefficient of fecundity in the Białowieża bison population (i.e., the number of calves born expressed as a percentage of the number of females of reproductive age) ranged between 25.9 and 100 %, the average figure being 47.1 % (Appendix 4, Fig. 10.4). The highest values characterised the period in which the population was taking shape (i.e., 1958–1966), the mean for these years being 70.3 % (Krasiński and Raczyński 1967). The figure declined subsequently to 42 %.

10.1.3 Mortality

Some 398 (233, 165) bison have died during the entire period of existence of the free-ranging population of the species (Appendix 5). The first fatality was noted in 1959, so it is for the period between that year and 2011 inclusive that coefficients of mortality (numbers of animals dying as a percentage of overall population size) have been calculated wherein the average mortality rate is 2.9 % per year. The greatest mortality—5.7 %—was noted in 1968, with values not much lower than this characterising both 1970 and 1996 (severe winters) and in 2010 (Appendix 5).

No distinct differences come to light when comparisons are made between percentage mortality in the first phase of development of the population (1959–1973) and in the phase in which the population had already been established (1974–2002). Mortality rates among bison have also been determined in relation to five age classes. Perhaps unsurprisingly, the highest mortality rates are to be noted among calves, while the group affected next most severely is that comprising

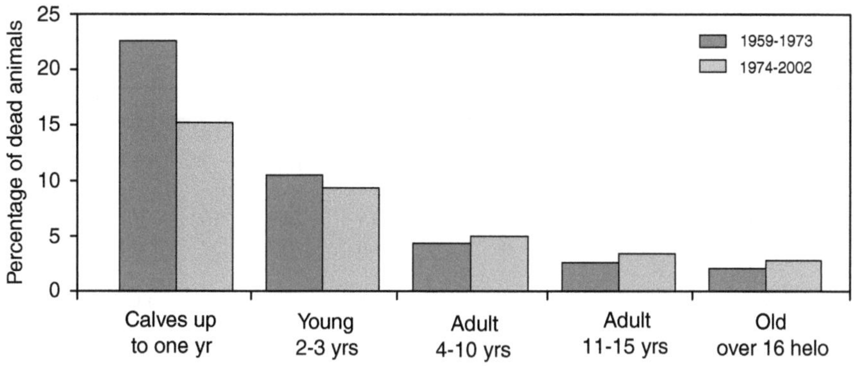

Fig. 10.5 Bison mortality in relation to age in the population inhabiting the Polish part of the Białowieża Forest in the years 1959–2002—calculated as the mean percentage of animals dying at a given age and in a given age class. *Source* Krasińska and Krasiński (2007)

2–3 year-olds. Older bison are characterised by much lower mortality rates (Appendix 6, Fig. 10.5). The number of bison died up to now does not in practice have a significant influence on population size, as birth rates exceed death rates.

In the conditions occurring in the Białowieża Forest, the loss of such a large animal as a bison is bound to be noted sooner or later. However, the discovery of a dead bison soon after its death increases the chances of the cause of mortality being established. Equally, as time passes, the ongoing process of decay limits those chances greatly. Thus, in around 30 % of cases, even an approximate cause of death may not be established (Table 10.3). Where causes are known, these are assigned to 14 categories (Table 10.3), although the process is hindered by the frequent existence of two simultaneous causes. Thus, where an emaciated specimen much infested by parasites was fatally injured by another bison, the primary cause of death—i.e., parasitisation in this case—was the one taken account of.

Causes of mortality among bison have been analysed in respect of the two phases: 1952–1973 (a 15-year period)—during which time the population was developing, and 1974–2008 (a 34-year period)—during which the population underwent stabilisation and assumed a developed structure in terms of age and sex. In the former period, the greatest percentage of losses (37.7 %) were brought about by different kinds of wounding, with other bison being the cause in 24.6 % of cases (Table 10.3). The setting-up of a hierarchical structure during the period of development of the population (and the accompanying injuries sustained) probably accounted for the numbers of deaths of bison induced by factors within the population (Krasiński 1976).

In the second period, in turn, the injuries sustained in the course of incidents of intraspecific competition only accounted for 5.9 % of deaths (Table 10.3). In contrast, harm inflicted by anthropogenic factors accounted for around 8 % of identified causes of death in each of the periods. The largest group here comprises fatalities arising from collisions with cars ortrains. Severe infestations by the lung-dwelling parasitic nematode *Dictyocaulus sp.* afflict younger bison and have been the cause of death

Table 10.3 Reasons for natural mortality among European bison in the Polish part of the Białowieża Forest, 1959–2008

No.	Reasons	1959–1973				1974–2008			
		M	F	Total	%	M	F	Total	%
1	Injuries:	15	8	23	**37.7**	18	20	38	**14.0**
	Injuries affected by different factors	5	3	8	13.1	7[a]	15	22	8.1
	Injuries affected by bison	10	5	15	**24.6**	10	6	16	5.9
2	Posthitis/balanoposthitis from:1980	0	0	0	0	27	0	27	**9.9**
3	Senility	0	0	0	**0**	7	11	18	**6.1**
4	Poaching	2	0	2	**3.3**	14	8	22	**8.1**
5	Parasitic diseases	1	2	3	**4.9**	7	13	20	**7.4**
6	Drowning	0	0	0	0	8	4	12	**4.4**
7	Diseases of the respiratory tract	2	0	2	**3.3**	2	3	5	**1.8**
8	At parturition	0	0	0	0	1	3	4	**1.5**
9	Underdevelopment	1	0	1	**1.6**	1	2	3	**1.1**
10	Diseases of the alimentary tract	3	2	5	**8.2**	2	0	2	**0.7**
11	Pathological changes the in the bone system	4	1	5	**8.2**	2	0	**2**	**0.7**
12	Killed by wolves	0	0	0	**0**	4	3	7	**2.6**
13	Different	2	1	3	**4.9**	12	5	17	**6.2**
14	Unknown, undetermined	13	4	17	**27.9**	54	41	88	**35.0**
	Total	**43**	**18**	**61**		**158**	**114**	**272**	
	Percentage	**70.5**	**29.5**			**68**	**42**		**100**

[a]*Balanoposthitis* was also observed in one bull killed by a train, *M*—males, *F*—females.
Source Krasińska and Krasiński (2007), updated
Data for 1952–1973 from: Krasiński (1978a)

in 4.9 % of cases in the first period, and 7.4 % in the second. Furthermore, parasite problems in general have a more negative influence upon the condition and development of younger—as opposed to older—individuals. As the population aged, the years 1974–2008 began to witness deaths due to old age (6.6 % of the total overall), although these were more likely to observed among females than males (Table 10.3).

The first cases of disease of the male urogenital organs were noted in 1980. All dead males found to be displaying symptoms of necrotic inflammation of the prepuce (posthitis) were included within this category (accounting for 9.9 % overall), notwithstanding the fact that the cause of death was often very complex in nature (Table 10.3).

The percentage of bison deaths brought about by poaching was twice as high in the later period as in the earlier, reaching 8.1 %. Almost all the animals in question fell victim to snares, primarily those set with other target game (like red and roe deer and wild boar) in mind. More often than not, bison succeed in pulling snares out of the ground, but the closed loop of steel wire around a limb or other body part induces permanent damage (Fig. 10.6). It is sometimes now possible to free a bison

Fig. 10.6 Snaring is a particularly cruel method of poaching. In this case, the snare has cut deep into the skin of the head beneath the orbits, producing swelling. *Photo* ZA Krasiński

Fig. 10.7 The steel wire of the snare has tightened around the horns of a bison that could fortunately be released once the snare was cut. *Photo* ZA Krasiński

with a snare caught around a limb or the horns (Fig. 10.7), but there remain cases in which animals die suffering.

In 4.4 % of cases (i.e., 12 animals) bison deaths were caused by drowning, notwithstanding the fact that there are no large rivers running within the Białowieża Forest. The problem lies rather with the heavy weight of bison and the boggy land, although there are even cases of animals drowning in drainage ditches (Fig. 10.8).

The consumption of fodder that ferments too readily (like potatoes, lucerne or rye) resulted in the deaths of 3 adults (Table 10.3). These cases were included amongst diseases of the alimentary tract. As the population grew, the years from 2002 onwards began to bring reports of bison being killed by wolves. Overall, such cases have now come to account for 2.6 % of all deaths (equating to 7 individuals) (Fig. 10.9). The category "miscellaneous" in turn includes a number of causes of death like peritonitis, extensive skin necrosis or a lighting strike (Fig. 10.10).

The issue of the seasonal breakdown to mortality was only addressed in relation to the whole period of 50 years over which the free-ranging population had been breeding (i.e., 1952–2002). This reflects the very limited level of mortality in any given year. Thus, the greatest numbers of bison deaths occur in the period from December to April inclusive (9–13 %; Fig. 10.11). Slightly higher mortality

Fig. 10.8 A bison passing through a muddy forest stream. Bison sometimes become bogged down. If the banks are steep, animals may be unable to escape and can even die. *Photo* ZA Krasiński

Fig. 10.9 In January 2002, an old cow (aged 20) fell victim to wolves, not far from the winter refuge of bison in forest compartment 422. The increase in the numbers of wolves in the Białowieża Forest may give rise to increased predation of bison, although there is no reason to expect a significant impact where the regulation of bison numbers in the Forest is concerned. This is a large and powerful beast that lives in herds, and is therefore just too difficult a prey to take often. *Photo* ZA Krasiński

Fig. 10.10 A dead female was found after a storm in September 1979, and post-mortem examination revealed that the individual had been struck by lightning. This kind of thing does not happen too often. *Photo* ZA Krasiński

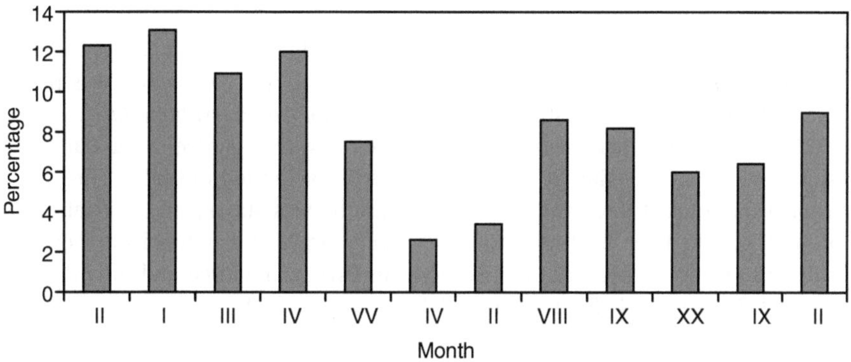

Fig. 10.11 Seasonal breakdown of bison mortality within the population of the Polish part of the Białowieża Forest, 1959–2002

characterises the period of the rut in August (8.6 % of cases) and September (8.2 %). The lowest mortality rates occur in June (just 2.6 %) and July (3.4 %) (Fig. 10.11). A deep cover of snow plus low winter temperatures result in increased mortality during winter, but not to such a degree that overall population dynamics are affected.

Jarocki (1830) defined the European bison as an animal of low fertility, but also limited mortality rates. Detailed information on mortality and the state of health of bison in the Białowieża Forest of the late 19th and early 20th centuries was provided by Wróblewski (1927). At the beginning of the 20th century, there was a widespread view that, among bison: "being a kind of archaic monument, elements leading the animal towards inevitable annihilation are simply inherent in it" (Wróblewski 1927). Büchner (1900) wrote"of the species's steady drift towards extinction". As Wróblewski spent two and a half years in the Białowieża Forest (1906–1909), and documented the animals' state of health, it is possible for us to compare the mortality and health of the 19th century and contemporary populations. What these have in common is a higher level of mortality among males than females. Wróblewski (1927) cites a ratio of 53:47 ($n = 88$), compared to our figures of 70:30 in the years 1952–1973, and 59:41 in the period 1974–2011 (Appendix 5 and 6).

Wróblewski carried out post mortem examinations on 81 bison, determining that 14 % of these died from wounds other bison had inflicted. The young individuals were the worst affected. Severe parasitic infection was the cause of death in 16 % of cases. That said, Wróblewski (1927) noted that only the odd animal was entirely free of flukes and/or worms on the lungs. Today, young bison eliminated from the population on account of their poor physical condition are usually found to be infested with the nematode *Dictyocaulus sp.* Wróblewski also noted the presence of whipworms (*Trichocephalus*), and these are also reported today. Particularly noteworthy are the records of trypanosomes *Trypanosoma* obtained by Wróblewski from bison blood. After consulting with specialists in St. Petersburg, he declared that a new species was involved, the name *Trypanosoma Wróblewski* being suggested for it. He further suggested that the trypanosomes were ultimately responsible for the deaths of five bison (6 % of the total). However, some specialists considered that the "new" species had actually been noted earlier in cattle. Dróżdż (2000) was of a different opinion, considering that the species was indeed typical of bison—thereby justifying the name *Trypanosoma wrublewskii.*

While it is true that foot-and-mouth disease was regarded as a danger to bison in Wróblewski's day, it was only the cause of limited mortality (just 5 %). Wróblewski noted it three times in bison while he was staying in the Białowieża Forest. It would later be found to pose a very grave threat to bison, however, resulting in some 35 deaths at the Reserves in Pszczyna, Gorce and Smardzewice in late 1953/early 1954 (Podgurniak 1967, Jaczewski 1959).

Fig. 10.12 Late 19th century, the Białowieża Forest. Villagers kill a female bison driven out on to the ice. Illustration by R Frenc. *Source* Karcov (1903)

Wróblewski reported 3 cases of death due to cancer which occur rarely today. What is noteworthy are two cases of suprarenal and hepatic adenoma in old (20+) female bison.

Poaching was a danger for bison back then, as now. Wróblewski wrote that c. 10 % of cases of mortality in bison needed to be ascribed to the activities of poachers (Fig. 10.12).

10.1.4 Selective Culling of Bison and Transfers Away

Regulation of numbers in the free-ranging population in the Białowieża Forest was commenced with in 1971, when 7 (2, 5) young bison were taken and then transferred to the Borecka Forest. Before that time, there was nothing more than sporadic interference with the population size. Only eight bison had been removed from the Polish population up to the end of the 1960s (Appendix 7). In the 1970s, the requirement to send animals elsewhere ensured on average the removal of 9.6 % of the population each year. In contrast, only 1.1 % qualified for culling. In 1986–2004, captures were a rare occurrence, while culling was the main measure applied to limit the population. Post-2005, there was a renewed rise in the incidence of captures and transfers, mainly of young animals (Appendix 7). Overall, the years 1955–2011 inclusive saw 339 captures and 892 cases of culling, together giving a removal of some 1,231 bison from the Białowieża population. Among these specimens, 25 % were bulls, 21 % cows, 27 % adolescents and 27 % calves (Appendix 8, Fig. 10.13). In the years 1971–2011, overall 559 males and 664 females were taken out of the population (Appendix 8). The 10.5 % removal annually in the years 1971–2011 was not great enough to bring about any limitation of population size.

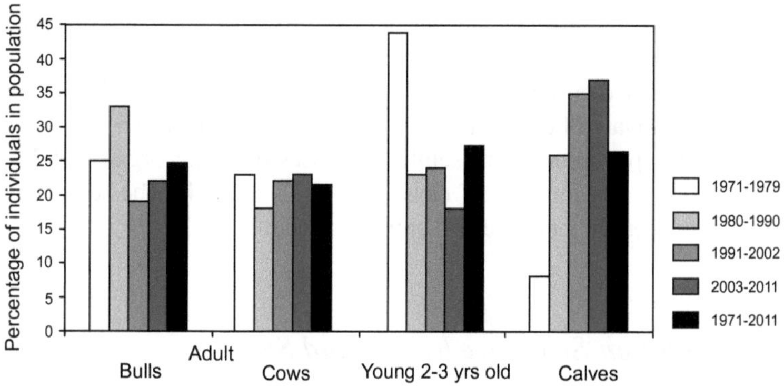

Fig. 10.13 The age and sex structure to the bison captured (for transfer) or culled from out of the Polish population in the Białowieża Forest, 1971–2010. *Source* Krasińska and Krasiński (2007), updated

Table 10.4 Reasons for the culling of European bison in the free-ranging population in the Białowieża Forest, 1971–2008

Reason	Number of culled bison			
	n	Males	Females	%
Reduction	354	62	292	**42,8**
Selection depend on:				
Posthitis/balanoposthitis from: 1980	202	202	0	**24,5**
Parasitic diseases	80	25	55	**9,7**
Entering fields	46	44	2	**5,6**
Diseases of motor organs	23	14	9	**2,8**
Late-born calves	36	14	22	**4,4**
Poaching	16	8	8	**1,9**
Injuries affected by anthropogenic factors	16	13	3	**1,9**
Senility	23	7	16	**2,8**
Poor condition, weak development	6	2	4	**0,7**
Blindness	6	2	4	**0,7**
Various	18	6	12	**2,2**
Total	**826**	**399**	**427**	**100**

Source Krasińska and Krasiński (2007), updated

The greatest reduction in numbers of males occurred in the years 1980–1990, following the appearance of the new disease of males known as necrotic posthitis/balanoposthitis. The practice has been for all ailing individuals to be killed. In the years 1971–1979, it was 2–3 year-olds that were removed in greatest numbers (accounting for 44 % of all instances). Young bison were captured and transferred to other centres. The years 1980–2007 in turn brought a steady intensification of the effort to reduce numbers of calves, though later (after 2008), the level declined once again to the point where it was less than half as high as it had been previously (Fig. 10.13).

Forty-three percent of the individuals eliminated from the population in the years 1971–2008 were healthy, though some had defects that were apparent externally (Table 10.4). Equally, the second largest group of removals involved cases of posthitis/balanoposthitis in males (accounting for 24.5 % of instances), while a third group comprised culling necessitated by parasitic diseases (9.7 %), continued straying on to cultivated fields with resultant damage to crops (5.6 %), poor condition of late-born specimens (4.4), senility, diseases of motor organs (at 2.8 each), poaching (1.9) and "various other" causes (2.2) (Table 10.4). The select culling of individuals has been the only method applied since mid-2000.

10.1.5 Population Structure by Age and Sex

The issue of the age/sex structure of the population of bison in the Białowieża Forest is addressed by reference to classes comprising adult bulls and cows (aged

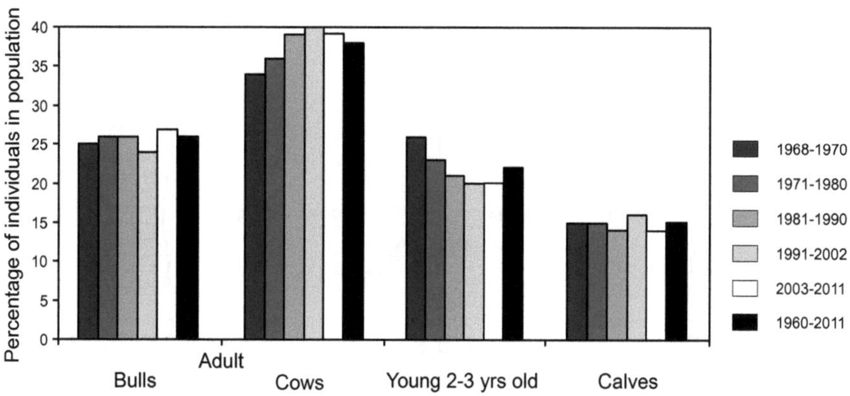

Fig. 10.14 The age and sex structure of the Polish population of bison in the Białowieża Forest, 1960–2010. *Source* Krasińska and Krasiński (2007), updated

4 and over), 2–3-year-old adolescents and calves aged up to 1 year old. This division into classes is a practical one, reflecting the possibilities for individuals to be aged in field conditions. Over the period 1960–2011 as a whole, the Białowieża population on average comprised 25 % adult bulls, 37 % adult cows, 23 % adolescents and 15 % calves (Appendix 3, Fig. 10.14). In analysing the population structure over the long term, we note a relative stability in different periods to the shares of the population comprising bulls aged ≥4 years, as well as calves. In contrast, there has been a steady increase in the contribution adult cows make to overall population size, the proportion exceeding the long-term average throughout the 1981–2011 period. The share of the population accounted for by 2–3 year-olds has in turn declined steadily (Fig. 10.14). These differences are associated with cows' longer lifespans (and the limited active effort to reduce their numbers), as well as with the specifically pursued policy of eliminating some adolescents from the population. In different years, the share bulls accounted for ranged from as little as 14.7 % in 1960 to as much as 30.5 % in 1985. The corresponding range for cows was from 29.4 % (in 1963) to 44.1 % (in 2000). While just 13.2 % of the 1987 population comprised adolescents, the figure was 38.2 % in 1960. Likewise, just 10 % of the 1986 population took the form of calves, compared to 21 % in 1994 (Appendix 3).

Careful registration of the bison population for the whole time since the free-ranging herd began to take shape has allowed for the construction of age pyramids characterising different periods in the population's development. Examples are those for 1973, which is to say a time just after the population had taken shape, then 1992 and 2002 (Fig. 10.15). Bison are in turn assigned to three categories in relation to their reproductive activity. Thus:

(1) **participation in breeding** characterises males aged 5–12 and females aged 4–20;

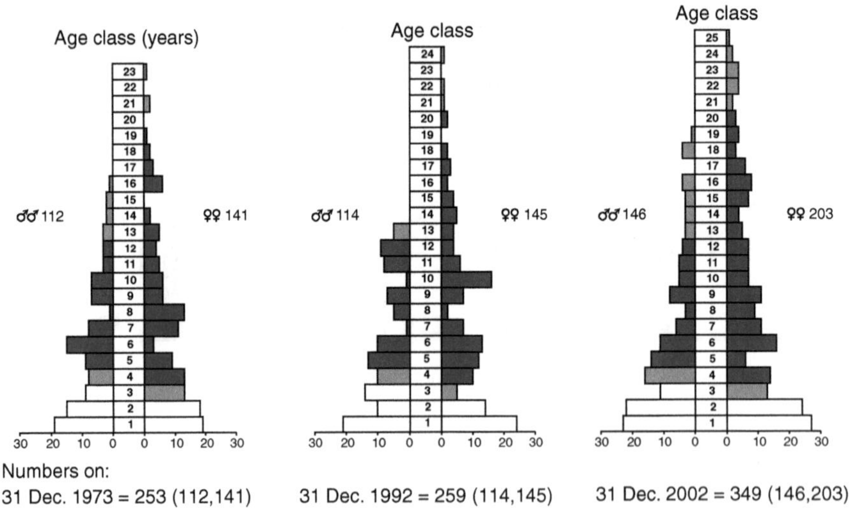

Fig. 10.15 The age pyramids for the European bison populations in 1971, 1992 and 2002. *White*—bison not reproducing, *pale grey*—limited reproductive role and *dark grey*—period of reproductive activity. *Source* Krasińska and Krasiński (2007)

Table 10.5 Percentage participation of European bison of different age-sexual classes in breeding in the Białowieża Forest

Category of breeding activity	Males			Females		
	1973 $n = 112$	1992 $n = 114$	2002 $n = 146$	1973 $n = 141$	1992 $n = 145$	2002 $n = 203$
Active participation in breeding	21.0	20.8	16.0	34.8	38.2	36.7
Limited participation in breeding	6.3	5.8	8.6	6.3	3.1	6.9
No participation in breeding	17.0	17.4	17.4	14.6	14.7	14.6

Percentage was calculated in relation to the numbers of population in given year (total sum of males and females)

n—number of individuals. *Source* Krasińska and Krasiński (2007)

(2) **limited participation in breeding** characterises males aged 4 or over 12 and females aged 3 and over 20;

(3) **non-breeding status** is assigned to males of ages 1–3 inclusive and females aged 1 or 2.

What is noteworthy when the population is described in these terms is the difference between the proportions of males playing an active part in breeding (16.0–21.8 %), compared with females (34.8–38.3 %) (Table 10.5). This reflects the much shorter duration of reproductive activity in the former than the latter. In turn, the proportions of both males and females playing a limited role in reproduction vary in the range 3.1–6.9 %. Finally, the proportions of non-participating males and females are similar, in the range 17.0–17.4 %.

Study of the age pyramids further confirms that males of the free-ranging population have shorter lives than females (up to 20 years, compared with up to 25). In turn, comparisons of the age pyramids for the three different years referred to above make it clear that the Białowieża population is of relatively stabilised structure (Fig. 10.15).

10.2 The Population in the Eastern (Belarussian) Part of the Forest

10.2.1 Year of Establishment, Origins of Bison and Founder Group

The periodic release of bison beyond the limits of the Belarussian Reserves was ongoing from 1953, involving bison of the Lowland-Caucasian line (Korochkina 1969, 1973a, 1974, Bunevich 2003b). The Pedigree Book first notes a free-ranging population of animals from this line among the entries for 1958. In turn, the first Biaowieża or Lowland line bison only began to be released in the eastern part of the Forest in 1961, the process being continued with until 1967. In line with what had been established at international level regarding the desirability of no other bison but those of the Białowieża (Lowland) line being raised in the Forest, all male Lowland-Caucasian bison aged 2 or more were removed from the wild herd in 1961, thus making it impossible for the two lines to mix further. The remaining Lowland-Caucasian bison were recaptured in 1964, before being transferred out of the area altogether via a stepwise process. The last Lowland-Caucasian bison left in 1968 (Korochkina 1973a, 1974, Bunevich 2003b).

The free-ranging population of Lowland bison began to develop in the Belarussian part of the Białowieża Forest in 1964. The founding group in this case comprised of 46 (26, 20) individuals, 43 of these having been brought in from other centres between 1949 and 1967. To reduce the level of inbreeding in the population, two males were transported here from Prioksko-Terrasnyjj Reserve in 1991. In turn, 23 bison (13, 10) had originated from the Białowieża breeding centre, 2 (2, 0) from the Pszczyna Reserve, and 18 (12, 6) from the said Prioksko-Terrasnyjj Reserve in Russia (Appendix 9, Bunevich et al. 2006). Alongside these are three further founders—the female BARCHATKA (pedigree no. 948), born on a Reserve in the Belarussian part of Forest and released in 1961 (with origins going back to the Pszczyna bison (M 720 PLUSZ and F 718 PLISZKA)) plus two calves—the female BANZAJKA (pedigree no. 1445) and the male BASZTAN (no. 1506), both born into the free-ranging herd to Reserve-covered mothers subsequently released (Bunevich 2003b, Bunevich et al. 2006). Once the introductions were over (in 1967), the herd of Lowland bison in the Byelorussian part of the Forest comprised 58 (31, 27) individuals.

10.2.2 Population Dynamics

By the beginning of 1970, the eastern population of Lowland European bison numbered 63 and was taken to have assumed a final shape not necessitating further human intervention (Appendix 10, Korochkina 1974, 1975, Korochkina and Kochko 1983, Bunevich and Kochko 1988). The population rose steadily subsequently, to reach 315 in 1990–1991, before declining a little thereafter (Figs. 10.16, 10.17, Bunevich and Kochko 1988, Kozlo et al. 1997, Bunevich 1994, 2003a). As of the end of the 2011, there were 428 bison living here (Appendix 10). The recommendation for the Belarussian part of the Forest is that the population be maintained at a level of 220–250. This determination is based on an analysis of the food base provided by the environment in which the animals live, the abundance at which maximal female fertility is to be noted and the assumed density of bison achievable where 8–10 individuals require an area of some 1000 ha (Kozlo et al. 1996, 1997, Kozlo and Bunevich 2009).

Values for the coefficient of fecundity in the first phase of population development (1964–1967) were very high, at 68.4 %. By the end of the 1960s, the figure had fallen to 45.1 %. 57 calves had been born, and the sex ratio was close to 1:1 (Korochkina 1974, 1975). In turn, over the entire period of existence of the Belarussian population of Lowland bison (1961–2005), some 1,112 calves have been born, giving a coefficient of births equal to 14.7 % on average. The sex ratio of

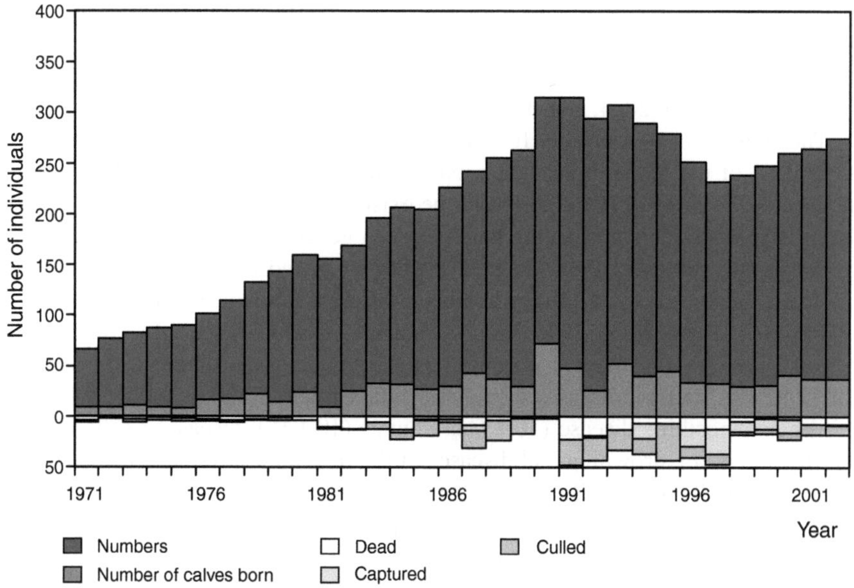

Fig. 10.16 Dynamics of the European bison population in the Belarussian part of the Forest in the years 1971–2002. Data from: Korochkina and Kochko (1982), Bunevich and Kochko (1988), Bunevich (1994, 1999) and Bunevich et al. 2006. *Source* Krasińska and Krasiński (2007)

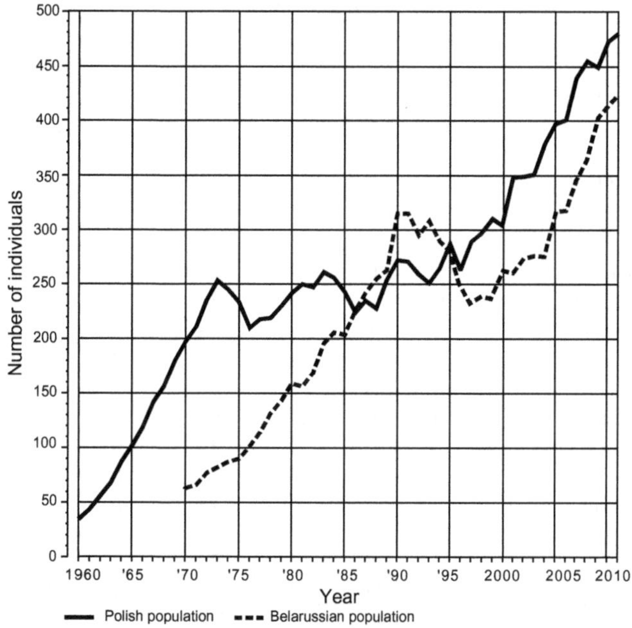

Fig. 10.17 Dynamics of the European bison populations in the Polish part of the Białowieża Forest (1960–2010) and the Belarussuan part (1970–2010). Data for the Belarussian population from Bunevich and Kochko (1988), Bunevich (1994, 1999b) and Bunevich et al. 2006, Kozlo and Bunevich 2009. *Source* Krasińska and Krasiński (2007), updated

calves at birth has not departed significantly from 1:1, though individual years have brought more major departures. The coefficients of fecundity were lower over the period 1971–2005 than they had been before, the average being 37.8 % (Appendix 11, Krasochko et al. 2004; Bunevich et al. 2006; Kozlo and Bunevich 2009).

The coefficient of mortality for the years 1961–2005 as a whole was of 3.6 %, with 58 % of the bison dying being males (Bunevich 1999b, 2003b; Bunevich et al. 2006). The most difficult period in the existence of the population came in the years 1980–1984, when breeding parameters in general diminished, while mortality was running at twice the average level (Bunevich and Kochko 1988). Pre-1980, the main cause of bison fatalities had been different kinds of wounding sustained (Table 10.6, Korochkina 1975; Korochkina and Kochko 1982). In contrast, the period 1981–2004 witnessed poaching as the major cause of death (accounting for 17.9 % of losses), diseases of the alimentary tract (12.3 %), followed by posthitis/balanoposthitis disease (10.1 %), wounds (10.1 %), diseases of the alimentary tract (9.8 %), drowning and diseases of the respiratory system (6.2 and 1.7 % respectively) (Table 10.6, Bunevich 1999b, d; Krasochko et al. 2004).

In the years 1971–1980, there were no captures of animals at all, while culling encompassed on average 2 % of the herd annually (Korochkina 1975; Korochkina

Table 10.6 Reasons for natural mortality among European bison in the Belarussian part of the Forest, 1971–2004, updated data for 1971–1980 from: Korochkina and Kochko (1982), for 1981–2004 from: Kozlo and Bunevich (2009)

Reasons	1971–1980				1981–2004			
	M	F	Total	%	M	F	Total	%
Poaching	1	0	1	**1.6**	17	15	32	**17.9**
Posthitis/Balanoposthitis	2	0	2	**3.2**	18	0	18	**10.1**
Diseases of the alimentary tract	2	4	6	**9.7**	7	15	22	**12.3**
Injuries affected by bison	7	6	13	**21.0**	10	8	18	**10.1**
Drowning	2	2	4	**6.4**	4	7	11	**6.2**
Pneumonia	6	2	8	**12.9**	0	3	3	**1.7**
Accidental injuries	8	0	8	**12.9**	3	4	7	**3.9**
Peritonitis, poisoning	0	2	2	**3.3**	0	0	0	**0**
Underdevelopment	0	1	1	**1.6**	2	4	6	**3.4**
Hypothermia	0	0	0	**0**	2	2	4	**2.2**
Lack of milk	0	0	0	**0**	1	1	2	**1.0**
After immobilisation	0	0	0	**0**	2	0	2	**1.0**
Cancer	2	0	2	**3.2**	0	1	1	**0.6**
Senility	0	1	1	**1.6**	0	0	0	**0**
Pathological changes in the bone system	0	1	1	**1.6**	0	0	0	**0**
Unknown	7	6	13	**21.0**	35	18	53	**29.6**
Total	**37**	**25**	**62**		**101**	**78**	**179**	
%	**60**	**40**		**100**	**56**	**44**		**100**

M—males, *F*—females
Source Krasińska and Krasiński (2007), updated

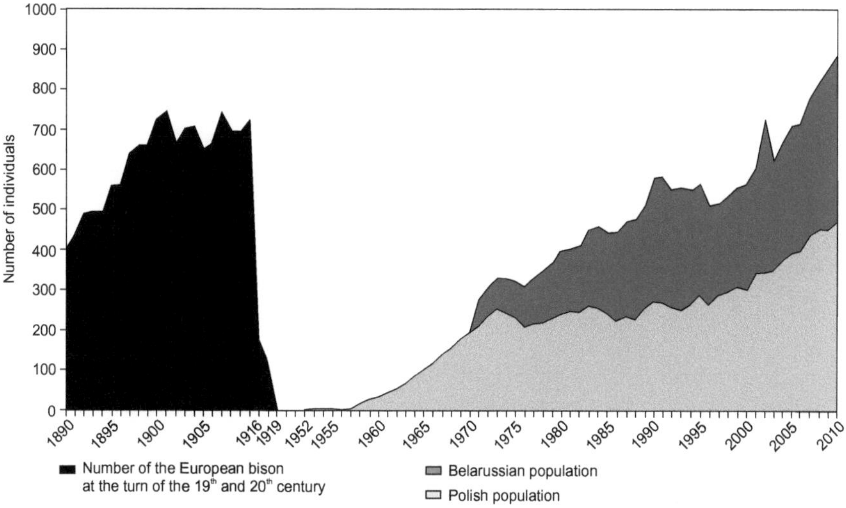

Fig. 10.18 Numbers of the European bison in the Białowieża Forest in the years 1890–2006. Data from Karcov (1903), Rörig (1918, after Pucek 1994), Sztolcman (1924), Wróblewski (1927, 1932), Krasiński and Krasińska (2007), Bunevich et al. 2006, Kozlo and Bunevich 2009; European bison Pedigree Book 2006-2010. *Source* Krasińska and Krasiński (2007), updated

and Kochko 1983). In turn, the years 1966–2005 taken together brought combined removals from the population (by culling or capture) equal to 4.5 % of the population each year on average. These data tend to disguise the fact that the last 15-year period for which figures are so far available (1991–2005) has been characterised by a much greater reduction effort (involving both culling and transfers), this taking in some 8 % of all specimens in the population each year (Bunevich et al. 2006). In the years 1981–2001, the most frequent justifications for culling involved the diseases posthitis/balanoposthitis in males (40.6 % of kills), old age (23.6 %) and wounds (14.9 %). Other reasons like diseases of the alimentary tract, aggression, dermatosis, deafness, or poor development accounted for between 0.4 and 4.7 % of specimens eliminated (Bunevich 1989a, 1999b; Krasochko et al. 2004).

Today the Białowieża Forest supports a population of European bison greater in size than the last wild population present here at the beginning of the 20th century (Fig. 10.18). This is the world's largest population of the species and is fortunately derived genetically from all seven founders of the world population of Lowland line bison (Grzegrzółka et al. 2004).

Chapter 11
The Rhythm to Bison Activity

11.1 The Annual Biological Cycle of the Bison

The annual cycle followed by European bison in the Białowieża Forest consists of two markedly different periods: a period of **free movement** throughout the period of the year without snow cover, in which bison make use of natural food; and a period of **winter concentration** around places at which supplementary food is made available (Fig. 11.1).

The natural environment of the European bison comprises extensive deciduous or mixed forests. These assure an abundance of food during the growing season, and their mosaic-like structure makes it possible for bison to find different foods in different seasons and phases of vegetational development. In early spring, the animals seek out the first green plants appearing on the still-unshaded forest floor of the oak-lime-hornbeam forest.

M. Krasińska and Z. A. Krasiński, *European Bison*,
DOI: 10.1007/978-3-642-36555-3_11, © Springer-Verlag Berlin Heidelberg 2013

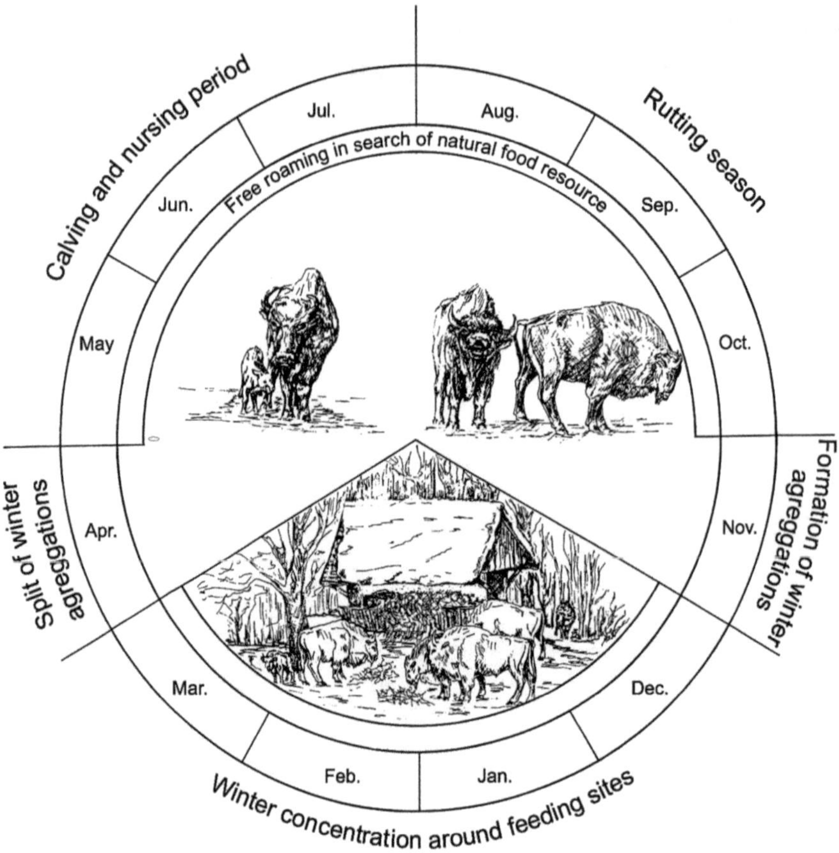

Fig. 11.1 The annual life cycle of a bison. *Source* Krasiński (1993), modified

Bison feed very intensively in the early spring, making up in terms of both quantity and quality for the shortages of food present over winter. The process is of particular importance for pregnant females, whose fertility is assuming its final phase of development at that time. The leaves developing on trees and shrubs at the beginning of May provide for the further enrichment of the diet, while the forest also provides better and better cover for the bison.

The calving and nursing period then begins. Pregnant females depart from the herd at this point, selecting a quiet place in which they can give birth. After several days, mothers with calves rejoin the group. Since food is abundant on the one hand, while young calves are present on the other, the groups move slowly, feeding intensively. At the age of about three months calves enter the period of the so-called juvenile moult, entailing a change of the light brown coat of hair for a darker one more similar in colour to the adults. Calves no longer stick so closely to their mothers at this point, often being noted in the company of their peers. Mothers in turn devote less and less time to their offspring, feeding them only rarely, even though calves try repeatedly to gain access to their udders.

The August appearance of groups of adult bulls is a sign that **the rut** is beginning. Bulls hoping to mate with females as soon as they come into season will not tolerate rivals, so younger males are wise to keep a safe distance from them in this season. Autumn marks the last time at which reserves of energy on which to overwinter can be accumulated. When the ground frosts of autumn begin to appear, animals also take the apical parts of the nettles Urtica dioica which are present in such abundance. The forest at this time echoes to the bellowing of red deer, while the herds of bison come steadily closer to their winter refuges.

November markes a **transitional period**, as bison approach those refuges in which reserves of hay have been gathered for them. The hay is laid out in racks that offer the bison unlimited access. Short movements away from the area in search of the last greenery do continue to take place, however. The first snow cover of more persistent duration makes it necessary for bison to make fuller use of the hay put out for them. The true period of **winter concentration** then begins, with hay accounting for virtually 100 % of the bisons' diet. More and more animals come into the winter refuges and large aggregations take shape. There is no fighting, however, as all bison know one another well, having come to the same places year after year. The adult bulls are the last to come in, a sign that winter proper has taken hold and that any last sources of natural food have now dried up. Males may sometimes move from one refuge to another, or even cover longer distances, living the life of loners or at most joining small groups. They sometimes make their way to other supplies of hay, set aside for livestock. Individuals of this kind—usually old bulls—are inclined to be troublesome, even becoming dangerous if disturbed or disquieted in some way. At the end of February they become the first to change their coats, rubbing intensively at their necks to shed the old coat and make way for the new—as a sure sign that winter is coming to an end. At that time bison are not really great lovers of hay, warming their brown bulk in the first warm rays of the sun. The final breakdown of the winter groupings usually takes place in April—the **transition period** between winter and spring. Animals move further and further away from their wintering grounds, and form smaller groups that vary in composition from year to year.

11.2 The 24-Hour Rhythm to Activity

Bison resemble ruminants in general in having a multi-phase 24 h rhythm to their activity. The phases in question consist of feeding, rest and movement. Feeding is interspersed with rest periods first and foremost dedicated to the chewing of a cud derived from food taken previously (Fig. 11.2) (Korochkina 1972a; Caboń-Raczyńska et al. 1983, 1987). Summer brings a high degree of synchronicity to the activity of all members of a group associated with the aforementioned, feeding, rest and movement.

It takes at most 15 min for all members of a group to shift from one form of activity to another. A group feeds actively while moving. While the decision to move is usually taken first by the leading cow, it has also been possible to observe situations in which the leading cow interrupts a long period of rest by actively initiating feeding in her group, raising adjacent cows from their prone position as necessary.

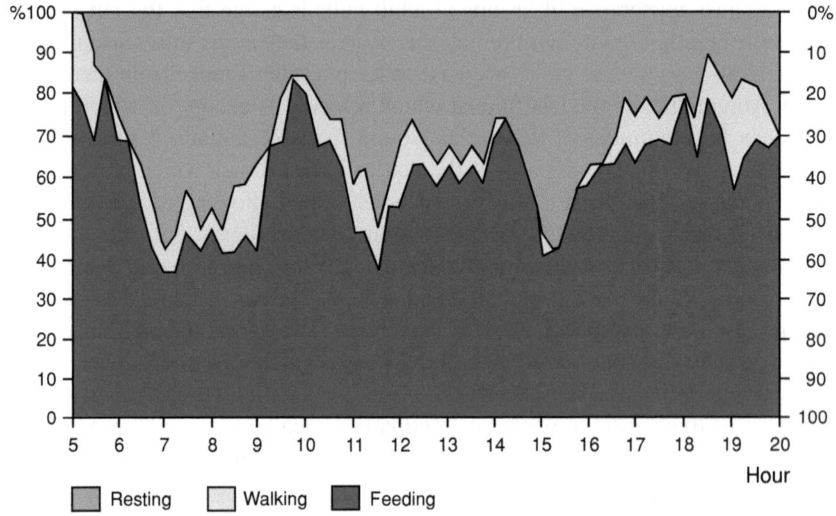

Fig. 11.2 The rhythm to the activity of a mixed group of bison during the growing season. *Source* Caboń-Raczyńska et al. (1987)

In the growing season, bison spend around 60 % of each 24 h period feeding. This leaves 30 % of the time available for resting, plus the remaining 10 % for periods of movement not involving the consumption of food. The reverse applies in winter, at which time bison take advantage of the supplementary food on offer, spending 30 % of each day taking hay, while 60 % is spent at rest (Caboń-Raczyńska et al. 1983, 1987) (Fig. 11.3). Animals in the Belarussian part devote more time to eating (80 % on average), probably as a consequence of the poorer food base that this part of the Forest has to offer. The length of time devoted to feeding obviously depends on both the nutritional value and the abundance of food available. Bison spend most time feeding in spring and autumn, when quantities of food available are more limited (Korochkina 1972a).

Observations of diurnal (dawn-dusk) activity among bison over the period of the year in which snow cover is absent has led to the identification of 4 feeding phases coinciding with sunrise/dawn, two periods during the day, and a fourth period associated with the approach of sunset (Caboń-Raczyńska et al. 1987). Studying the 24 h activity of bison in the Belarussian part of the Forest, Korochkina (1972a) reported that bison feed at night too, devoting just 3–3.5 h to rest during the hours of darkness. Radiotelemetry confirms that, while night-time activity does take place, it is of shorter duration than that during the day. In particular, the period between 23.00 and 02.00 seems to be set aside for minimal activity or movement (Rouys 2000).

While the feeding periods around dawn and dusk are something of a constant, the remaining sessions of feeding during the day are subject to variable timing, depending inter alia on the weather, insect numbers, the state of the forage base and the degree to which bison are being disturbed by people. Just as now,

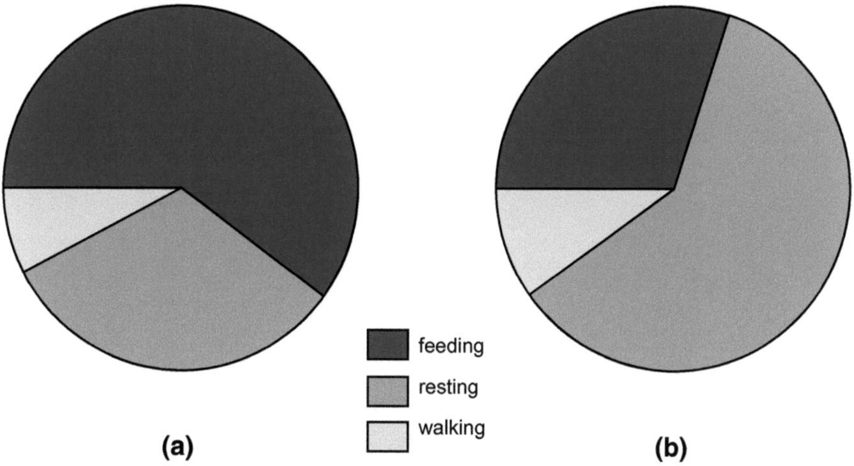

Fig. 11.3 A comparison of percentage diurnal activity among bison in the vegetative period (**a**), and during winter (**b**). *Source* Caboń-Raczyńska et al. (1987)

Wróblewski (1927) observed early-morning feeding by bison as a constant feature, as well as continued instances of feeding through to dusk and the evening. During heatwaves, bison feed in the morning, in the evening and at night, spending much of the day immobile (Korochkina 1972a). Activity is particularly curtailed when temperatures exceed 30 °C (Rouys 2000).

On the Belarussian part, the arrival of autumn sees some bison head out of the Forest for adjacent cultivated fields offering attractive food. Since people are usually present at such areas during the hours of daylight, bison tend to come out on to them at night, in the early morning or in the late evening (Korochkina 1972a). Unit foraging times range from 15 min through to 5 h and 15 min, while periods of repose also last between 15 min and in excess of 4 h (Caboń-Raczyńska et al. 1987).

During the season in which animals move around freely, there is no particular time set aside for arrival at sources of drinking water. Natural watercourses are present throughout the Forest, along with the stagnant water present in depressions, fens and bogs, as well as artificial watering places. Bison living in groups generally drink from permanent bodies of water or watercourses. Lone males on the move are inclined to make use of puddles they meet along the way. On rainy days or days following rain, bison may not bother to move over to bodies of water at all. The commonest time for animals to drink is in the early hours of the afternoon, although they also come to watering places in the middle of the day in heatwave conditions. In winter, bison make water available by breaking the ice along streams, or by trampling frozen ground in alder forests so as to take the water gathering in the depressions created in this way. The consumption of snow is another way of gaining a drink of water in winter, being observed 1–4 times a day (Fig. 11.4).

Fig. 11.4 Thirst is only partially satisfied by the consumption of snow. *Photo* ZA Krasiński

Other forms of activity (grooming, the suckling of calves and play) are engaged in individually and take place at different times among different bison. They mostly occur at times when a herd is resting or feeding.

Bison move around without simultaneous feeding when they are making their way to other pastures or watering places, or else in search of places to rest. They may also change localities if disquieted by the presence of people or large numbers of insects. The time designated for movement usually represents around 10 % of the day, although a maxima of 27 % can be noted. A herd does not cover greater distances every day, and may spend time over several days in the same place. The distances covered in the course of a 24 h period tend not to be great, although the search for entirely new places to forage can take them up to 12 km a day (Korochkina 1972a, Krasińska et al. 1987). Bison equipped with radiocollars have been checked for the distances covered, although with no capacity to determine what proportion of this time is spent feeding simultaneously. In periods with no snow cover, the distances covered by such bison averaged 5 km per day (Rouys 2000).

In winter, it is possible to observe three peaks of feeding activity over a 24 h period, although there is little synchrony in activity between individuals. Even when most members of a herd are resting, there are always some individuals feeding. Winter herds display two periods of rest, however, in the evening between 18.00 and 21.00, and again between midnight and the early hours of the morning. Resting is in fact the most highly synchronised winter activity, along with journeying to watering places (Caboń-Raczyńska et al. 1983).

During winter, radio-tracked individuals spending time around supplementary feeding sites cover just 1.25 km a day on average (maximally 2.3 km). Those specimens not making use of the hay provided during the winter in contrast travel distances of between 3.3 and 13.1 km daily (Rouys and Krasińska 2001). Equally, the presence of lying snow combines with low temperatures to limit the mobility of bison greatly (Krasińska and Krasiński 1994, 1995; Krasińska et al. 2000).

Chapter 12
Reproduction

12.1 Sexual Maturity and Fertility

Male bison can reach sexual maturity at ages of 15–20 months, although research suggests that such cases are rare (Mohr 1952; Zablocky 1949; Jaczewski 1958). Histological studies of the testes of Reserve-bred males or individuals from the free-ranging population in the Białowieża Forest make it clear that some males begin maturing sexually in their third year of life. However, by the age of 4 all males studied had fully developed spermatogenesis, and were thus mature sexually (Czykier et al. 1999). Recent studies have revealed an increasing number of cases of spermatogenesis among young males. Out of 45 males aged 3 months to 3 years, 8 of those in their second or third years of life were capable of full

M. Krasińska and Z. A. Krasiński, *European Bison*,
DOI: 10.1007/978-3-642-36555-3_12,

spermatogenesis. Early spermatogenesis was present in males with the greatest body masses in their given age classes, and was observed in Reserve-bred males (Czykier and Krasińska 2004). Wróblewski (1927) held that males of the world's last free-ranging European bison population at Białowieża in the early twentieth century would only mature in their fifth year of life. Today, the young 4–5-year-old males in the wild population are sexually mature, but do not usually participate in reproduction for behavioural reasons, older bulls seeing to it that they are kept well away from cows on rut (Fig. 12.1). Bulls aged 4–12 were found to have fully developed spermatogenesis and produced mature sperm (Czykier et al. 1999). Males aged 12+ years in turn manifest an impaired process of spermatogenesis, perhaps indicative of curtailed reproductive function (Czykier et al. 1999). In today's free-ranging population, old bulls over 12 years old are driven away from cows on rut by younger males (Krasiński 1967, 1978a, b; Krasiński and Raczyński 1967; Krasińska and Krasiński 1995). In the wild populations, an active role in reproduction is generally played by males aged 6–12. It is also among such animals that the highest levels of free testosterone are to be noted in blood serum (41 pg/ml) (Czykier and Krasińska 2006). In contrast, in Breeding Reserve conditions, males engage successfully in breeding upto the greater age of 15. Indeed, the bull M 980 PORANEK became a father at the Białowieża Reserve aged 17 (Fig. 12.2).

Testis weight in bison increases with age. Among sexually mature males, the mean value for left and right testes is lowest among individuals aged 4 (at 137 g), while bulls aged 6–22 years are characterised by a value of 222 (67–390) g.

Fig. 12.1 Older and stronger animals prevent a young (4-year-old) male from breeding, even though he is in fact sexually mature. *Photo*: ZA Krasiński

Fig. 12.2 Bull M 980 PORANEK, who was born on July 14, 1955 at the Białowieża Breeding Reserve, was released into the wild aged 3, spending some 6 years at liberty in the Forest. He was recaptured for the reserve in 1964, however, siring numerous offspring in both the free-ranging and Breeding Reserve populations. He died on April 4, 1975, aged 20. *Photo*: ZA Krasiński

The length of the testis among the youngest sexually mature males is 88 mm, the width 48 mm and the circumference 137 mm. This compares with corresponding figures for bulls aged 6–22 of average length 116 (54–146) mm, width 64 (48–75) mm and circumference 178 (152–220) mm (Krasińska et al. 2007).

In captive breeding conditions, females become pregnant when aged between 24 and 28 months (Jaczewski 1958). However, in the majority of cases, they mature sexually at the age of 3, the first calves being born in their fourth year of life. Such a situation has been found to apply to 54 % of captive breeding cows (Krasiński and Raczyński 1967, 1969; Krasiński 1978b; Belousova 1986). At the end of the 19th century, females of the free-ranging population in the Białowieża

Fig. 12.3 KANIA F 1131 is one of the most fertile cows at the Białowieża Breeding Reserve, having given birth to calves in 14 consecutive years: 1961–1974. *Photo*: ZA Krasiński

Forest mated first in their third and fourth years of life (Karcov 1903; Wróblewski 1927), which is much the same as they do today.

Cows remain fertile into old age. Analysis of the oldest data from the captive breeding centres (years 1891–1946) shows that females can give birth to calves when aged 19–22 (Koch 1956). Zablocky (1956) relates the case of a calf being born to a female aged 24. Even in wild conditions, a well-known cow living in the Białowieża Forest gave birth to and successfully reared her last calf when aged 23. In turn, in Reserve conditions, it is normal for females to keep giving birth until they reach 20 years of age. Furthermore, between 63 and 91 % of the Reserve females capable of reproducing do give birth to calves in a given year. On average, therefore, a cow produces 9 calves in a lifetime. Beyond this, the famous case of the cow F 1131 KANIA (Fig. 12.3) involved the production of offspring annually over a 14-year period (Krasiński and Raczyński 1967; Krasiński 1978b). A high level of fertility is also to be noted among the cows on the Russian Reserves (Zablocky 1957). However, Kiseleva (1974, 1986) noted that the Oka Reserve females displayed highest fertility (with 85 % of cows capable of breeding) in the first 10 years of that centre's existence, the figure declining subsequently to 44 % due to a deterioration in the quality of the dietary base. There is thus no doubt that the higher fertility of Reserve females reflects the abundance of food on offer to animals throughout the year.

The bison population of the early 20th century was characterised by females producing calves every other year and thus having 7–8 calves in a lifetime (Wróblewski 1927). The currently existing free-ranging population in the Białowieża Forest had a high (60–100 %) coefficient of fecundity in its first 10 years of development (Krasiński and Raczyński 1967; Krasiński 1978a), but the 60 years of existence of the population have been associated with a decline to 47 % of the females in the population on average (although the range is still 28–100 %). The figure for the Belarussian population is somewhat lower, at 40 % on average (range 15.7–86 %) (Bunevich and Kochko 1988; Bunevich 1999b; Krasiński et al. 1999). Females experiencing the natural conditions of the Białowieża Forest thus give birth every other year on average, just as did their forebears living wild at the beginning of the 20th century (Wróblewski 1927). In contrast, Jarocki (1830) has it that breeding took place still less often in his day—one calf even being born every 3 years.

Another population of Lowland bison present in Poland's Borecka Primaeval Forest is characterised by higher coefficients of fecundity than Białowieża, for example of 55 and 66 %, respectively, in the years 2005 and 2006 (Żoch 2007). This phenomenon may reflect extensive supplementary feeding of the population in winter.

12.2 The Rutting Period

Bison are characterised by a polygynous mating system whereby one male may mate with many females. Sexually mature males aged 6–12 are generally only present in mixed groups during the breeding season. At other times, they join small bachelor herds or remain alone. At the end of July, just prior to the breeding season, bulls make more and more frequent visits to mixed groups to check if any of the cows there are receptive. Those few bulls preferring the margins of the Forest in winter and spring will definitely return to the forest proper in July, where the mixed herds occur.

The breeding season is a period of intensified activity on the part of adult bulls. At this time they occupy larger home ranges and cover greater distances. It is often possible to see lone males wandering the Forest in search of cows. Most mixed groups at this time include adult males (Krasiński 1976, 1978a, b; Krasiński et al. 1994a; Krasińska and Krasiński 1995; Krasińska et al. 2000; Daleszczyk 2002). In the course of a single breeding season, an adult male over 6-years old will spend between 1 and 29 days (usually 2–6) in the same mixed group, visiting 1–8 such groups on average (most often 2–5) (Krasińska and Krasiński 1995; Daleszczyk 2002). Young (4–5 year-old) sexually mature males are also said by Daleszczyk (2002) to apply another breeding tactic

Fig. 12.4 A rare example of a winter rut at the breeding reserve in Białowieża (on January 11, 1971) which resulted in autumn parturition. Such births are in fact a rarity in nature, being mainly characteristic of animals in reserves and zoos. *Photo*: ZA Krasiński

whereby their participation in reproduction is assured. Specifically, they intensify their activity prior to the true start of the breeding season, then visiting a larger number of mixed groups, spending more time there than do the older bulls.

According to Jarocki (1830), the rutting season in the early 19th century Białowieża population began in August and lasted for c. 14 days. In contrast, according to Karcov (1903), the late 19th and early 20th century breeding period began in late July and continued through to the beginnning of September. Wróblewski (1927) in turn notes August and September as the breeding period for bison. Today's season seems rather more protracted, lasting as it does from August through to October, although the greatest intensity of activity definitely falls in August and September. Cows of the free-ranging population will rarely engage in mating at the end of July, or in November–December, while breeding activity is still less frequent in January and February (Caboń-Raczyńska et al. 1983; Daleszczyk and Czykier 2010).

Receptivity outside the true breeding season (i.e. in autumn or winter) arises much more frequently in captive breeding (Fig. 12.4), the high-protein diet available year-round tends to disrupt the normal seasonality of reproductive activity. Thus, the breeding season in the Białowieża Reserves begins in July and runs on until March, albeit with 70–80 % of matings taking place in the August–October period (Krasiński and Raczyński 1967; Krasiński 1978b). A similar situation occurs in Russia's Okskjj Reserve, at which—in the years 1960–1970—the mating season extended between the first half of July and January, sometimes even to April (Kiseleva 1974).

Females are in oestrus for 1–3 days, which is associated with a short-lived relationship with a bull in which each partner displays characteristic behaviour (Krasiński and Raczyński 1967; Krasiński 1978b; Daleszczyk 2002).

12.3 Sexual Behaviour

Sexual behaviour on the part of bulls is not confined to the breeding period, being noted at other times of the year also. The young (2–4-year old) males present in mixed groups sniff the outer parts of the female genitalia from time to time, adopting the so-called *flehmen* attitude, whereby the head and neck are raised, and the lips parted with the upper lip folded back. Flehmen serves to allow the male to detect the hormone concentrations in the females urine, and thereby assess when she is sexually receptive. Mounting of cows also sometimes takes place.

Breeding season sexual behaviour among adult males is associated with intensified aggression, targeted at both other bulls in the same group and at any person who makes the herd nervous. In captivity, breeding aggression even extends to cases of injuring or even killing calves by bulls (Krasiński and Raczyński 1967). In the free-ranging population, lone bulls sniff out the trails of bison that have passed through an area and then pursuing mixed groups. Adult

Fig. 12.5 A bull covers a cow, assuming the so-called "*flehmen*" attitude. *Photo*: ZA Krasiński

males seek to demonstrate their strength when aroused, pawing impatiently at the ground, snapping young trees, bending down to the ground or even breaking taller spruces, and even using their horns to uproot trees. Males also quite frequently issue the croaking vocalisation, known as grunt, during the breeding season.

The most fundamental sign that a female is in oestrus is its being covered by a bull. A bull accompanies a cow uninterruptedly for 1–2 days, rarely even bothering to eat or rest. The bull in this way seeks to ensure isolation of the cow from the herd, remaining with her on its edge and engaging in a series of ritual activities. The male frequently licks or sniffs the area around the vulva, mounts the cow fully or partially, adopts the aforementioned flehmen position repeatedly (Fig. 12.5), and frequently grunts and snorts, pawing repeatedly at the ground and sometimes scattering soil around. From time to time a male wallows, frequently urinating in the wallow. An incidence of mounting ending in actual copulation is a short-lived affair (Krasiński and Raczyński 1967; Krasiński 1978b; Caboń-Raczyńska et al. 1987). Young, 4–5-year-old bulls present in the same group also become aroused at this point, brushing against one another with their horns, breaking branches, perhaps wallowing on the ground and trying to manoeuvre themselves into the immediate vicinity of a cow in heat, whenever the dominant male becomes slightly more separated from her. However, they are immediately repelled by the male intent on covering the female.

12.4 Fights Between Bulls

R. Frenc

Battles between red deer stags during the rut have been described in the literature and captured on canvas or celluloid many, many times. This is nothing surprising, since nature has equipped stags with impressive antlers that sometimes achieve huge dimensions. Battles between stags are also accompanied by noticeable sound effects, the roar of a male often being enough to dissuade a rival, attesting as they do to the power of the owner and his capacity to hold total sway over his territory.

In comparison, the courtship period in bison is an altogether more discreet matter. For a start, a bull bison does not bellow out, producing little more than a quiet if resonant 'grunt'. It is at the end of July that adult bulls appear in the herd, alongside females and adolescents. Two or more bulls can then be observed in any given herd, and they stand out from the rest of the group in their powerful lines. Only rarely, however, does an actual battle between bulls ensue. Indeed, a bull taken up with the matter of his female partner has neither the time nor the desire to fight.

Thus, the free-ranging populations only rarely bear witness to rutting season battles between bulls. Nevertheless, wounded bulls are noted from time to time, bearing traces of a blow from another's horns. Males occasionally even die from the wounds they have sustained in battle. In reality, though, fights are a rarity precisely because it is rather unusual for evenly matched males (of the same size and strength) to come into contact with one another, and that is when a trial of strength is most likely to start. Each such trial is different, the course and duration depending on the character, temperament and strength of the rivals pitted against each other. There is a repetition of the behaviours characteristic for fighting bulls in different phases of their contests (see colour insert). Prior to any actual conflict, the rivals show just how strong they are by breaking and/or uprooting young trees (mostly shallow-rooted spruce in the case of the Białowieża Forest). The ground is hoofed repeatedly, urine is passed and the places then wallowed about in.

One such battle was observed in August 1984. A herd in which a large, 8-year-old bull was mating with a cow in oestrus was infiltrated by another male of similar age and condition. The newcomer quickly made its way into the centre of the group, sniffed one of the cows, raised its head to assume the flehmen position, and immediately set off in the direction of the mating bull. He circled the latter, prior to the two of them standing alongside each other with their heads turned in opposite directions (Fig. 12.6a). Further preparations for the fight lasted some 10 min. During this time, the bulls came head to head with each other several times, butting and pushing. Each raising of the head by one of the bulls immediately encouraged the assuming of a head-on position on the part of the rival.

In the next phase, the bulls stood alongside each other, turning their heads to come face to face (Fig. 12.6b). In the course of 5 min, each several times hit the rival in the back or head with its horns. The next, most fundamental, phase of the conflict lasted 15 min. The bulls stood head to head (cf. Fig. 12.6c), regularly locking horns and heaving powerfully against each other (Fig. 12.6d). The attack was renewed constantly, the males striking each other with their horns or slamming the horns into the head of their opponent. This behaviour was accompanied by powerful flicks of the tail. The horn-wrestling bulls trampled down the ground cover to the point where bare soil was exposed. Such a struggle required a huge amount of effort, such that the males became more and more tired, snorting and steadily slowing down the pace of their activity. The site of the fight was redolent of that very strong and characteristic musky scent of bison. At a certain moment, the defeated bull turned and slowly made off. The other animals in the group had in the meantime paid not the slightest attention to the conflict, continuing to feed or rest. The contest had lasted around 30 min in all, and was ultimately won by the bull who had been mating in the first place. Following all his exertions, he went to lie down in the bushes, only rejoining his herd after two hours of repose. Shortly thereafter, he had to take another lie down. When another hour had passed, he rose, uprooted a young spruce by digging it out with his horns and began to feed.

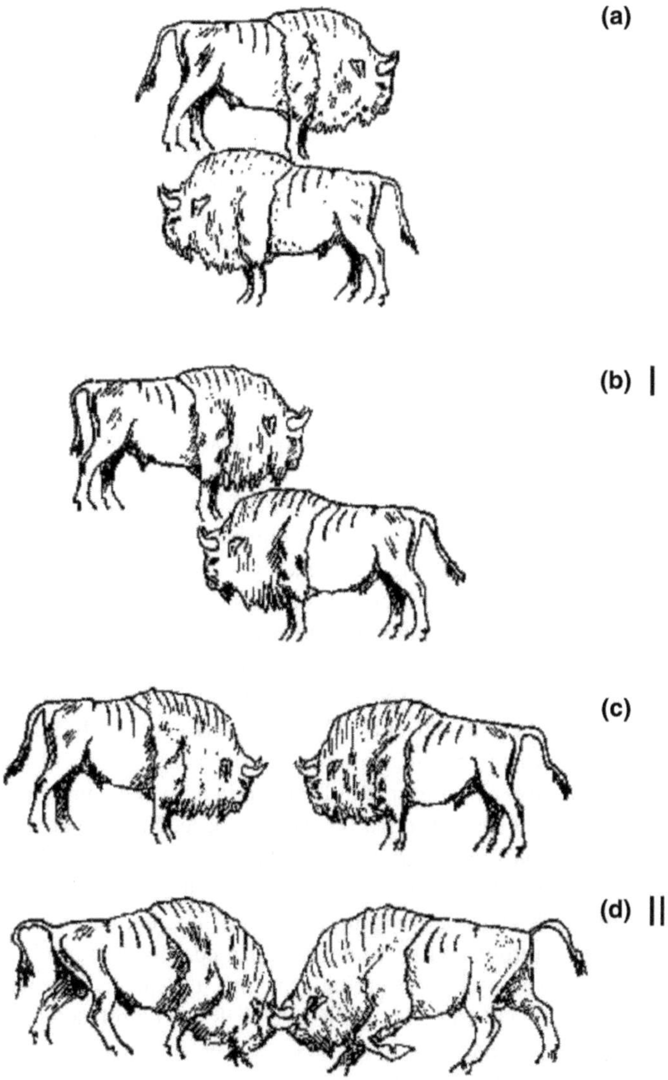

Fig. 12.6 The positions bulls adopt as they fight during the rut: (I-**a**, **b**, **c**)—preparation for the fight, and (II-**d**) the true fight itself. *Source* Caboń-Raczyńska et al. (1987)

During the evening, however, the group under observation linked up with another, in which there was also an adult male in similar condition to the first. The new bull closed in on his rival, circled him, and shook his head several times—but these behaviours drew no reaction. After ten or more minutes the bull that had won the battle in the middle of the day simply walked off and left the new, enlarged group.

In captivity, rutting season bulls kept in adjacent enclosures have been known to break down a fence in order to fight, notwithstanding the fact that the fences in question are made of 1.5-inch boards. The overcoming of such obstacles represents no more major problem for a bull, especially if it has caught the scent of a female in oestrus. The stubbornness of rivals is further attested to by an event that took place at the Białowieża Breeding Reserve. A very prolonged struggle between bulls was observed there one autumn. At one particular moment, one of the rivals suddenly moved off. Reserve employees witnessing the event breathed a sigh of belief, but before long it emerged that the bull had only gone off to his trough for a drink and was now seeking to renew his effort. After a certain length of time, the tired bulls gave up fighting of their own accord. They then rested for the two following days, lying in different corners of the enclosure and not going off to eat.

Battles between bulls may occasionally take place outside the breeding season. However, the first such conflict between bulls in the wild state took place, not in the breeding season, but in February (of 1964). In the first years after the release of bison from the Reserve, all the bulls were known by name. The adult males gathered by one of the supplementary feeding sites. Several times, the powerful bull PORANEK floored a male who was of similar age, but in a weaker condition. After the fight, POCZET was only able to rise with difficulty, before heading off into the young forest. From that time onwards, he only ever went to a feeding site when there was no sign of PORANEK anywhere in the vicinity (Fig. 12.7).

The case of the 11-year-old bull PONURY convinced us that the release into the wild of adult males from the breeding centre was not a particularly good idea. PONURY left the Breeding Reserve at Białowieża at the beginning of April 1964. After a couple of days he had found his way to the feeding site at which the free-ranging herd of bison took their supplementary fodder. PONURY, despite being in fine condition, was feeling somehow unsure of himself in the new situation, standing to one side and not going up to the feeding site, and instead observing his surroundings carefully. When 15 min had passed, the oldest and strongest bull POLAS began to wallow on the ground, making it clear that his blood was up. He then rose rapidly and came up against PONURY, who refused to give ground. Short tussles took place several more times, but no longer fight ensued. POLAS then headed off, demonstrating his strength by breaking young spruces. At that time, PONURY was accosted by the young bulls, who formed a circle around him. Seeing that the ring was closing, PONURY attacked one of his would-be captors and broke through the cordon. He was finally able to make his escape and never came back to the same feeding site. In November of the same year, having broken the fence, he returned to the Reserve, where he spent the whole winter calmly and safely. In spring, he was released once more, but was never again met with in the free-ranging herd, always choosing to pursue a life of solitude, and never moving too far from the Reserve. He died 4 years later, close by the Reserve fence.

In July 1983, a bull from the wild population elected to eschew the search for the herd of females with young, instead seeking to gain access to females living in the Breeding Reserve. Separated off by a fence at that time, they met with two

Fig. 12.7 The bull POCZET following its fight with PORANEK. In the groin area, there is a clear white scar caused by being gored with horns. *Photo*: ZA Krasiński

males, the 8-year-old POETA and the bull from the free-ranging herd. The two bulls were of similar age, although that from the wild herd had thick, blunt-ended horns, leucoma in the right eye and an estimated weight of 850 kg. POETA was in a slightly worse condition, weighing some 750 kg, and having long, spreading and sharp-tipped horns. The breaking down of the wooden fence was a joint effort by the two bulls. Once the obstacle had been overcome, the two came head to head with great force. The ground in the area of the contest looked as if it had been ploughed, while the thin trees and shrubs were all damaged. POETA, as the more agile and faster of the two bison, was able to bring old "One-eye" down, hitting him several times with his horns. The battled ended in a triumph for him, but when all was done both bulls were very tired and breathing in a laboured fashion. "One-eye" began to move off, but POETA followed him step for step. The two silent brown colossuses quietly circled each other and did not stray far from the place of battle. "One-eye", having been exhausted by the effort, moved around with some effort and stumbling over a tree trunk. However, the proximity of the rival ensured that there would be no rest for him. After four hours of such wandering POETA had become less aware and had started to browse at hornbeam shoots. The defeated rival took advantage of this and made good his getaway unseen. POETA lost sight of him and—disquieted in that way—began to look around, vocalise and move off at the trot, making straight for the hole in the fence via which he gained access to the females. "One-eye" needed a long rest after that, and spent his time lying in the young forest thicket (Krasiński 1994c).

12.5 Pregnancy and Parturition

European bison have an average gestation period of 264 (254–270) days (Krasiński and Raczyński 1967). Other authors state similar periods of 266 (256–279; Jaczewski 1958) and 267 (259–279) days (Kiseleva 1974), however, there are few visible outward signs of pregnancy among female bison, a fact associated with the build of cows and the limited size of the foetus.

Parturition is of brief duration. In four observed cases on the Białowieża Reserves, parturition lasted between 1 and 2 h 11 min. Cows calve both day and night, while standing or lying. Licking of the calf follows immediately from delivery (Fig. 12.8), and the placenta is eaten within 4–5 h. The calf begins to stand 20–45 min after birth (Figs. 12.9 and 12.10) and will start suckling within the first hour of life (Daleszczyk and Krasiński 2001). In the free-ranging population, a cow about to give birth leaves the herd to deliver the calf in a safe place. After several days the mother and calf rejoin the herd. According to Wróblewski (1927), it was most common for births in the early 20th century bison population at Białowieża to take place at night or in the early morning.

Cows give birth to a single calf. While twins are noted occasionally. In 1945, the cow F 219 PLANARIE gave birth to dead female twins (F 681 PLAX and F 682 PLAY), while in 1960 a female at the Prioksko-Terrasnyjj Reserve near Moscow produced the male twins M 1314 MOŁODEC and M 1315 MOŁODCZIK. A third case of twins being born is known from Vienna Zoo, two females again being produced in 1971 (and named F 2601 SCHOLLE and F 2602 SCOTTIN). Yet more recently, a fourth case arose on the reserve at Germany's Damerower Werden Breeding Centre (Kelterborn et al. 2009). In the case of a free-ranging population, it is usually hard to arrive at unambiguous proof that twins have been born. In turn, the free-ranging (LC line) population at UST'KUBENSKOE OKHOTKHOZJAJJSTVO in the Vologda region of Russia has twice reported the birth of twins (Tyapougin and Gusarov 2004; Gusarov 2011), though the veracity of this information is not certain.

Calves are born in spring, which is the period most favourable to their subsequent development. According to the oldest captive breeding data (for the years 1891–1946), the peak of calving (66 % of instances) falls in May, June and July. Remaining births take place throughout the year, although only occasionally in January and February (Koch 1956). At the Białowieża Breeding Reserve, the calving season is of shorter duration, extending from April to December (Fig. 12.11), although there is a peak of births (80 %) in the May–July period (Krasiński and Raczyński 1967). In the Russian Breeding Reserves, bison births most often occur between May and September, although they are also known from November and December (Kiseleva 1969, 1974; Belousova 1986).

In the last free-ranging population present in the Białowieża Forest in the late 19th and early 20th centuries, the earliest births occurred in April and May (Karcov 1903), although cases were also known from late summer and winter (Wróblewski 1927). In today's Białowieża population, most bison calves

Fig. 12.8 A female licks a newly born calf. *Photo*: ZA Krasiński

Fig. 12.9 Within around 20 min of being born, a calf is making its first attempts to stand up. *Photo*: ZA Krasiński

Fig. 12.10 The first steps are not very successful. The mother moves barely a single pace away from the calf, and is quite capable of mounting a fierce attack on any intruder coming within her zone of security. *Photo*: ZA Krasiński

Fig. 12.11 Calves born in June, July and August at the Białowieża Breeding Reserve. *Photo*: ZA Krasiński

(70–80 %) are born between May and July inclusive (Krasiński and Raczyński 1967; Krasiński 1978a). It is rare for births to occur at the end of April. Equally, in the first 10–20 years of the development of the Białowieża population, 20–30 % of all calvings took place in late autumn, something that can be explained by reference to the influence of many years of captivity (Krasiński 1978a). In the last 20 years, the numbers of such calves born late in the wild herd have declined—today representing just 15 % of all the births taking place in a given year.

12.6 The Sex Ratio at Birth

Births in both Reserve and free-ranging populations of bison are characterised by non-significant departures from the 1:1 male:female ratio, although particular years inevitably bring certain departures from this (Zablocky 1957; Jaczewski 1958; Krasiński and Raczyński 1967; Kiseleva 1974; Krasiński 1978b; Bunevich and Kochko 1988; Krasiński et al. 1994a; Krasochko et al. 2004). From the time of the first bison birth back in the wild in 1957 up to 2011, a total of 2,070 calves had been born in the Polish part of the Białowieża Forest, 982 of these being male and 1,088 female.

Insert II. 1, 2 Bison live in herds whose size depends on the season and the habitat. Thus, groups in the forest are smaller than those of meadows and clearings. *Photo*: ZA Krasiński

Insert II. 3, 4 Old bulls lead loners' lives, or else go around in pairs. *Photo*: ZA Krasiński

Insert II. 5, 6 In the moments immediately after giving birth, a cow will not move a step away from her calf. At this time, the urge to defend the young one entails dramatic attacks upon any intruder. Such thorough care extends to a calf for the first several months of its life. *Photo*: ZA Krasiński

Insert II. 7, 8 Clouds of insects surround bison during the summer months, and efforts to be rid of them for even a short while entail animals wallowing and covering oneself in sand. Wallowing on to the side is a particular kind of skin massage, though the ritual is often performed with a view to emotional stress being relieved. *Photo*: ZA Krasiński. A bison takes advantage of the deep cover of snow to look out beyond the Reserve fence—however, this is more in the nature of curiosity than any nostalgia for life in the wild. *Photo*: ZA Krasiński

The sex ratio at birth for the Polish population did not depart significantly from 1:1 over the 1957–2002 period. In contrast, the years 2003–2011 brought some skewing of the ratio in favour of female calves ($chi^2 = 10.59$, $p = 0.001$). Furthermore, this was a period associated with bison being in less favourable condition, this being in line with recent research suggesting that the condition of cows may influence the sex ratio of calves brought to term, inasmuch as that more female calves are born in these circumstances. The corollary also applies, since years with an abundance of nutritional food (like mast years for oaks) are associated with the subsequent birth of a higher proportion of males (Hayward et al. 2011).

Chapter 13
Behaviour Patterns

B. Bańka

13.1 Feeding Behaviour

Feeding bison pass through different types of forest and non-forest habitat (the latter including meadows and forest clearings). As they feed, different individuals within the group are only separated by a couple of metres and are guided in an overall direction by the movement of the group as a whole. Smaller calves stick close to their mothers. The dominant cow may move to the front of the group in certain situations, such as when there is a forest road to cross at the end of a longer period of feeding. However, if feeding is continuous, even such visible boundaries will be crossed by the herd without any obvious leadership.

M. Krasińska and Z. A. Krasiński, *European Bison*,
DOI: 10.1007/978-3-642-36555-3_13, © Springer-Verlag Berlin Heidelberg 2013

13.2 Excretion

Defecating bison come to a halt, raise their tails and lean back slightly on their hind limbs. Urinating females behave in a similar way, while males urinate while standing and raising their tails slightly. Excretory behaviour most often takes place (on 63 % of observations) as bison rise, having rested. In only 14 % of instances does it take place prior to or in the course of a rest period. The behaviour is still rarer when feeding is taking place, while urination is extremely rarely observed while animals are on the move. Calves urinate immediately after suckling, while adults do it just 2–3 times a day. It is likewise usual for adult bison to engage in defecation just once a day.

13.3 Behaviours Associated with Rest

Bison go down to the ground by kneeling first, then lowering their forequarters and finally their rear ends. They in turn stand up by coming into the kneeling position and raising their hindquarters first (Fig. 13.1). After standing up, they mostly shake themselves and their coats. A typical rest position for a bison entails the adoption of the physiological position, on the sternum with tucked-in forelegs (Fig. 13.2). Only rarely do bison lie on their sides with legs stretched out. Bison also rest in the standing position. In the course of a rest period lasting a couple of hours, individuals are seen to rise every now and again, stretch and lie back down

Fig. 13.1 Old bison can only rise with difficulty. Animals getting up first raise their rumps, and lean on their carpi. *Photo* ZA Krasiński

Fig. 13.2 After feeding on hay, bison lie down not far from the feeders in order to chew the cud, usually in a sparsely-forested area or clearing. *Photo* ZA Krasiński

on the other side. A change of position is often also associated with wallowing. Sleep with closed eyes is only rarely observed during the day, and when it does take place it lasts for between 5 and 20 min (Caboń-Raczyńska et al. 1983).

When a mixed group is at rest, the small calves lie close to their mothers, while older ones rest close to one another, quite often forming a sub-group of calves. Adults lie at distances several metres apart. Resting bison do not make physical contact with one another. The shift from active feeding by all members of a group to a restful state takes longer than the rise after resting.

Bison have favourite resting places within the parts of the Forest they frequent. These are mostly slight elevation (always dry), quite often located close to forest roads or in felled areas. In periods of heightened insect activity, bison seek out places less frequented by the latter, like thick young stands of spruce or pine, or open areas. For example, on one hot day in June 1979, it was possible to observe a group of no fewer than 48 bison resting on a sandy and fully sunlit cleared area of forest (Fig. 13.3), at which the herd spent the whole time from 07.00 to 18.00, with nothing more than a single break to drink around midday. Some animals lay on their sides with extended legs, panting away (Fig. 13.4). Thus bison would apparently prefer to put up with the hot sun than to expose themselves to the torment imposed by swarms of insects.

13.4 The Behaviour of Bison on the Move

Bison move around to find new feeding sites, places to rest or watering places. The dominant female usually takes up position at the head of the group in these cases, other bison following on one after the other, not far behind. Young individuals and older calves are usually located in the middle of the group. Should the

Fig. 13.3 A herd lies in the full sunshine of a clearing for many hours, notwithstanding what must have been fierce heat. The behaviour at the time (early June 1979) might be accounted for by the mass emergence appearance of mosquitoes. Only in the evening did the bison rise and head off to water for a drink. *Photo* ZA Krasiński

Fig. 13.4 It is possible for bison to take a nap, especially when warmed by the sun. A very close approach to the dozing individual is then possible, and there are even known cases of a beast being considered dead by an observer. However, a slight touch resulted in this scared cow getting very rapidly to her feet. *Photo* ZA Krasiński

group include an adult bull, he will most likely bring up the rear. Wandering bison make use of the tracks and paths trodden by various game animals. In 50 % of observed cases, such bison were seen to choose ways already identified as typical for them. Unlike animals in groups, lone bulls have no reservations about making use of forest roads, even over distances of several kilometres (Caboń-Raczyńska et al. 1987). Korochkina (1972a) notes that bison are not very ready to push their way through dense young stands of trees. They nevertheless take the shortest route when heading for a feeding ground known to them from previous years.

13.5 Antagonistic Behaviour

Active aggressive behaviour is rare among cows, adolescents of either sex, or calves. In the course of more than 2,000 observations of mixed groups, only single instances of females attacking other members of a group were noted. In the cases that have been observed, cows were most likely to prod their horns into the hindquarters of the second individual. A cow may also react in this way to her own calf, if the latter is too persistent in demanding to suckle. Groups of bison are in no way of constant composition, mergers taking place frequently, with subsequent re-division into smaller herds. Outside the rutting season, 99 % of such mergers that have been observed took place without any visible reaction. Observations suggest that the hierarchical structure of groups is maintained in the course of the process, and is recognised by newcomers. More severe aggression between two females was only observed once—a couple of minutes after two herds had joined together. A solution to the problem had clearly been arrived at within 15 min of the onset of this aggression, as both cows then proceeded to feed calmly.

In contrast, the rutting period is quite frequently characterised by aggressive interactions between adult bulls as groups merge. Where older bulls (aged 6–12) are present in each, these may assume the characteristic pose intended to determine which male is dominant. Such a bull may, for example, close in on the rival, circling him slowly, coming to a halt, hoofing the ground, wallowing on it and damaging young spruce trees. If the other male does not adopt a submissive attitude, or else demonstrates symptoms of aggression for its own part, then a battle will take place (see colour insert).

Aggressive behaviour toward younger rivals is also displayed by bulls covering females on rut. Young males keeping a safe distance between themselves and the mating pair often engage in pushing behaviour, trials of strength, etc., although these are only make-believe conflicts. A bull attending a receptive cow may also show aggression towards her calf, propelling it out of the way should it try to suckle. However, cases of calves being killed by bulls during the breeding period have not been noted in the wild herd, notwithstanding the fact that this behaviour does arise where captive animals are kept in small enclosures.

In stressful situations (such as those induced when bison are disturbed repeatedly by people), it is possible to observe an intensification of aggressive interactions between different members of a group. What are displayed most often are passive forms of aggression and mock attacks, with no direct contact taking place. Behaviour such as shaking the head by a female is usually enough to bring other members of the group into line, and to encourage them to move off to a safe distance.

13.6 Relations Between Mother and Calf

Observation of parturition and of the subsequent development of relationships between mothers and calves is only realistically possible in the conditions of the Breeding Reserve, in which the identification of individuals is possible, as well as the constant observation of the same pairs of animals. Only once did it prove possible to observe the arrival at one of the free-ranging groups of a female with newborn (still-wet) calf. When that herd moved off, the cow encouraged the calf to follow on by nudging it with her head.

Small calves generally lie close to their feeding mothers. One-month-old calves are already very mobile, following close to their mothers when groups are on the move, while their several-month-old counterparts assume positions in the middle of the group.

The first week in the life of a calf is the period of the closest relationship between mother and offspring (Daleszczyk 2002, 2004). The calf remains very close to its mother, no more than 3 m away. At this time the female limits the time spent feeding, in the interests of fuller engagement in maternal behaviour, including licking and feeding. It is she who most often initiates contact with the calf, and it is in this period that the latter comes to recognise the scent, call and appearance of its mother (Fig. 13.5).

Fig. 13.5 A calf shortly after birth. Olfactory and visual stimuli become imprinted, allowing the calf to recognise its mother unfailingly within the herd, and vice versa. *Photo* ZA Krasiński

Calves move around with their mothers from the earliest times, not being left hidden at all—as is the case with some other ungulates (like roe deer). The tendency to follow the mother as opposed to remaining in a place of concealment is characteristic of animals inhabiting open areas and, since this strategy is pursued by both the forest-dwelling European bison and the prairie-inhabiting American bison, it is considered by Daleszczyk (2002) to have evolved in a common ancestor that also inhabited open areas.

In captive conditions, the duration of the suckling period is dependent on the sex and age of the calf. Compared with females, young males suckle for longer, but less often during the day. The first 3 months in the life of a male see it suckle for an average of 7 min at a time, cf. 5–6 min in heifers. Actual times noted vary from several seconds to 22 min (Daleszczyk 2002, 2004).

The pattern to maternal and calf behaviour in the first 3 months of life was described by Daleszczyk (2002), and remains true for the species irrespective of the conditions in which the bison are living. Like Reserve-dwelling specimens, free-ranging bison engage in suckling behaviour at the initiative of either the mother or the calf. By grunting, the mother calls the calf up to feed. The young one begins feeding by nudging its head against the udder several times. In the course of suckling, a calf most often stands alongside its mother's body, sucking off to the side and wagging its tail rapidly at the same time. When a mother tries to move away, the calf runs neck and neck with her, encouraging her to stop. If she does so, milk will be taken from the other side as well. Larger calves, from the previous year, most often suckle from behind, approaching the udder from between the mother's back legs. In the case of free-ranging bison, all that are known from observation are unit suckling times. Between May and July, calves aged from several days to 3 months suck for shorter periods and more frequently, on average for 4 min (range 2–6). Those aged 3 months and over in turn feed for longer (on average for 7 min—range 4–19) but less often (Caboń-Raczyńska et al. 1987). If a cow has no new calf, she may go on feeding last year's calf even up to October, although suckling most often ends in spring.

Care for the calf in turn entails thorough licking, generally during feeding or after the young one has defecated. Such behaviour is generally observed in the first month of life. The first cud-chewing (rumination) is observed among calves aged 2–3 weeks, suggesting that they have begun to take their first solid food. Equally, it is from the age of 6 weeks on that time spent on feeding begins to increase markedly (Daleszczyk 2005).

Calves remain with their mothers for the whole first year of life, and it is also usual for them to stay in the same group as the mother in the second year of life. There are nevertheless cases in which a young individual barely more than a year old may leave the herd of its mother (Krasińska et al. 1987).

By the time winter arrives, most of the calves born in May–July are already a few months old and do not require any special care or attention from their mothers. Calves eat hay by this time, although they are still being fed by their mothers. Calves that are between 5 and 10-months-old suckle for 8 min on average (3–24), with feeding only taking place from 1 to 3 times a day. Only a small percentage of the calves enduring their first winter were late (August–October) births, their development being weaker such that they enter winter poorly prepared. Calves have also been seen suckling at night in winter, and there are numerous observations of two calves being suckled simultaneously by a cow at a place of winter assembly. At these times, there is one calf taking the udder from the side, the other from between the back legs, suggesting that the latter calf is most often last year's, or rarely the current year's calf of a different cow. There is even one observation of a late-born cow being suckled by a non-maternal cow for a period of as much as 15 min (Caboń-Raczyńska et al. 1987).

The strength of the maternal instinct and role of scent in mutual contacts is attested to by the story of the calf POLARKA (described in Figs. 13.6 and 13.7). The tender loving care extended to a small calf is also confirmed by observations of a marked female who separated off from her group and spent 7 days in a small area with a sick, very lame calf plus a 2-year-old bull. A female usually stays with a dead calf for up to 3 days.

Fig. 13.6 POLARKA was born on August 17, 1985 at the Białowieża European bison Breeding Reserve. Like all other European bison in Poland, she obtained a name beginning with the letters PO. Her mother was the 18-year-old cow POLARNA, and she was sired by the 5-year-old bull PODKOMORZY. In the course of the frosty night of January 29, 1986, the 5-month-old POLARKA became trapped beneath a lying tree, from which she was unable to extract herself using her own strength. The low temperature and lack of food quickly left the calf exhausted. By the time people had freed her, she was too weak to rise. The mother's constant nosing to encourage her to stand up were thus to no avail. While the mother continually sought to bar access to the young one, it did eventually prove possible to remove her from the paddock. By this stage the calf was in a state of extreme exhaustion. However, once her initial resistance was overcome, POLARKA quickly learnt how to drink milk from a bottle. Nevertheless, it was only after 2 weeks had passed that the calf finally managed to get up. She would drink 3 litres of milk with added oat flakes at one go. By mid-April it was possible to see clear signs of the changeover from the winter to the summer coat. *Photo* ZA Krasiński

Fig. 13.7 POLARKA grew quickly and became ever stronger, and it was decided to take her back to the Reserve. She was thus released into the same enclosure from which she had been removed three and a half months previously. After a couple of minutes, her mother emerged from the forest and POLARKA began to approach her slowly and cautiously. First she sniffed her side, then the two sniffed each other. After that, POLARKA instinctively came up to her mother's udder. The following day it was possible to observe how she suckled from her mother, who had continued lactating thanks to her supplementary feeding of other females' calves. While she no longer wanted to approach those who had cared for her, POLARKA would react to their calls for a long time subsequently. *Photo* ZA Krasiński

13.7 Play Behaviour

This type of behaviour is not particularly common in bison, being mainly observed in calves and adolescents. In groups, play can be observed at most once or twice a day, although it is not characteristic for all young animals. Calves most often engage in play when the herd comes out into an open area, clearing, clearcuts or meadow. They run, raise their tails in the air and kick out with their back legs. Sometimes a single calf may play in an open space, using back kicks to raise its hindquarters clear of the ground. Young males in turn engage in mock fights, locking horns and pushing against one another (Fig. 13.8). Such periods of play do not

Fig. 13.8 This is not a serious fight between bulls, but the pushing and shoving of young males at play. *Photo* ZA Krasiński

normally exceed 30 min in length. Play behaviour has also been seen to take place at the same time as feeding or rest, the participants being 2–3 pairs of young males (Caboń-Raczyńska et al. 1987). During winter, play behaviour is a rarity among bison, although young males are sometimes seen to flick their horns into piles of fluffy snow. In the warmer part of the year calves head-butt one another in groups of two or three for perhaps 5–10 min. Young females also nudge against young males, while males may also practice-mount the females.

13.8 Grooming Behaviour

This group of activities includes rubbing, licking of its coat and wallowing on the ground. These behaviours are observed with similar regularity among bison of different sexes and ages. Rubbing takes place against tree stumps, fallen trunks and their legs and the crowns of lying trees. Sometimes bison also scratch their skin with the hoofs of their hind legs. Grooming activity is observed equally often while animals are at rest or feeding. Rubbing continues for between 1 and 25 min and may be engaged in several times a day. The maximum amount of time bison spend in cleaning themselves is in turn 1 h. Rubbing behaviour intensifies in summer when insects swarm, as well as during the spring moult, at which time the activity facilitates the removal of the old winter coat.

Wallowing is most often observed among adults, and only rarely so among calves and young specimens. As many as 73 % of all observed cases of wallowing among bulls occur in the breeding season. Furthermore, if a given herd includes several adult bulls simultaneously, the frequency with which the behaviour is engaged in increases, especially in the case of the dominant bull. In turn, when a bull is mating with a cow in oestrus, the subordinate bulls also engage in more frequent rolling. An equivalent situation applies among females, inasmuch as that the dominant cow wallows more often than other cows. Wallowing by females and young most often takes place when a group is resting. The favourite places for wallowing are sandy hillocks, gravelly or sandy forest roads, roadsides and areas of exposed soil in felled areas or gravel pits. The places in question are known as "dust baths" (Polish – "kąpieliska", although they once went by the name of "kupryska"). Before beginning to wallow, a bull may paw the ground with its fore legs, scattering sand or soil over itself in the process. Urine is sometimes passed at the site in advance of the wallow. In wallowing, a bison lies on its side and kicks out its straightened legs, but does not turn on to its back (see colour insert). Thus, a bison wishing to wallow the ground with its other side, will not roll over on to it, but stand up and then lie down once more. Among 253 observations of wallowing, there are only six observed instances of any synchronisation of the behaviour among all members of a group. A rare observation concerns cows that lined up to take it in turns to use a sandy hollow in a felled area of forest, wallowing on its sloping, sandy sides. As Wróblewski (1927) rightly noted, bison never wallow in mud.

Wallowing does also take place in winter, however, most often in the case of adult specimens (in 76 % of instances), only more rarely with adolescents being involved (16 %), and very rarely with calves (8 %). The behaviour takes place in slight elevations. Prior to a wallow, the place chosen for the activity is subjected to a pawing of snow using the hoofs. The wallowing of a series of bison (up to 8) in a given place has also been observed.

13.9 Vocalisation

It is rather rare for bison to communicate vocally. A mother uses grunt to call a calf over to feed, while only rarely does a calf grunt to bring its mother. A mother also uses this means to summon her calf when she is feeling unsure of the situation. Bulls also grunt when they feel threatened. Their voice is lower and harsher than that of cows. During the rut, the call of a bull may be described as threat grunt where it serves to repel or discourage potential rivals. At this time, bulls also produce other sounds that are sometimes termed blowing and puffing.

13.10 Human–Bison Interactions

A herd of bison encountered in the forest usually reacts to the sight of a person by escaping. However, the behaviour animals indulge in depends on the composition of the herd, the reaction of the lead female and the degree of anxiety the animals are feeling. Where bison become severely irritated by insects, they are inclined to flee even more quickly at the sight of a person. Startled bison gallop over short distances, but after covering 100–150 m at speed they usually come to a halt, and often crowd together. The most nervous herds are those with a large share of adolescents. Long-term observations of selected herds have raised animal's reaction thresholds (habituated them), reducing the distance at which flight takes place, and making it possible for a herd to be followed from dawn until dusk without any more major disturbance being caused.

Bison may react aggressively to a human presence if they feel they are threatened. The presence of calves seems to provoke females into "courageous" behaviours. However, if people keep a safe distance from the herd, then bison will attempt to keep their distance without resorting to aggressive behaviour.

It is usual for adult bulls to feel more confident than cows, while the largest and strongest of all behave like true lords of the Forest. They thus feel at liberty to treat people with an ostentatious lack of concern. Indeed, the reaction of bulls in the rutting season to an approaching person is positively alarming: they shake their heads, grunting threateningly, pawing the ground and flicking their tails. If the person involved fails to recognise the signs and to treat discretion as the better part of valour, then a bull may go on to the attack. In contrast, outside the season, bulls generally depart quietly enough, although they are not particularly keen to make way. Bulls are known for their tendency to enter settlements, and if driven off, barked at by dogs, or even shot at, they are again inclined to attack. It is particularly difficult to discourage bulls from approaching stored hay or from going out on to cultivated fields. Even the capture and transfer of such specimens to other parts of the Forest may not in many cases suffice, since the animals frequently return to their favourite places. Jarocki (1830) recalls how a bison enamoured of a peasant's haystack could not be kept away from it by any kind of fencing.

It should be stressed that an encounter with bison from the free-ranging population in the forest is not hazardous, providing that a safe distance is maintained and a person remains calm. Bison living in captivity can show greater aggression towards people than those living free.

In the winter period, the behaviour of bison towards people changes markedly. Animals are less nervous, since the sight of a person is first and foremost associated with a supplementary food supply.

It has never yet proved possible to bring an adult bison to a state of complete tameness, while an excessively familiar relationship with this powerful ruminant is likely to meet with a drastic and unfavourable response.

13.11 Interactions Between Bison and Other Wild or Domestic Animals

The authors of old were inclined to stress the antagonistic behaviour bison displayed towards horses, cows and dogs (Jarocki 1830; Karcov 1903; Wróblewski 1927). Jarocki underlined just how scared of bison horses were, while Wróblewski noted that "bison cannot stand horses, and there is no year in which several of the latter do not lose their lives as a result".

All documented cases of bison-induced wounding of horses (11) or cows (5) are presented in Table 13.1. The most recent such case relates to 1996. In this respect, declines in the numbers of head of cattle and of horses in the Białowieża Forest area go a long way to explaining the lack of such incidents in later years. In two cases, the wounding of horses took place in forest, while the remaining instances relate to fields and meadows close to Pogorzelce and Teremiski villages, which are located in the middle of the Forest or else in Chrabostówka, Łosinka and Wiluki (all on the Forest edge). In three cases, the incidents concerned took place in a typical circumstances and places: along a narrow road fenced off on both sides, a passing bison hit a horse pulling a cart. In two cases, bulls from the free-ranging herd wounded working horses in the Reserves they had entered. In each case, the incidents involved a one-off impact of the horns in the lower part of the body, inflicted by an adult male. Such cases generally occur in the rutting season, although this is not always so. Horses bringing in hay have not been subjected to aggressive behaviour on the part of bison. Young bison and tarpan-type horses—for long kept side by side in the same enclosure—proved well able to get along together (Fig. 13.9). Likewise, bison coming out on to meadows and cultivated fields beyond the Forest will tend to graze peacefully within herds of domestic cows (Fig. 13.10).

Table 13.1 Cases of the wounding of horses and cattle by bison in the Białowieża forest and its surroundings

No.	Date of event	Location	Remarks
		HORSES	
1.	14.08.1967	Meadow along the Łutownia river	
2.	31.12.1967	Teremiski village	Wounding of a horse pulling a cart along a narrow road
3.	25.11.1969	Forest	Przewłoka Forest District
4.	19.09.1971	Meadow and fields at Białowieża village	
5.	8.06.1975	Forest	Compartment 272D
6.	9.02.1981	Experimental Reserve No 4	Bull from the free-ranging population entered the reserve
7.	6.10.1983	Meadow along the Łutownia river and Teremiski village	
8.	5.10.1989	Łosinka village	On the outskirts of the Forest
9.	12.10.1989	Łosinka village	On the outskirts of the Forest
10.	5.06.1992	Wiluki village	
11.	4.06.1994	Reserve nr 1	A bull from: the free-ranging population aged 12 entered the reserve
12.	13.09.2010	Wiluki village	Marked bull on the outskirts of the Forest
		DOMESTIC CATTLE	
1.	29.08.1967	Meadow along the Łutownia river	
2.	18.09.1985	Meadow at Chrabostówka village	On the outskirts of the Forest
3.	11.10.1985	Meadow at Pogorzelce village	
4.	21.10.1985	Meadow at Pogorzelce village	
5.	30.09.1996	Meadow at Chrabostówka village	On the outskirts of the Forest

Source Krasińska and Krasiński (2007), updated

Fig. 13.9 A young bison in an enclosure along with a working horse, each showing interest in the other. *Photo* ZA Krasiński

Fig. 13.10 A bison and a domestic cow feed calmly side by side on a cultivated field (village of Grabowiec, September 1988). *Photo* ZA Krasiński

Fig. 13.11 Deep snow cover and frost make it necessary for wild boar to keep an eye open for bison leftovers. Winter 2002/2003, Breeding Reserve in the Białowieża Forest. *Photo* ZA Krasiński

Wróblewski (1927) wrote that, as bison were unnerved by the very sight of dogs, the latter should not be taken into areas where they occur so as to avoid being attacked. Equally, if a scared dog runs back to its owner, that owner may also then be put in danger. Our observations suggest that bison pay no attention to dogs that bark or jump at them. However, a dog that approaches too close risks being kicked out at. There have also been several cases of dogs chained to haystacks or sheds being killed by bison.

Fig. 13.12 In the 19th century in the Białowieża forest bison and red deer gathered at feeding sites during winter. Illustration by A. S. Hronov, *Source* Karcov (1903)

There are many reported observations of wild boar feeding among bison in both summer and winter, although again an excessively close approach may provoke an attack (Fig. 13.11).

19th century observations refer to red deer approaching bison feeding sites (Fig. 13.12). There is at present no confirmation of the observation from Wróblewski (1927) that bison avoid those places in which red deer dung is present.

Chapter 14
Food and Use of the Environment

B. Bańka

14.1 Dietary Selectivity of Wild Ruminants

The Białowieża Forest is inhabited by four species of large wild ruminants: the European bison, moose, red deer and roe deer. The wild boar is another ungulate (Table 14.1). The most abundant of these species is the red deer, which accounted for 36 % of the overall population of ungulates in the 1980s and 1990s. Bison accounted for 3 % of that population (Jędrzejewska and Jędrzejewski 1998). According to data from winter 2002/2003, bison's share of all ungulates reached 8.5 %, due to the culling or hunting of the remaining species and an increase in bison numbers. By early 2012, this situation had changed again, with bison accounting for just 5 % of the ungulate population, and account of increases in numbers of red deer and wild boar. In the face of the disproportions in size

M. Krasińska and Z. A. Krasiński, *European Bison*,
DOI: 10.1007/978-3-642-36555-3_14, © Springer-Verlag Berlin Heidelberg 2013

Table 14.1 Percentage community structure of ungulates in the Białowieża Forest, 1991–1995

	European bison	Moose	Red deer	Roe deer	Wild boar	Total %
Percentage of all ungulates number	3	1	36	33	27	100
Percentage of total biomass of ungulates	17	3	46	9	25	100

Data from: Jędrzejewska and Jędrzejewski (1998). *Source* Krasińska and Krasiński (2007)

Table 14.2 Percentage composition of the food of ungulates living in the Białowieża Forest

Type of food	European bison	Roe deer	Red deer	Moose
Herbaceous vegetation	90	80	40	10
Shoots of shrubs and trees, and trees-bark	10	20	60	90

Data from: Jędrzejewska and Jędrzejewski (1998). *Source* Krasińska and Krasiński (2007)

characterising the different species, it is more reliable to offer biomass-based comparisons. In this context, the share accounted for by bison is higher—at 17 % of ungulate biomass, while red deer account for no less than 46 % of the total (Table 14.1) (Jędrzejewska and Jędrzejewski 1998). In the 19th century, the Forest area was also inhabited by the introduced fallow deer (Dama dama).

Each of these ruminants has different dietary preferences (Table 14.2). Bison are among the species manifesting limited selectivity: their sheer size makes it imperative that they simply take in large amounts of food. At the other end of the scale is the roe deer, a highly selective feeder, while the position of the red deer is somewhat intermediate.

The structure of an animal's alimentary tract predisposes it to specific types of food (Hofmann 1978). Bison resemble domestic cattle in having a 20:1 ratio for gut length as compared with body length (Pytel 1969; Gill 1999). The stomach comprises four chambers, the three pre-stomach compartments of the rumen, reticulum and omasum, as well as the abomasum serving as the true stomach. The bison rumen is particularly impressive, its volume sometimes exceeding 100 L (Pytel 1969). It serves as a fermentation vessel, and food from it is regurgitated for second chewing before being remixed. Symbiosis with specialised bacteria ensures that plant food is digested effectively and may be used as a construction material and energy source. Some of the products of metabolism (the volatile fatty acids) are absorbed from the rumen itself, while the mass of food is next transferred to the reticulum and omasum, the excess water being removed at this stage. Finally, the abomasum is the site of the ultimate breakdown of food to liberate nutritional components.

The fermentation processes in the rumen of a bison continue uninterrupted, so constant augmentation with food is essential. This fact finds its reflection in the rhythm of bison feeding and general activity. Intensive feeding prior to sunrise allows bison to fill their huge pre-stomach once again, following a several-hour break during the night. Bison then lie down to rest and chew the cud of the selected food. Such a sequence of events is repeated several times a day.

The food of the roe deer is composed first and foremost of leaves, herbs, flowers, stems and fruits. Bulk food like hay is not suitable for this species, as the rumen has too few bacteria capable of digesting cellulose (Hofmann 1978). In the winter, it is even possible to come across roe deer that have died of hunger, notwithstanding the fact that their rumens were full of twigs. In turn, in the Białowieża Reserve, there are observations of fatal atony of the rumen in roe deer following on from the consumption of excessive amounts of food concentrates (cereal mixes).

The rumen of a red deer, with a volume of some 25 L (Hofmann 1978), is still only one-quarter as large as that of bison. Some 40 % of a red deer's diet is in the form of green plants (Table 14.1). Supplying individuals of this species with concentrated food in winter leads to an unsafe increase in acidity levels capable of modifying the bacterial and protist flora in the gut. For its part, the European bison has two kinds of protist in its pre-stomach, representing the families *Ophryoscolecidae* and *Isotrichidae*, the former always accounting for some 70–90 % of the total (Gill 1999).

While the poor 19th century development of the population of European bison was traditionally ascribed to competition for food with red deer (Wróblewski 1927), the above remarks would seem to support the idea that both bison and red deer populations wwere suffering first and foremost from an overall shortage of food caused by over-dense populations of each, rather than competition. The situation was made worse by cattle grazing, as Jędrzejewska and Jędrzejewski (1998) noted. In a situation of ecosystem balance, the large herbivores resident in the Białowieża Forest will not be in competition for food with one another, because each has its own dietary preferences shaped over evolutionary time. In this respect, Hofmann (1978) reminded people how inappropriate it was to consider "red deer units", whereby one red deer corresponds to five roe deer, while one bison is equivalent to 1.5 red deer.

14.2 Diet Composition

The European bison is a herbivorous ruminant dwelling in a forest or forest-field environment. From one season to another, the animals select such habitats that will assure them of suitable food. Bison utilise the vegetation of the forest floor, clearings or meadows and of the shrub layer, as well as food originating from trees, i.e., leaves, sprouts and even bark.

There is no significant information available by which we could learn more about the diet of the bison living in the Białowieża Forest centuries ago. Rather, the data that are usable already derive from a period in which human beings had exerted a powerful influence on the environment of the once-wild Forest. By even the 16th century, people had begun to cut the grass of the meadows along the rivers here, putting the hay into service to meet their own needs and the supplementary-feeding requirements of their stock (Hedemann 1939). However, the authors of old agree that the primary food of Białowieża Forest bison comprised plants of the forest floor. Karcov (1903) wrote that the bison resident in the Forest in the 19th century showed a preference for herbaceous plants in both spring and summer. Finally, with the arrival of August and the tendency for grass in the forest to dry out completely, they turned to the young shoots of deciduous species. Wróblewski (1927) in turn studied the last free-ranging population of the early 20th century, and concluded—on the basis of rumen-content analysis—that grassy food made up 80 % of a bison's diet, leaving 20 % that originated with trees. It was his view that the structure of both teeth and stomach predisposed bison for the consumption of large amounts of soft food. It was for this very reason that the species inhabited lowland forest with a diverse herb vegetation (Fig. 14.1). In studying the feeding of bison on trial plots while at the same time observing the Polish part of the Białowieża Forest directly, Borowski and Kossak (1972) were able to establish that the diets of bison included 137 plant species, of which 27 were of trees and shrubs, compared with 14 of grass and sedge and 96 of herb. The food of bison was 33 % tree and shrub matter, 67 % of grasses, sedges or herb plants. Tree species of significant importance in forestry made up 19 % of what was consumed.

Again according to Borowski and Kossak (1972), the preferred trees and shrubs species of which shoots are eaten are hornbeam *Carpinus betulus*, sallow *Salix caprea*, ash *Fraxinus excelsior* and raspberry *Rubus idaeus*. Where grasses and sedges are concerned, the species eaten most commonly are rough small-reed *Calamagrostis arundinacea*, wood sedge *Carex sylvatica* and hairy sedge *Carex hirta*. In turn, the herbaceous plants taken most willingly are ground elder *Aegopodium podagraria*, common nettle *Urtica dioica*, woolly buttercup *Ranunculus lanuginosus* and cabbage thistle *Cirsium oleraceum*. Holy grass

Fig. 14.1 Herbaceous vegetation forms the main food of European bison. *Photo* ZA Krasiński

Hierochloe odorata is a rarity in the Białowieża Forest and as such plays a limited role in the diet of the European bison, notwithstanding the fact that its popular Polish name is *żubrówka*, i.e., bison grass (Fig. 14.2).

Where debarking by bison takes place, the species most often involved are the pedunculate oak *Quercus robur*, hornbeam *Carpinus betulus*, ash *Fraxinus excelsior* and Norway spruce *Picea abies*. Acorns are in turn a favorite food of bison, although they are only present in abundance in the masts occurring every 6–9 years or so (Fig. 14.3). Years following mast years in the Białowieża area are

Fig. 14.2 Sweet grass or holy grass *Hierochloe odorata* is known in Polish as *żubrówka* (or bison-grass). It is in fact rare and protected in the Białowieża Forest and has not played an important role in the diets of bison since written records were made. On drying, the grass gives off a strong scent of cumarin. The undersides of the leaves are covered with a whitish waxy coating that is easy to rub off. A second species more often met with in the Forest is the southern holy grass *Hierochloe australis*. *Source* the BNP archives

associated with more effective reproduction in the bison population (Mysterud et al. 2007). Likewise, in the submontane environment of Poland's Bieszczady Mountains, bison remain longer in the more elevated beech stands in mast years, where they feast on beech nuts (Paszkiewicz 2004). The so-called mountain bison (hybrids between the American and European species) of the Caucasus Biosphere Reserve devote the early spring (March and the beginning of April) to the consumption of spring-flowering herbs, as well as the stems, twigs and bark of deciduous trees plus firs. In summer, they graze on the subalpine meadows existing above the treeline. They also willingly take fruit from abandoned orchards (Nemtsev et al. 2003).

Research on the dietary composition of bison living in the Polish part of the Forest has also been based on analysis of rumen contents. This reveals that grasses, sedges and herbs represent the main food of bison (accounting for 87–93 % of the total), while food originating from trees accounts for between 7 and 13 % of the contents of the rumen (Gębczyńska et al. 1991). Seeking to identify seeds in the faces of bison, Jaroszewicz and Pirożnikow (2008) found that the diet extended to no fewer than 454 vascular plant species, of which 409 were herbaceous and 45 woody.

Fig. 14.3 A mixed group eating acorns beneath oak trees in autumn. Acorns are a favorite food of bison, although they do not digest them fully—undigested parts can be found in the dung. Masting of oak trees happens once every 6–9 years on average in the Białowieża Forest. *Photo* ZA Krasiński

Beyond that, the preliminary results of dietary studies based around the direct observation of radio-collared animals feeding has confirmed the proportions accounted for by herbs, trees and shrubs. However, certain changes in diet compared with earlier times are indicated when it comes to choices made at the level of the individual species. Among the plants consumed most often contemporarily are: stinging neetle (*Urtica dioica*), meadowsweet (*Filipendula ulmaria*), tussock grass (*Deschampsia caespitosa*), rough small-red (*Calomagrostis arundeacea*) and red raspberry (*Rubus ideus*). The strongest preferences are shown for raspberry and American millet grass (*Milium efusum*), while wood anemone (*Anemone nemorosa*) and creeping buttercup (*Ranunculus repens*) are clearly avoided. It is also now clear that cows consume more grass and sedges, while bulls eat more plants of woody origin. However, the share of the diets accounted for by green plants is similar in the two sexes (Kamiński et al. 2010).

In turn, work done in the Belarussian part of the Forest by Korochkina (1969c, d, 1971a, 1972b) shows that the diets of bison extend to 331 species of green plant from 42 plant families, among which there are some 85 species that can be considered main food (being eaten willingly and present in the diet frequently), 85 that constitute supplementary items in the diet (readily taken but not encountered very often), and a further 90 that may be considered of secondary importance only. The final group of 71 species are regarded as chance food only. On the list of species most often consumed by bison and mentioned by Borowski and Kossak (1972), all items except nettles are also present among the species representing main food in the Belarussian part of the Forest. Korochkina (1969d) could only regard nettles as a supplementary food. Among the 15 species of green plant readily taken in the Polish part of the Forest are 11 considered main food by Korochkina, as well as the remaining four regarded merely as supplementary food items. Bison living in the Belarussian part also feed on 38 species of trees and shrubs (Korochkina 1969c, 1974)—which is more than in the Polish part. The greatest role is nevertheless played by ash, maple, aspen, service tree, sallow and elm. Food from woody plants plays the most important role in the diet in spring and autumn. According to Korochkina (1969b, 1974), bison debark trees of diameters less than or equal to 10 cm. In turn, Bunevich (1999a), studying the natural foods bison take in winter, came to the conclusion that the species most afflicted by debarking are hornbeam, lime, hazel and spruce, primarily those with diameters of up to 15 cm, although also of up to 30 cm in the cases of ash and lime.

The fact that the populations living in the two parts of the Forest differ in the numbers and shares of species present in their diets may reflect the failure on the part of Borowski and Kossak (1972) to distinguish between species of grass. It may also be due to genuine habitat differences between the Belarussian and Polish areas. However, all the studies make it clear that herbaceous plants are the primary component of the diet of bison (Fig. 14.1), while food derived from trees is, at most, a supplementary item (Korochkina 1969c, d, 1972b, 1974; Borowski and Kossak 1972; Dzięciołowski 1991; Dzięciołowski et al. 1975; Gębczyńska et al. 1991). This fact is also confirmed by work on bison activity, inasmuch as that—during the Forest's growing season—bison devote as much as 95 % of their

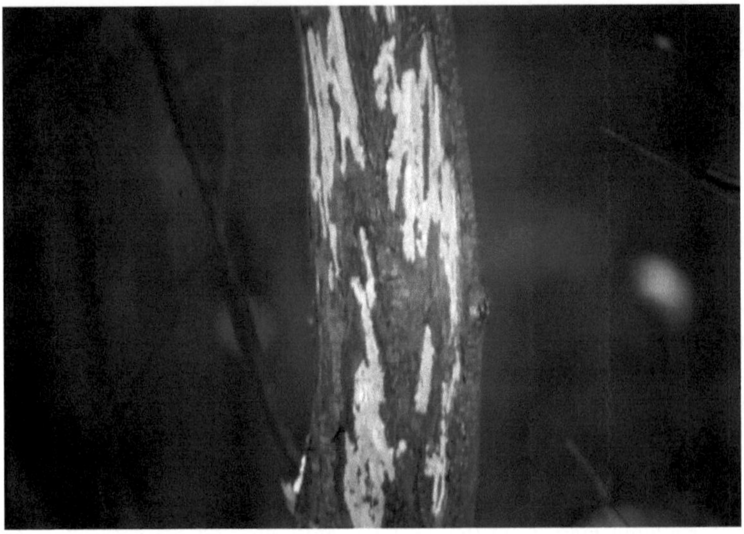

Fig. 14.4 Woody food constitutes around 10 % of a bison's diet. The chewing off of bark (debarking) is at its most intensive in early spring, as animals start to supplement their diets with natural food. Foresters are not very well-disposed towards this behaviour when it concerns important forest-forming tree species in production forests (in this case—an ash). *Photo* ZA Krasiński

feeding activity to feeding on forest-floor plants. Whatever the type of forest, the herb layer is the primary source of food for bison. The chewing off of young leaves and stems happens quite commonly, but does not provide a more major addition to the diet (accounting on average for just 3 % of the time set aside for feeding). The selection of this kind of food takes place most intensively in June and July, mainly in deciduous or fresh mixed/coniferous forest, which at the time of the study accounted together for 12 % of time assigned to feeding. The greater share of sprouts in the diets of bison thus coincides with trees being in full leaf, the nutritional value being greater at this stage (Drożdż et al. 1975).

The rubbing or chewing off of bark becomes more frequent in late winter and early spring, and most especially in April, when there is not much forest-floor vegetation (Fig. 14.4). In this month, it is possible to come across whole groups of bison debarking trees. In other months of the growing season, bison only debark the odd oak, ash, spruce, lime, hornbeam or occasionally hazel. The only species among these that are of greater economic importance are the oak and ash. In the conditions of the growing season, bison devote just 2 % on average of the time set aside for feeding to bark. However, this average conceals a higher figure for April, namely 18.5 %—making bark a significant component of the diet. In winter, the bison living in the Polish part of the Forest intensively debark oak, hornbeam, ash, lime, spruce and different species of willow. The causes underlying debarking behaviour have not yet been fully identified. In the winter period, bark may simply represent one of the few forms of food available, but when the snow cover has retreated debarking may rather reflect a need to take in more coarse-fibrous food.

The Lowland-Caucasian line bison living in the Vologda region of northern European Russia (and thus the southern taiga sub-zone) are found to consume a variety of herbaceous plants during the growing season. However, as summer goes on, animals begin a steady increase in the amounts of tree-stem material browsed upon, only for this to become the main item of the diet when winter arrives (Tyapougin and Gusarov 2004; Gusarov 2011).

When winters are mild and with little snow, European bison continue to display a preference for their natural food, as opposed to what they might receive via supplementary feeding. Bison can be observed digging up snow in forest clearings to reach grass (Figs. 14.5, 14.6). In the eastern part of the Forest, the grass consumed by bison is found to lie beneath snow that is not more than 10–25 cm deep (Korochkina 1974, 1978). Thus, in severe winters, bison rely mainly on the hay put out for them (Fig. 14.7). They are also ready to wander off to places where trees are being felled. A long-practised custom in the Białowieża Forest involves the cutting of aspen especially as food for game animals. In the 19th century, some 7,000 trees per year were cut with that aim in mind (Karcov 1903). Today likewise, aspen are cut at the end of winter, in order to that the diet of the bison might benefit from the diversification natural food can bring. The trees in question are entirely stripped of their smooth bark, while sprouts of diameters up to two cm are also consumed (see color insert).

Not all the bison in free-ranging populations make use of the supplementary fodder on offer in winter. For example, the animals of the Bieszczady Mountains

Fig. 14.5 This bull feeding on a meadow in winter had moved the snow aside with his muzzle and eaten the dry grass beneath. This means of obtaining food is only rarely observed. *Photo* ZA Krasiński

Fig. 14.6 A group of young bulls in search of natural food during winter. *Photo* ZA Krasiński

Fig. 14.7 Good hay disappears quickly in the powerful stomachs of bison. The hay is scattered across the snow so that all animals can gain access to it. *Photo* ZA Krasiński

(of south-east Poland) make little use of extra food made available in winter—mainly silage and root crops. The Lowland line bison of the Nalibokskaya Pushcha (western Belarus) mainly take natural food, even in winter, with browse accounting for 64 % of the diet in this season, the tree species selected most often being pine, birch, buckthorn and spruce (Shakun 2011).

In contrast, bison spending the winter on farmland beyond the limits of Poland's Białowieża and Knyszyn Primaeval Forests go out on to the fields in search of winter cereals and rape hidden under the snow. In the fields around the Białowieża Forest, 51 % of all the damage done by bison affects cereals, as followed by hay (34 %) and rape (13 %). In turn, on the fields around the Knyszyn Forest, 76 % of cases of damage involve cereals, followed by rape (19 %) (Hofman 2008; Hofman-Kamińska and Kowalczyk 2010, 2012).

Dackiewicz (2000) made use of sample plots in determining the influence bison exerted on stands close to two places of supplementary feeding within the Polish part of the Białowieża Forest, where two largest winter groups were concentrated. The most intensively browsed there was oak (accounting for an average of 62 % of the trees damaged), as followed by ash (51 %), lime (42 %) and hornbeam (30 %). In turn, the least intensive browsing affected species like spruce, birch and alder. In terms of actual severity of impact, the trees most affected by debarking were found to be hornbeam and lime, while oak, ash and spruce suffered less, and birch were not touched at all.

Kowalczyk et al. (2011) applied a novel DNA-based method of herbivore diet analysis to determine the degree to which supplementary winter feeding modified the diet of radio-collared bison in the Polish part of the Białowieża Forest. It was found that, where the regularity with which supplementary feeding took place was lower, the amount of food of woody origin (from trees and shrubs) was greater. Where there was intensive use of extra food put out, woody elements accounted for just 16 % of the diet, whereas in animals not taking advantage of supplementary feeding at all, the corresponding figure was 65 %. In turn, a level of consumption of herbs, grasses and sedges amounting to 82 % in the case of animals making intensive use of supplementary feeding compared with a figures of just 32 % where only natural food was eaten.

Recently, the work by Kowalczyk et al. published in 2011 revealed that, in winter, the bison in Poland's Białowieża Forest most often eat such tree species as *Carpinus/Coryllus,* followed by *Betula* sp. *Salix* sp., *Pinus silvestris* and *Quecus robur. Carpinus/Coryllus*, *Quecus* and *Alnus* are selected in line with their level of availability in the habitat, while there is active avoidance of *Tilia cordata*, *Acer platanoides* and *F. excelsior* compared with the degree to which these tree species are abundant in the forest. Browse most often derived from *Carpinus/Coryllus* (which account for 34–49 % of food from trees). Since these results are markedly different from those of earlier studiem, confirmation should be sought in research on the content of bison rumen in winter. It may emerge that the increase in the Polish (Białowieża Forest) population of bison has resulted in a change of preference where species of tree consumed in winter are concerned.

The Lowland-Caucasian line bison of Russia's Vologda region in turn make only minimal use of supplementary feeding (Tyapougin and Gusarov 2004; Gusarov 2011). For their part, the bison of that same line within the free-ranging (CEJSKIJJ ZAKAZNIK) population, inhabiting the Central Caucasus Mountains (North Osetia region, South Russia) do not receive any supplementary food during the winter.

At the beginning of that season, when snow cover is still shallow, bison diet mainly comprises of blackberry and herbaceous plants (59 %), while bark accounts for 39 %. However, in the second half of winter—in conditions of deep lying snow—the share of bark increases to 79 %, while blackberry and green plants account for just 6 % of what is consumed. The bark chewed there at that time comes from eight species of tree, mainly lime, maple and elm (Kaz'min and Smirnov 1992). During winter, the main food of the bison hybrids living in the Caucasus Biosphere Reserve are blackberry and red fescue, along with the bark of trees and shrubs. The mosses growing on tree bark are also consumed readily (Nemtsev et al. 2003). In contrast, the Western Bieszczady Mountains (of Poland) are inhabited by bison whose main winter food is blackberry, this representing a calorific and highly digestible diet option (Paszkiewicz 2004; Aleksandrowicz et al. 2009).

14.3 Food Requirements

The fresh mass of food eaten over a 24 h period by an adult bison was determined by Kalugin (1968) at between 30 and 45 kg. This compares with a figure of 39–60 kg obtained by Aleksandrov and Golgovskaya (1965). Holodova and Belousova (1989) write that 2–3 year-old bison on average take 27 (25–30) kg of fresh food per day, as compared with 48 (40–50) kg in the case of 4–5 year-olds. Gębczyńska and Krasińska (1972) reported that 5-year-old bison took 23–32 kg of fresh fodder each day. However, the food supplied to the latter was rich in materials originating from trees and shrubs (it was 39 % tree leaves and branches). Holodova and Belousova (1989) noted that, during winter, the demand for food components and energy in bison fell 1.4–1.5 fold in relation to the situation applying in the growing season.

14.4 The Digestibility of Dietary Components

Experimental research into the digestibility of food comprising meadow hay (70 % of the diet) and pearl barley (30 %) revealed that the digestibility coefficients as regards total protein, ether extract, lignin and ash contents were greater in bison than in domestic cattle (Gębczyńska et al. 1974). The ability to digest lignin points to a specific adaptation among bison to forest conditions and the possibility of food rich in that component being made use of. Also subject to study was the digestibility of natural food comprising plants easily accessible to (and willingly consumed by) bison in different parts of the growing season. The results obtained point to the bison's effective use of the natural food available to it, and the high degree to which components of that food prove digestible for the species. The coefficients of digestibility were lowest in the autumn, as was the utilisation of energy from food (Kowalczyk et al. 1976).

14.5 Habitat Use

Studies of the evolution of *Bison bonasus* have yet to make it clear exactly when what is essentially a grass-eating animal began to make use of the forest environment. Bison have in fact retained their selenodont dentition, though a series of adaptations to life in the forest have developed, these concerning both overall build, the structure of the intestines and the capacity to digest certain components of food (notably lignin) (Pytel 1969; Hofmann 1978; Gill 1999; Kobryńczuk 2000; Gębczyńska et al. 1994).

Kuemmerle et al. (2011) put a time-calibrated niche model together with a model for the dynamics of vegetation to conclude that the bison of the Holocene period inhabited either semi-open areas, or else broadleaved, mixed and coniferous forests, in habitats that represented a mosaic of woodland and grassland vegetation. Since the last 8,000 years have brought a decline and fragmentation in habitats suitable for bison (thanks to increased numbers of human beings and the associated development of agriculture) the effect has inevitably a steady decline in numbers of bison and a restriction of the species' range.

In the view of several authors (Heptner et al. 1966; Kaplan et al. 2009; Kuemmerle et al. 2011), it was not by chance that *B. bonasus* lingered on longest in north-central and north-eastern Europe, since until even the 19th century, these areas retained forests of a suitable nature, did not experience agricultural development until rather late on, and continued to support rather low-density human populations. Kerley et al. (2012) are in turn of the view that the grass-eating European bison may be regarded as a refugee species forced to seek shelter in forest that took the place of open steppe during the last postglacial period. This leads to the presumption that forest actually represents a sub-optimal environment for the animals.

The main factor shaping population densities among ungulates in Polish forests is food (Jędrzejewska et al. 1994). The degree to which an environment offers attractive opportunities for foraging on the part of large herbivores is determined by the luxuriance of the herb-layer vegetation (Krupka 1990).

Bison resemble other large herbivores in being animals of limited dietary selectivity. As they wander around the Forest in search of food, they select the places in which most of the plants they prefer are present in abundance, the aim being to satisfy their energetic demands rapidly. Indeed, it is the high level of demand for food that represents the main factor underpinning seasonal movements among bison in the first place. The main food items taken are herbs and grasses growing on the forest floor or in areas deprived of their forest cover artificially or naturally (e.g., after gales), as well as in mid-forest clearings or on meadows. There is thus marked variation in the abundance of plant food from season to season.

Borowski and Kossak (1972) carried out research on the food of bison between 1966 and 1971, and noted that the forest communities most often used by bison are deciduous and mixed/deciduous forests (in c. 75 % of observations). The work done between 1969 and 1973 shows that bison most often penetrated mixed/deciduous forests (accounting for 44 % of summer observations), as well as deciduous ones (35 %). Only further down the list do we find alderwoods (17 %). Coniferous forests were only visited rarely (Krasiński 1978a).

Up to the early 1970s, mixed groups concentrated in the central area of the Polish part of the Forest, where they mainly fed in mixed, fresh and moist deciduous forests. They had at their disposal relatively numerous clearcuts. In later years, deciduous forests still took the leading position among the forest habitats environments utilised by bison, although by then it was possible to observe increased penetration of fresh mixed coniferous forest. The consequence of the increase in population was the spread of bison into the southern Polish part of the Forest, an area with a high proportion of fresh mixed coniferous forest. In the years 1976–1995, the main environment inhabited by bison was still deciduous forest, which accounted for 54 % of observations of the species. This compared with the second-placed fresh mixed coniferous forest (42 %) (Fig. 14.8). From the mid-1990s onwards the Polish part of the Białowieża Forest witnessed a renewed intensification of deciduous forest use (to 73 %), as well as the utilisation of and alderwoods (15 %), in parallel with a reduction in the use made of fresh mixed coniferous forest (to 11 %) (Fig. 14.8). This would seem to be associated with the higher average daytime temperatures and lower annual precipitation totals to be noted in this period (Daleszczyk et al. 2007).

In bison populations in the north-east of Poland, deciduous forests are also the main environment in which bison live. For example, in the Borecka Forest, 87 % of observations of bison relate to fresh deciduous forest (Krasiński and Krasińska 1994b). No work on bison habitat preferences has been done in the Knyszyn Forest, so all that can be noted is that the areas in which the population actually lives have a prevalence of mixed/coniferous forest, with deciduous forests in second place (Krasiński and Krasińska 1994b). In the vegetative season, the

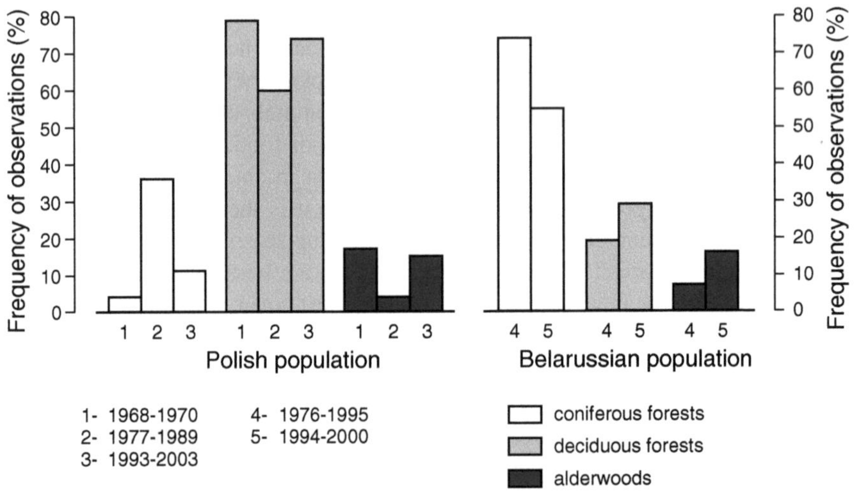

Fig. 14.8 Use made by bison of different types of forest in the Polish and Belarussian parts of Forest, in the course of the plant growing seasons of years 1976–2003. *Source* Krasiński et al. (1999) modified and Daleszczyk et al. (2007)

population of bison living in the Western Bieszczady Mountains also most readily utilises beech stands (noted in 57 % of observations), followed by stands with fir (20 %), pine and alder (10 % each) and spruce (just 3 %) (Paszkiewicz 2004).

As has been noted, bison use of habitats varies with the seasons. Spring (April–May) sees bison concentrating their feeding in deciduous forests (51 % of observations). In contrast, mixed/coniferous forest is increasingly visited in June and July (71 % of observations), when less use is being made of deciduous forest (22 %). However, in years in which June and July are hot and dry, the preference for deciduous stands continues to be shown. Penetration of mixed/deciduous forest increases steadily from May to August. In the Polish part of the Forest there was no intensive use of alder forest up until the mid-1990s (such stands only accounting for 4.5 % of observations). Alder forest is most often visited in early spring, only rarely in summer (and then when there are droughts), but once again more frequently in autumn. In contrast, the use of alderwoods in the last decade has increased (to account for 15 % of observed situations), the probable underlying cause again being climatic. In dry and warm years, alder woods become more accessible to bison as they dry out (Daleszczyk et al. 2007). The August to late autumn period again brings increased penetration of deciduous forests (this habitat accounting for 58 % of observations at this time), while the use of mixed/coniferous forests goes down (to just 15 % of observed instances). Finally, the summer (July–August) period sees a similar degree of penetration of both deciduous and mixed/coniferous forest (Krasińska et al. 1987).

In the last wild population of bison inhabiting the Białowieża Forest at the beginning of the 20th century, the seasonality to the use of different types of forest was different to what we record now. In spring, the departure of the last snow was followed by a tendency for bison to feed first on herbaceous plants in coniferous forests. They would then move on to deciduous (oak-lime-hornbeam) forest, feeding in alderwoods between mid-summer and the end of autumn (Wróblewski 1927). The maximal development of the forest floor in alderwoods takes place in summer (Falińska 1973); hence the decision by bison to remain in this area which can supply enough food from summer through to autumn.

In the years 1966–1971, bison occurred in equal frequency in forests of all ages (Borowski and Kossak 1972). However, up to the beginning of the 1970s, more than 40 % of encounters with feeding bison took place on open areas (Fig. 14.9) (Krasiński 1978a). In the years 1976–1982, bison were most often noted in stands aged 20 and over, while they were only rarely present in thickets. In contrast, only 20 % of observations involved encounters in open areas (Fig. 14.9). Young plantations of deciduous species were most markedly penetrated in early spring and again in autumn, while those comprising coniferous and deciduous species are more inclined to be visited in spring.

Phenological differences to the vegetation in different forest habitats of the Forest (Falińska 1973) influence seasonal changes in bison's environmental selectivity. Deciduous forests and mixed forests are habitat types in which feeding bison can be observed throughout the growing season. The environment in question includes plants of both short and long growing seasons, the latter lasting into

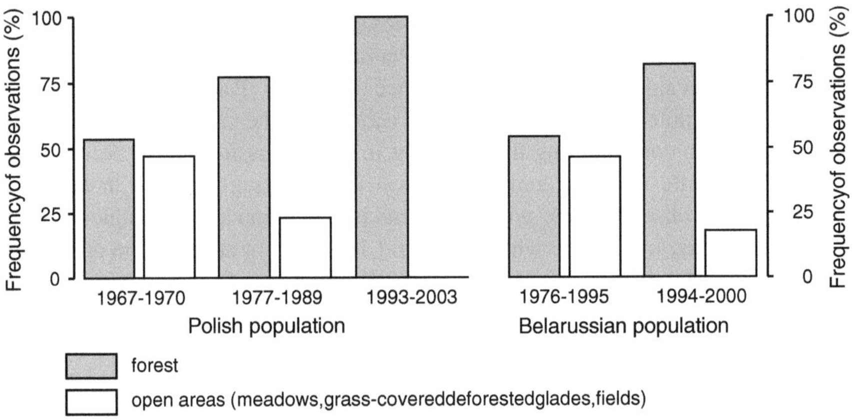

Fig. 14.9 Comparison of the use bison make of forest biotopes and open areas in the Polish and Belarussian parts of the Forest. *Source* Krasiński et al. (1999) modified and Daleszczyk et al. (2007)

late autumn. Forest-floor vegetation appears first in this environment (Falińska 1973). Among the species with a long growing season, there are four species (ground-elder *Aegopodium podagraria*, hedge woundwort *Stachys silvatica*, wolly buttercup *Ranunculus lanuginosus* and nettle *Urtica dioica* that are readily consumed by bison. In August and September, several rhizomatous plants are present in the deciduous forest herb layer, e.g., yellow archangel *Galeobdolon luteum* and wood sorrel *Oxalis acetosella*, these offering an autumn supplement to the normal bison diet. Fresh mixed deciduous forest is a habitat that bison remain keen on throughout the growing season. Several places in the Forest are known where bison come specifically for this habitat (Krasińska et al. 1987).

Another habitat very frequently visited by bison is fresh mixed coniferous forest. Vegetation grows here later than in deciduous forests (Falińska 1973). The reed-grass *Calamagrostis arundinacea* flowers in June, and is one of the four plants that European bison include most willingly in their diet (Borowski and Kossak 1972). There is thus a ready explanation for the species' frequent use of mixed coniferous forest habitat during summer.

Equally, radiotelemetry studies into the behaviour of individuals confirms earlier observations regarding the rotational use of habitats by bison. Herds are continually on the move, if at a leisurely pace, describing some kind of loop or circuit around the terrain—in such a way that, when between several and several tens of days have passed, they find themselves back in areas they have visited previously (Fig. 14.10) (Krasińska et al. 1987; Korochkina 1973).

Observations made in the Belarussian part of the Forest in the late 1960s and early 1970s made it clear that bison spent the early spring feeding in alderwoods, before turning to the deciduous and mixed deciduous forest stands in late spring and summer. Late summer and autumn were in turn spent in coniferous forests, as well as on mown meadows or in forest clearings. As the amount of available

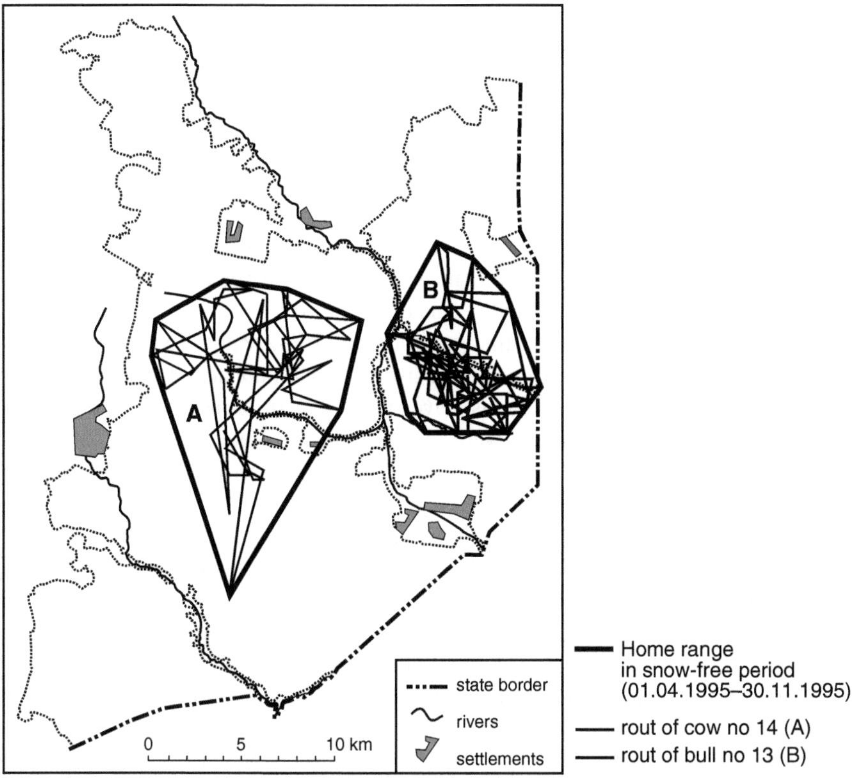

Fig. 14.10 The routes taken within the home range of an adult bull and a cow during the 1995 vegetative period. *Source* Krasińska and Krasiński (2007)

vegetation remaining on the forest floor diminished with the advancing season, bison began using more open habitats (Korochkina 1973).

The Belarussian part of the Białowieża Forest has more coniferous forest than the Polish side, so bison in Belarus make fuller use of this habitat than do their counterparts over the border (Fig. 14.8). Indeed, the food base of fresh and wet coniferous forests is utilised, as well as that of fresh mixed coniferous stands (Bunevich 1989c). Nevertheless, bison have increased their use of deciduous forests and alderwoods since the mid-1990s (Daleszczyk et al. 2007).

In the growing season, the Lowland line bison present in Poland's Puszcza Borecka most often make use of the so-called fresh deciduous forest habitat, seasonally also the mixed fresh deciduous forest and alderwood (accounting for 65 % of observations). In autumn, they also venture out on to cultivated fields (Żoch 2007). In contrast, the bison of the same line present in the Belarussian forest known as Nalibokskaya Pushcha were found to have cultivated meadows as their primary summer habitat, as well as mown meadows along rivers. Beyond that, the animals in this population also feed in mixed and deciduous forests. However, the use of different types of habitat by the above population of bison does change from season to season, with managed meadows being grazed in spring and the middle of summer, while birch, alder and mixed spruce-deciduous forests are favoured (Shakun 2011)

The home range of the small population of Lowland-Caucasian line bison existing in the Poloniny National Park area of northern Slovakia has 72.6 % of its area within forests, while open areas account for the remaining 27.4 %. The *Fagetum typicum* association has most intensive use made of it (hence accounting for 83.3 % of observations). Abandoned meadows and orchards are in turn the open-area habitats most often frequented (Pčola et al. 2006).

The bison—again of the same line present in the Bieszczady Mountains of south-eastern Poland do indeed make principal use of forest habitat (which accounts for no fewer than 96.6 % locations). This leaves just 3.4 % of occasions on which bison were located in open areas. They were most inclined to frequent (on 61.5 % of all locations) the kind of forest dominated by beech (*Fagus sylvatica*). This being followed far behind by spruce (*Picea abies*) (10 %), mixed broadleaved/coniferous stands in turn account for 12 % of locations (Perzanowski et al. 2011).

Historically, the river valleys of the Białowieża Forest supported meadows that were mown to obtain hay for bison. The haymaking was done by farmers living close to the Forest. In turn, game animals showed a tendency to concentrate in the river valley areas (Fig. 14.11). A related policy in the 19th century was to sustain the increasing populations of game animals through the creation of large numbers of forest clearings, which were often sown with grass or cereals, potatoes, lupins and Jerusalem artichokes. Unsurprisingly, the hay to be gained from the wet riparian meadows was of poor quality, but was nevertheless stored and brought out to the bison in winter. Local farmers were still utilising the meadows after World War II, and bison were also feeding on them at that time. Likewise, there were some 3,000 domestic cows then grazing within the Forest boundaries, although a steady reduction in the number of cattle kept on local farms then ensued, so the area of exploited meadows in the Polish part of the Forest declined. The last meadows continuing to be cut were those along the Łutownia River, and it was in this area that bison had their main winter strongholds as recently as in 1973 (Fig. 14.12). However, here too, the meadows began to go wild with the onset of the 1980s, becoming overgrown with alders and birches as succession got underway. Thus, as of the second half of the twentieth century, the phenomenon of grazing by cattle in the Forest has largely been a feature of its Belarussian side (Fig. 14.13), and even there, this land use has become marginal since the early 2000s. Furthermore, the resultant change in the exploitation of forest with much more limited large clearcuts has led to a further reduction in open areas in the Forest. The 1980s brought severe feeding pressure by ungulates to bear upon the young plantations and thickets, making restocking problematic. In consequence, some 500 ha of these young stands of the Forest were fenced off between 1993 and 1995, which further limited the availability of food for bison. Indeed, the overall effect of these various land-use changes was to reduce drastically what had been an abundance of herbaceous plants and grasses associated with open habitats.

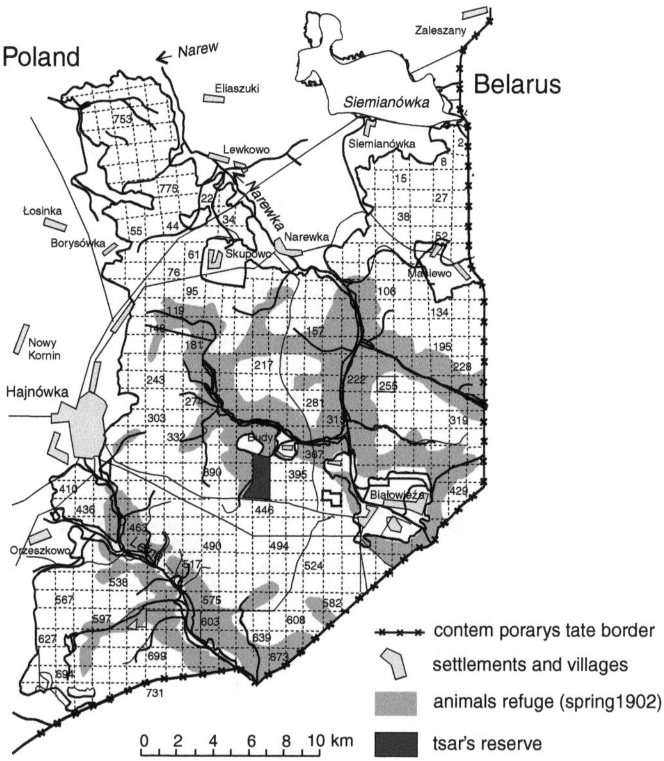

Fig. 14.11 The refuge for game animals in the Białowieża Forest at the beginning of the 20th century. Animals concentrated close to watercourses and riverside meadows. The Tsar's menagerie was an enclosure for keeping bison, red deer and roe deer. It was built prior to Tsar Alexander II's hunting trip of 1860 and removed during World War I. Data from Karcov (1903), figure by T. Samojlik. *Source* Krasińska and Krasiński (2007)

With these issues in mind, the Białowieża National Park sought (in the years 1994–1995) to re-enrich the feeding base available to game animals by bringing 50 ha of meadows and clearings back under exploitation. It was borne in mind that European bison much prefer large clearings with a good cover of grass subject to mowing (Krasiński 1996b), while clearings covering less than 1 ha are used by single individuals only, most often bulls (Fig. 14.14).

The importance of meadows is confirmed by the intensive use made of them by bison in the Belarussian part of the Forest. The large (361 ha) complex of meadows known as Dziki Nikar is at times visited by herds of 100 or more individuals (Fig. 14.15).

The level of use of open areas (clearcuts, young plantations, clearings and meadows) in the Białowieża Forest by bison depends on the degree to which these sites are under management. Clearings and mown meadows are visited by the animals most willingly in spring and autumn. Up until the 1980s, open areas were used as much as forest in the parts of the Forest on either side of the border

Fig. 14.12 Mown meadows along the Łutownia were a regular haunt of bison. However, they have only rarely been seen here since cutting of the grass stopped in the early 1970s. *Photo* ZA Krasiński

Fig. 14.13 Cattle put out to graze in the Białowieża Forest are a threat to free-ranging animals in the sense that they may pass on diseases. They also deplete the Forest's food resources. On the other hand, the meadows mown by farmers offer a good feeding base for bison and other wild ruminants. In fact, designated meadows in the Belarussian part of the Forest are still subject to cattle grazing. *Photo* ZA Krasiński

Fig. 14.14 A bull bison grazing in a small forest clearing. *Photo* ZA Krasiński

Fig. 14.15 A large herd of bison on the "Dziki Nikor" meadows of the Belarussian part of Forest. *Photo* ZA Krasiński

(Korochkina 1973; Krasiński 1978a; Krasiński et al. 1994a). However, a decline in use lasting for many years ensued as mowing ceased on the Polish part, while limitations on access came in on the Belarussian part (Daleszczyk et al. 2007; Krasińska and Krasiński 2007). Renewed management of forest clearings and meadows in 2006 resulted in increased use of those areas by bison. Moreover, in some seasons, bison had tended to stray beyond the edges of the Forest in some seasons. The results of telemetric studies carried out between 2006 and 2010 made clear the animals' preference for open areas both within the Forest complex and beyond it (Hofman-Kamińska and Kowalczyk 2010, 2012; Kowalczyk 2010).

Work in the Borecka Forest also reveals the major role open areas overgrown with luxuriant herbaceous vegetation play as a source of food for bison. An increase in the open area of that Forest brought about by the passage of severe gales through the area in 2002 (and the consequent felling of a belt of forest about 2 km wide along the complex's entire length) ensured that the area occupied by the population in 2003–2006 declined, as did the distances wandered during the summer (Żoch 2007). Meadows in Puszcza Borecka are also used seasonally—in May, as well as following mowing in August (Żoch 2007). On the other hand, in the Knyszyn Forest, the period from September through to early spring sees bison out feeding on cultivated fields located beyond the limits of the forest complex (Hofman-Kamińska and Kowalczyk 2010, 2012). In turn, in the so-called STADO ZACHODNIOPOMORSKIE (Western Pomeranian herd) present in north-western Poland, one part spends the entire year in the forest, while the other is mainly out on agricultural land (Tracz and Tracz 2010).

All the free-ranging populations of bison living in open areas (cultivated meadows, grassy forest clearings, clearcuts and young plantations) make almost as much use of these habitats in feeding as they do of the forest itself (Dzięciołowski 1991; Krasiński and Krasińska 1992, 1994b; Krasińska and Krasiński 1997a; Krasiński et al. 1994b, 1999). In Lithuania, the wild bison population occupies a forest-field landscape, where the buffer zone comprising mixed deciduous forest covers some 50 km^2. The bison feeding there make full use of fields and managed meadows (Balčiauskas 1999).

Korochkina and Bunevich (1980) estimated the plant biomass of natural clearings at 1,330 kg d.m./ha. They found that some 17 % of this was taken by ungulates. The biomass of the grassy vegetation presents on meadows sown with perennial grasses and mown in turn reached 2,796 kg d.m./ha, its level of utilisation by ungulates being 39 %. Still more biomass was available at managed meadows—some 3,600–7,800 kg d.m./ha depending on the species present (Podkówka et al. 1984). In comparison, the plant biomass of the forest floor in fresh deciduous forest is of just some 430–543 kg (Faliński 1973; Aulak 1976).

14.6 The Impact of Bison on the Forest's Natural Environment

Assessments of the biomass of herbaceous plants taken by animals usually involve work on trial plots. However, in the Białowieża Forest, an attempt was made to estimate the overall level of consumption of plants by bison over the 6 months growing season. To this end, it was assumed that the population of bison was 250, of which 62 animals were bulls, 87 cows, 63 young aged 2–3 and 38 calves. The 24 h consumption of herbaceous vegetation (in fresh mass) was taken to amount to 50 kg in the case of bulls, 40 kg for cows and 30 kg for adolescents. This gives a calculated daily consumption of 8.5 tonnes of fresh mass by the population as a whole, converting to 1,530 tonnes over the 6 months period. In line with data from Faliński (1973), it was further accepted that in the deciduous forests the fresh mass: dry mass ratio for forest-floor vegetation is of 8–1, suggesting that the bison population take a tonne of dry mass each day, or 191 tonnes in the growing season as a whole. These estimates were then set against the production of biomass by herb-layer plants in the Białowieża Forest. In the case of fresh mixed coniferous forest, the mean is 519 kg d.m./ha, compared to 536 kg for fresh mixed deciduous forest (Dvorak and Tolkach 1980), as well as 430 or 543 kg d.m./ha in fresh deciduous forest (Faliński 1973; Aulak 1976). The average figure for all the forest environments utilised by bison was thus assumed to be 500 kg d.m./ha. As the Polish part of the Forest utilised by bison represents some 76.5 % of the overall area, there is some 430 km^2 of land available. Very roughly, this area can be thought to generate 21,500 tonnes by dry mass of herb-layer vegetation, suggesting that—over the whole growing season—the bison population consumes something like 1 % of the herb-layer biomass produced by mixed coniferous, mixed deciduous and deciduous forests. Overall then, consumer of ground cover vegetation in forest as it may be, the European bison present in a population of 250 cannot be said to be disturbing the dynamics to forest-floor communities of the Białowieża Forest in any significant way.

As has been noted, rumen-content studies suggest that food matter originating from trees and shrubs represents some 7–13 % of the overall diet of bison (Gębczyńska et al. 1991). From the forestry point of view, the least favourable activity entailed by this is the stripping bark from younger trees, especially economically-important species like oak, ash and spruce. Effective mitigation of this is one reason for the supplementary feeding of bison in winter, although there is no more recent research to determine what the current impact of hoofed mammals on the environment in the Forest might truly be.

The fact that herbivorous mammals move from ecosystem to ecosystem in line with increases in their food resources was noted by Faliński (1986), who also stressed that—as a dispersed population living in small groups—the bison could not be exerting any intensive influence on the structure and dynamics of ecosystems in the Białowieża Forest as a whole.

Chapter 15
Social Organisation of the Population

The early authors concurred that the European bison is a gregarious animal. Bojanus (1827) wrote that the early nineteenth century population of bison in the Białowieża Forest went around in herds of 10–40 individuals, although lone animals (mostly old bulls) also occurred. Jarocki (1830) in turn gave group sizes of 5–15 individuals, noting that old bulls left groups once the rut was over, to live in pairs or occasionally small groups, or else alone. Karcov (1903) referred to family herds of 15–18 individuals, whose composition changed for natural reasons. Old males of 10 and over would, he noted, remain in small groups or stay alone, although large groups would form in winter.

At the beginning of the twentieth century, the bison were living in family groups of varying compositions in terms of age and sex, and sizes of between 7 and 18 individuals. Such groups would include, not only cows, calves and adolescents, but also young bulls aged 4–5. Outside the rutting season, adult bulls would most often form separate groups of 3–4. Such bulls would join the family herds for the period of the rut, but not always the same ones each time (Wróblewski 1927).

M. Krasińska and Z. A. Krasiński, *European Bison*,
DOI: 10.1007/978-3-642-36555-3_15, © Springer-Verlag Berlin Heidelberg 2013

15.1 Group Composition and Size

The population of European bison now living in the Białowieża Forest have manifestly different social structures in the winter period compared to the growing season.

15.1.1 In the Growing Season

In the period from spring through to autumn (i.e. more or less the whole growing season in the Forest), bison predominately occur as mixed group and smaller groups of bulls. A mixed group comprises cows, 2–3-year-old adolescents, calves and—from time to time—adult bulls (aged 4 and over). Group sizes and compositions vary seasonally, as calves are born, and as bulls join the mixed groups for the rut. Other changes reflect circumstances at the level of the population as a whole, inter alia involving the overlap of home ranges and consequent merging of groups.

Mergers between mixed groups occur frequently, and they subsequently break apart just as frequently—mostly with exchanges of individuals having taken place. Early spring groups are the most stable, persisting with the same composition for the longest period (17 days on average, 30 days at most). In turn, group composition changes most in the parturition period from mid-May through to the end of June when 39 % of all changes in group composition take place. However, there is a reintensification of exchanges of group members in August and September, when 36 % of all observed changes occur. Stabilisation sets in once more from the end of September and into October.

At all times, it is adolescents and bulls that form the most variable components to group composition, while adult cows are most consistently present in given groups (Krasińska et al. 1987). Calves begin to appear in mixed groups in May, and by June they account for 10 % of all mixed-group members (15 % in the August–October period).

The rutting season brings a doubling in the share of mixed groups taken by adult bulls. On average, 14 % of group members in August and September are such bulls. Nevertheless, 61 % of mixed groups have just a single adult bull present. Twenty-seven percent of groups have two bulls, 8.2 % three and just 4 % of groups have as many as four. There are only two known instances of the Polish population featuring groups with more bulls (6 and 8) (Fig. 15.1) (Krasiński 1978a; Krasiński et al. 1994a). While some mixed groups retain adult males for the whole spring–autumn period, the bulls in question are mostly younger (aged 4–6). Otherwise, in May and June only 23–48 % of groups include any bulls at all, as compared with a higher percentage in July and a peak figure of 80–90 % at the height of the rutting season in August and September (Fig. 15.2) (Krasiński 1978a; Bunevich and Kochko 1988; Krasiński et al. 1994a).

In the first years of development of the Polish population (1963–1970), the mean size of mixed groups was 12, the maximum 40. Average group size in fact

Fig. 15.1 Numbers of adult bulls living in mixed groups. *Source* Krasiński et al. (1994a)

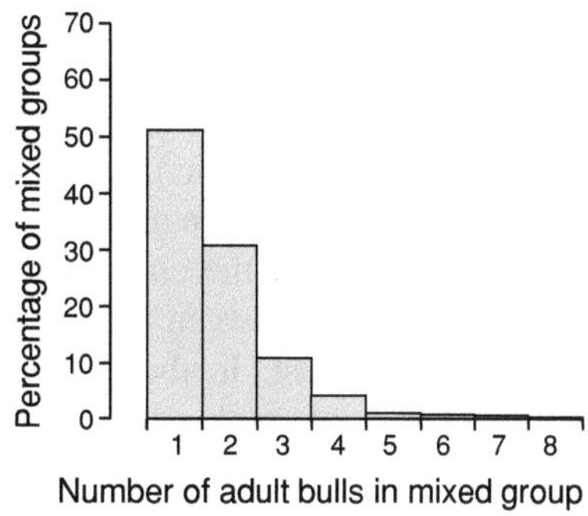

Fig. 15.2 Percentage of mixed groups including adult bulls. *Source* Krasiński et al. (1994a)

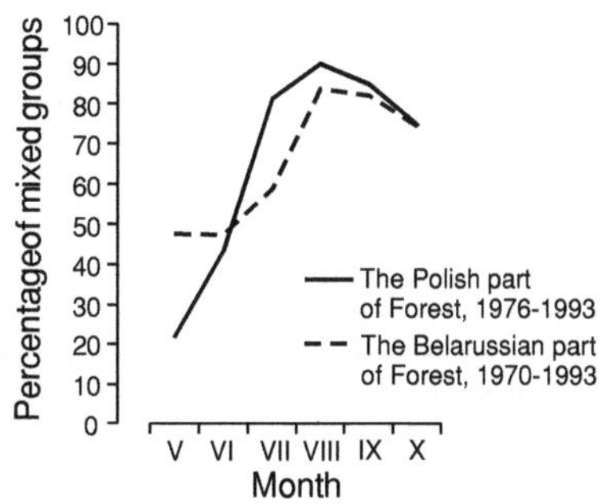

Table 15.1 Seasonal changes in mixed-group size among European bison in the Polish part of the Białowieża Forest during the snow-free seasons of years 1963–1993

Month	1963–1970					1976–1993				
	n	Min	Max	Mean	±SD	n	Min	Max	Mean	±SD
May	221	1	40	12.1	7.5	324	1	50	13.2	8.7
June	167	1	35	11.2	6.5	329	1	48	13.7	9.3
July	181	1	30	12.6	6.5	372	1	60	13.6	9.5
August	235	2	40	12.7	6.9	369	2	92	15.1	9.3
September	187	1	38	12.9	6.9	316	2	43	13.0	7.4
October	213	1	40	12.9	6.9	344	1	40	11.9	7.4
Total	**1204**	**1**	**40**	**11.9**	**7.0**	**2054**	**1**	**92**	**13.4**	**8.6**

Data for years 1963–1970 from: Krasiński (1978a), for 1976–1993 from: Krasiński et al. (1994a). n: number of observations. *Source* Krasińska and Krasiński (2007)

remained steady in this period, notwithstanding a 4-fold increase in the population. In contrast, the years 1975–1995 brought a small increase in the sizes of mixed groups, to 13 individuals on average (Table 15.1). Maximum group size in turn rose from 40 to as many as 92 individuals, although the latter large group was only actually maintained for a couple of hours in the Polish population in August

1992, as bison assembled at a large forest clearing. The group in question had disintegrated into several smaller ones before the day was out. Indeed, the Polish part of the Forest most often has groups of 20 or fewer individuals (these accounting for 84 % of observations). Just 15 % of groups number 21–40 animals (15 %), while only 1 % of known groups have been any larger than that (Fig. 15.2).

The influence of the environment on group size in European bison is attested to by differences between the mixed groups present in the Polish and Belarussian parts of the Forest. In the Polish part in the period during which bison spent most of their time within the forest complex, which only has small clearings, the groups were markedly smaller (Table 15.2). In contrast, the Belarussian part has large meadows and clearings for grazing, and so the average size of a group is

Table 15.2 Comparison of mixed-group sizes in different habitats of the Białowieża Forest, 1975–1995

Population	Habitat	Size of mixed groups				Percent of groups by size:		
		n	Mean	Min	Max	Up to 20 indiv	21–40 indiv	>40 indiv
Belarussian	Open area	628	23.0	2	140	40	24	36
Belarussian	Forest	930	10.8	1	74	89	9	2
Polish	Forest	1974	13.4	1	92	85	14	1

Data from: Krasińska and Krasiński (1997). n: number of observations, indiv: individuals. *Source* Krasińska and Krasiński (2007)

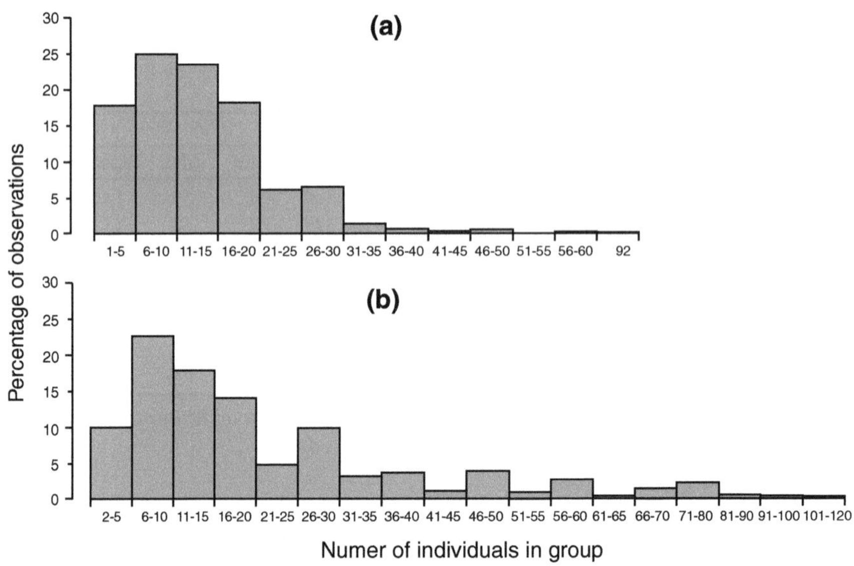

Fig. 15.3 Frequency distributions of the size of mixed groups in the Polish (**a**) and Belarussian (**b**) bison populations of the Białowieża Forest, within the vegetative period of the years 1976–1993 inclusive. *Source* Krasiński et al. (1994a)

Table 15.3 Seasonal changes in groups size among European bison in the Belarussian part of the Białowieża Forest during the snow-free seasons of years 1971–1993

Month	Mixed groups				Groups of adult cows				Bull groups			
	n	Min	Max	Mean	n	Min	Max	Mean	n	Min	Max	Mean
May	264	2	140	18.3	30	2	9	4.2	91	1	11	2.7
June	232	2	111	22.0	23	2	9	3.0	81	1	7	2.1
July	306	2	70	19.7	27	2	6	3.8	100	1	7	1.7
August	281	2	90	21.3	8	2	6	3.3	97	1	3	1.5
September	235	2	101	25.0	16	2	5	2.0	73	1	7	1.5
October	142	3	80	20.2	10	2	3	2.4	66	1	8	2.0
Total	**1460**	**2**	**140**	**21.0**	**117**	**2**	**9**	**3.5**	**508**	**1**	**11**	**1.9**

Source Krasińska and Krasiński (2007)
Data from: Bunevich and Kochko (1988). Krasiński et al. (1994a). n: number of observations

21 individuals (from 2 to 140 individuals) (Table 15.2) (Bunevich and Kochko 1988; Krasiński et al. 1994a). Similarly, larger groups of 40–50 individuals were indeed observed on agricultural areas beyond the Polish part of the Białowieża Forest from early 2000 onwards. In Belarus, mixed groups of up to 20 individuals are most common (accounting for 65 % of groups observed). There are also many medium-sized groups of 21–40 individuals (21 % of the total), although large groups (of 61–140 bison) are again rare; representing just 5 % of the total (Fig. 15.3). Groups numbering more than 40 occur in spring and autumn on meadows and in forest clearings (Bunevich and Kochko 1988; Krasiński et al. 1994a; Krasińska et al. 1997a). Likewise, within the Belarussian population itself, there are differences in size between the groups inhabiting forest areas and those mainly residing in open areas. The groups observed in the former environment only number 11 bison on average, while those in the open areas have 23.

Since 1975, a phenomenon unknown in the Polish population has come to the fore in the Belarussian part of the Forest, that is the establishment of groups composed entirely of cows. These have 2–9 individuals (on average 3.5) (Bunevich and Kochko 1988) and are observed in all months of the growing season, although most frequently in spring (Table 15.3). The fact that such groups form may be linked with the large share of the Belarussian population accounted for by old cows. The groups in question nevertheless represent just 3 % of all the mixed groups observed in the Belarussian part of the Forest.

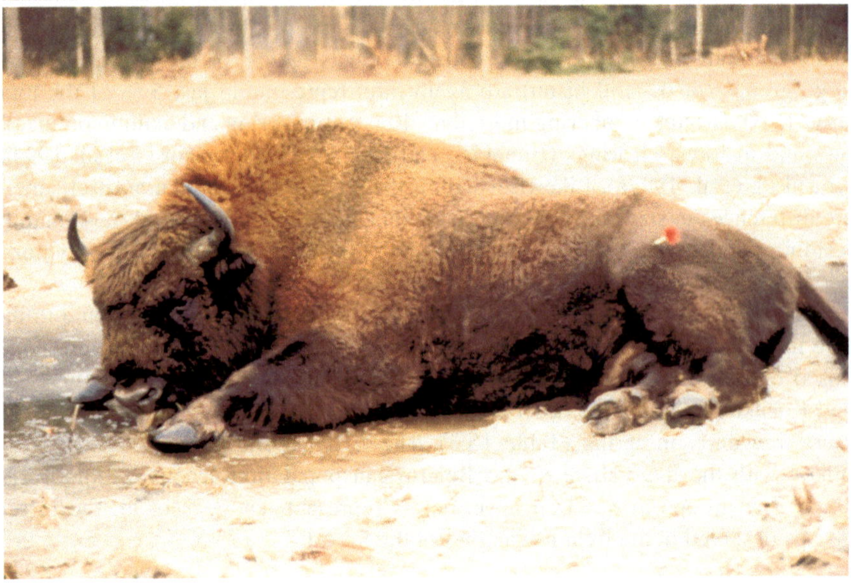

Insert III. 1, 2 Bison can be immobilized using drugs, or by catching them in special cages. *Photo* ZA Krasiński

Insert III. 3, 4 Telemetric studies using radio-collars placed around the necks of bison have allowed many aspects of the ecology of the free-living animals to be determined. *Photo* ZA Krasiński. Windthrown aspen trees represent a very important source of food in late winter and early spring. Both the bark from apical parts and the stems are chewed very thoroughly by both bison and red deer. *Photo* ZA Krasiński

Insert III. 5, 6 During winter 2002, an old bull comes to the "Zwierzyniec" car park at the centre of the Białowieża Forest, in search of hay and carrots. *Photo* A. Wajrak/AG. It is much rarer for groups of bison with young to enter human settlements. Here a herd has reached the village of Olchówka, to the north of the Forest. *Photo*: ZA Krasiński

Insert I. 7, 8 Bulls return stubbornly to farmers' fields. Sometimes they are captured and taken off deep into the forest, but in 90 % of cases they come back. *Photo* A. Wajrak/AG

The sizes of mixed groups vary with the season. The largest groups are present in the breeding season, that is in August in the Polish population and September in the Belarussian one (Tables 15.1 and 15.3) (Krasiński 1978a; Bunevich and Kochko 1988; Krasiński et al. 1994a).

Populations occurring in the forests of north-eastern Poland are most often characterised by the presence of mixed groups of not more than 20 individuals, allowing it to be suggested that this is the optimal size for a group of bison when it comes to life in the conditions of lowland forests (Krasiński 1978a; Krasińska et al. 1987; Krasiński and Krasińska 1992; 1994a; Krasiński et al. 1994a; Żoch 2007).

In turn, in the conditions pertaining to mountain forests in Poland (in the Bieszczady region), most observations concern mixed groups of up to 20 individuals. Larger groups (of 31–50) are encountered but rarely (in just 2.5 % of observed cases) (Perzanowski et al. 2005). The population of so-called "mountain bison" (i.e. European bison with some American bison blood as well) that are present in the Caucasus Biosphere Reserve includes mixed groups numbering between 13 and 35 individuals (Nemtsev et al. 2003).

Groups of bulls are to be met within all months of the growing season. Young, 1 and 2-year-old males spend most of their time in mixed groups. In turn, bulls aged 3–5 are as likely to be found in a mixed group as in a bull-only group.

Groups of bulls are most often made up of younger (4–6 year-old) specimens, only more rarely 3-year-olds (Fig. 15.4). Young bulls usually link up into bachelor

Fig. 15.4 Young bulls often stick together in small "bachelor" groups. *Photo*: ZA Krasiński

groups of 2–7 individuals, while adult or old bulls generally go around on their own or in pairs. Through the entire, 50-year period of development of the Polish population, the average size of groups of bulls was 2. Most often encountered are groups of 2–3 males (accounting for 32 % of male-only groups), while groups of 4–9 bulls are a rarity (accounting for just 6 % of cases). Between 52 and 62 % of bulls live alone (Fig. 15.5). The numbers of lone bulls vary from season to season, starting to rise prior to the rut in July, such that 70–80 % of all males observed in August and September are alone (Fig. 15.6) (Krasiński 1978a; Krasiński et al. 1994a; Krasińska and Krasiński 1995).

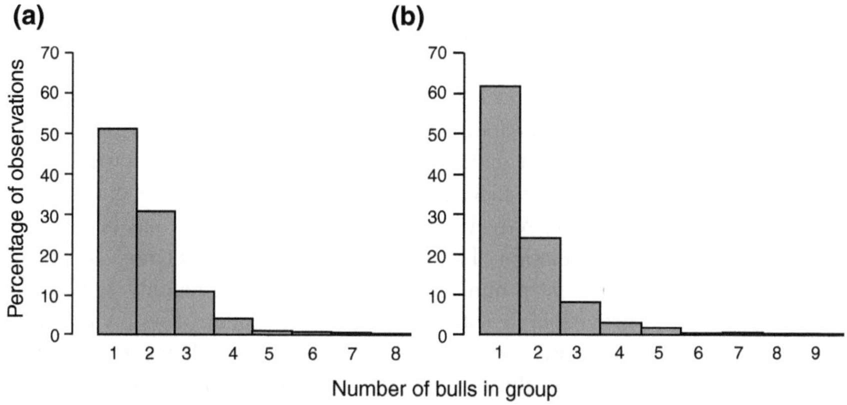

Fig. 15.5 Sizes of groups of bulls in the Polish (**a**) and Belarussian (**b**) bison populations of the Białowieża Forest, within the vegetative period of years 1976–1993 inclusive. *Source* Krasiński et al. (1994a)

Fig. 15.6 Seasonal changes in the frequency of observation of lone bulls and those present in bachelor groups during the growing seasons for plants of years 1976–1990 inclusive. *Source* Krasiński et al. (1994a)

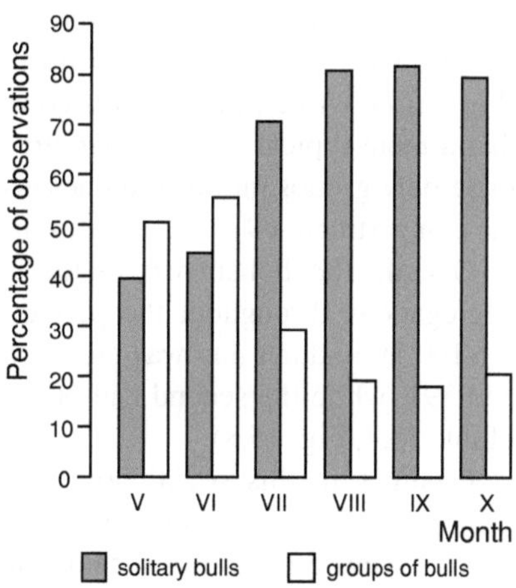

Within the Belarussian population, the average size of groups of bulls is also 2, again with changes being noted from season to season. More than half of all males (63 %) seen were loners (Fig. 14.5) (Bunevich and Kochko 1988; Krasiński et al. 1994a; 1999). In turn, in the Bieszczady Mountain bison population, lone bulls occur year round (Perzanowski et al. 2005), while in the Caucasus (Russia) males aged 3–20 from today's "mountain bison" population groups of males numbering 3–6 individuals (maximally 10) are most often present (Nemtsev 2003).

15.1.2 In the Winter Season

No information on the abundance and habits of wintering bison has come through to us from earlier centuries. The late nineteenth and early twentieth century population of bison in the Białowieża Forest was under special protection, however, and we know that supplementary fodder was on offer in winter. We also know that this generated the same kind of gatherings around feeding places as occur now. The bison spent 5–6 months in such aggregations, mostly between mid-October and the first days of April (Karcov 1903; Wróblewski 1927). As has been noted, the current Białowieża population also pursues such habits. A change of organisation can result in increases in numbers around permanent feeding sites, while any abandonment of provisioning with hay has consequences of its own. Indeed, mild winters of limited snow cover are never associated with the formation of larger aggregations. Likewise, throughout the whole period in which the populations reintroduced into the two halves of the Forest were developing, it was always possible to observe bison gathering at feeding sites in large mixed aggregations, as well as smaller aggregations of bulls only (Krasiński 1967, 1968a, 1978a; Korochkina 1972a; Korochkina and Kochko 1982; Bunevich and Kochko 1988). The size of the winter mixed aggregations has increased, in line with the re-establishment of the population, and the founding of permanent feeding sites has not prevented this. However, in the first years of development of the Polish population, the winter aggregations were not yet of constant structure. For example, the feeding sites in forest compartment 422 attracted nothing but bulls until 1962 (Fig. 15.7), while in later years cows with young began to come in as well, such that a more typical mixed group gradually took shape. Similar transformations of male-only groups into mixed ones also took place in later years, as in the 1990s at the compartment 697 feeding sites in Hajnówka Forest District.

Overall, the Polish part of the Forest continued to feature two main mixed aggregations throughout the period until the late 1980s. The first of these had 90–100 individuals and was present around the feeding sites in compartments 390 and 391, while the second had 60–75 bison and was present in compartment 422 (Table 15.4, Fig. 15.8).

Changes in this social organisation were slow but sure, accompanying as they did the gradual spread of bison across ever-greater areas of the Białowieża Forest. New aggregations came into being as a result of this process, with

Fig. 15.7 An adult bull of the first winter aggregation of males at the place of permanent supplementary feeding within forest compartment 422 of Białowieża Forest District. *Photo* from 1964 by ZA Krasiński

Table 15.4 Sizes of winter aggregations of European bison in the Polish part of the Białowieża Forest

Aggregation	I	II	III	IV	V	VI*	VII
Comp.	391	422	Starzyna 697	Hwoźna 162/193	Postołowo 211	Nieznany Bór 441	Siemianówka (beyond forest)
Period				Number of animals			
1971–1989	90–100	60–75	0	0	0	0	0
1990–1993	70–80	51–77	12	15–38	0	0	0
1994–2002	90–126	45–85	10–33	15–47	0	0	28–52
2003–2011	49–104	50–77	32–55	38–59	25–51	15–18	46–112

Data from: Krasiński (1978a), Krasiński et al. (1994a). Comp: No. of forest compartment, *from 2010. *Source* Krasińska and Krasiński (2007), updated

Fig. 15.8 The winter aggregation in forest compartment 391 of Hajnówka Forest District has numbered around 100 individuals (range 80–126). Such a high density at the place of supplementary feeding is not conducive to the animals' health. *Photo* ZA Krasiński

early 1990 bringing the first establishment of a third mixed aggregation (of 15–47 individuals) in compartments 163 and 193 within the National Park's Hwoźna sub-district. A fourth aggregation, of up to 33 bison, also appeared in the mid-1990s in compartment 697 (Starzyna sub-district) (Table 15.4). Two new mixed winter groupings took shape in the 2000s. In compartment 211 ("Postołowo") there are periodically between 25 and 51 bison, while in compartment 441 ("Nieznany Bór") there are around 18. A seventh mixed grouping is of limited stability, since it numbers between 46 and 112 individuals, but regularly splits off to 2–3 smaller groupings. Furthermore, it is not associated with any more permanent feeding area, but does go out on to the meadows around the villages of Olchówka, Zabrody, Osowe, Siemianówki and Babia Góra, which are situated beyond the northern most border of the Polish Białowieża Forest, and are able to supply hay accumulated on several haystacks owned by farmers contracted specially to ensure supplementary food for the bison (Table 15.4).

If supplementary feeding sites are close to one another, bison move from one to the next with ease and can in general be considered to form a single aggregation that might split into two from time to time. Such a situation applies to the bison gathering around the compartment 390 and 391 feeding sites in Hajnówka FD, as well as the compartment 162 and 193 sites within the Białowieża National Park (Fig. 15.9). In contrast, mergers between the compartment 391 and 422 aggregations do not take place, even though the two feeding sites are only 4 km apart. At most, only occasional exchanges of group members are to be observed there, and these primarily involving bulls.

In the Polish part of the Forest the winter aggregations have consisted of a large mixed herd in compartment 391, plus five medium-sized ones ("422", "Hwoźna", "Postołowo", "Starzyna" and Siemianówka) and one small herd ("Nieznany Bór") (Table 15.4).

In the Polish part of the Forest, 10–50 % of bulls existing in the years up to the early 1970s were in mixed winter aggregations, which also included cows and

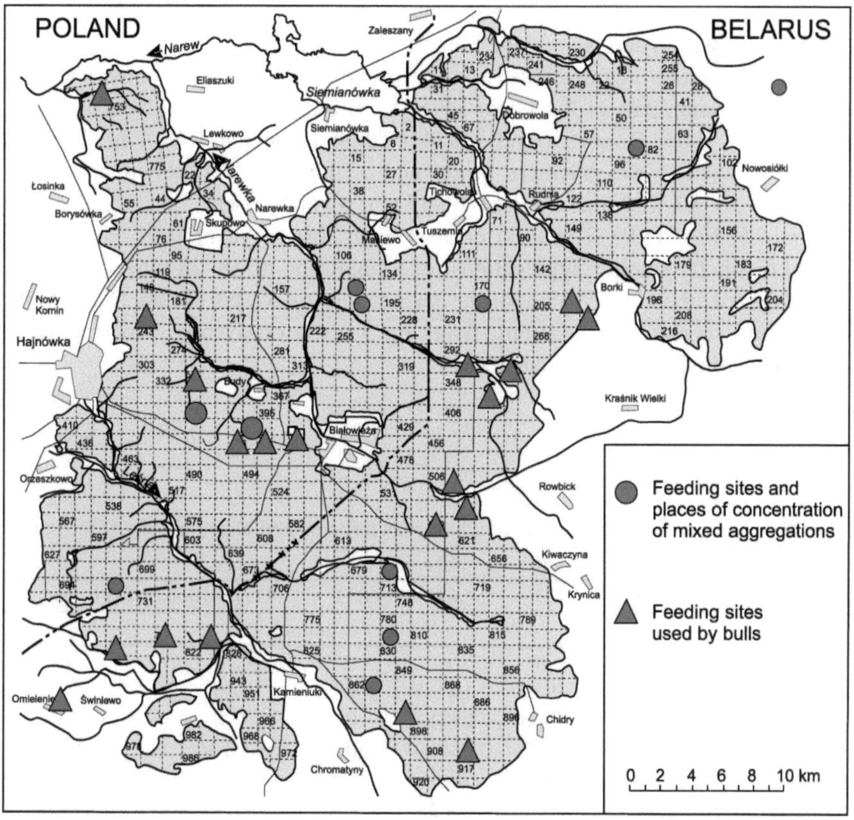

Fig. 15.9 The distribution of supplementary feeding sites for bison in the Polish and Belarussian parts of the Forest. *Source* Krasiński et al. 1994a, modified

young. The remaining bulls created 5–9 separate aggregations of up to 15 individuals (Krasiński 1978a). In contrast, in the years 1976–1992, the mixed winter aggregations accounted for approximately 35 % of adult bulls, while the remainder formed aggregations among themselves or stayed alone (Krasińska et al. 1987; Krasińska and Krasiński 1995). There were then seven separate winter aggregations of males, four in the central part of the Forest and three on its margins. Their size was of three males on average (at most nine individuals).

Some bulls present on the Forest edge went out onto cultivated fields in the period of snow-free winters. The animals involved did so both singly and in groups of average size four individuals (maximally 12; Krasiński and Krasińska 1995). Bulls are always ready to take advantage of the hay put out for deer in hunters' feeding racks in the exploited Forest, but the end of this practice on a regular basis in 1996 encouraged some to join up with the winter aggregations. The years 1997–2003 thus brought an increase to 42 % in the proportion of all bulls spending the winter in mixed aggregations, while the figure for the years 2003–2011 was still higher at 59 %. Nevertheless, mild winters continued to see males living a solitary life or wandering around the whole Forest in small groups. There was also ever-greater penetration of cultivated fields and meadows in hamlets and villages both within and beyond the Forest.

Four main winter aggregations form up around places of supplementary feeding in the Belarussian part of the Forest (Fig. 15.9) (Korochkina 1972a; Korochkina and Kochko 1983; Bunevich and Kochko 1988; Bunevich 1994). In the years 1986–1993 there was a southern group numbering 100–180 individuals, a central one of 30–80, a northern one of 20–45 and a north-eastern one of c. 10 bison (Table 15.5, Fig. 15.9) (Krasiński et al. 1994a). In the years 2001–2005, bison formed distinctly smaller aggregations: one small one of 18–23 individuals, four medium-sized ones of 25–47 and one largest one of 57–60 (Table 15.5). Males formed into separate aggregations, featuring 3 bison on average and at most 13 individuals (Kozlo and Bunevich 2009).

Table 15.5 Sizes of winter aggregations of European bison in the Belarussian part of the Forest

Aggregation	I	II	III	IV	V	VI
Comp	680–713	201	81	832	FD Novosiolki	FD Belyanskoe
Period	Number of animals					
1978–1981	102–113	0	0	0	0	0
1982–1985	130–140	9–30	16–20	0	0	0
1986–1990	140–150	32–40	23–45	0	0	0
1991–1995	105–114	56–78	33–45	0	3–9	8–13
1996–2000	32–87	65–75	26–31	7–21	10–20	17–26
2001–2005	33–47	57–60	23–27	25–30	21–32	18–23

Comp: No. of Forest compartment, FD: forest district. *Source* Kozlo and Bunevich (2009), modified

Over the whole Forest, with its populations of about 900 bison, there are mixed aggregations of three sizes: small ones of 30 individuals or less, medium-sized ones of 30–80 and large ones with more than 80 bison (Tables 15.4 and 15.5).

15.2 Hierarchy and Leadership

Dominance relationships were studied in bison living in zoos and reserves, and hence in groups founded artificially in which there was an adult male, several cows and their offspring. In such groups, the dominant animal was always the bull (Vajner 1980, van den Brink 1980, Krischke 1983), as would also be the case at the Białowieża Reserves.

In turn, observations of free-ranging groups of bison with marked individuals make it clear that adults in mixed groups are subject to rather persistent linkages. In one group, the same leading cow for three consecutive growing seasons had several of the same females gathered around her, in spite of the fact that the composition and size of the group went unchanged (Krasińska et al. 1987).

It would seem that European bison resemble the American bison (Rutberg 1983) in that an individual's age is a factor determining dominance. The young (4–6 year old) bulls present in mixed groups do not occupy positions at the top of the hierarchy. Likewise, the older bulls of six and over joining a group in the breeding season do not disturb the existing structure, as their presence is entirely connected with the task of mating. Only in relation to younger bulls can these males be said to occupy a superior social position (Caboń-Raczyńska et al. 1987). Old bulls of 15 years and over in turn seek to avoid contact with younger and stronger males by going around on their own. If they do join a mixed group this is for a short time only.

A hierarchy is also in operation in the large winter aggregations. The hay that has been put out is first taken by cows and bulls, and only later by adolescents and calves. When the food provided is a very attractive item (like fodder beets) present in just a small amount, the dominant cows and bulls will simply prevent younger animals from gaining any access to it at all. In contrast, the spreading out of hay over a larger area is a measure sufficient to allow subordinated animals to also take necessary advantage of supplementary feeding.

Mixed groups of European bison are headed by cows, and the position of the leader is generally held by an older female with a calf in tow. Leaders (dominant cows) may be mild-mannered, as in the case of the identified cow "Babula" (Fig. 15.10), which showed a great deal of tolerance of observers. Equally, some herd-leading females prove to be exceptionally alert and vigilant, making a closer approach on the part of an observed a near-impossibility—especially when small calves are present in a group. Such behaviour was demonstrated by the female called "Pocieszna", who was recognisable by the white tips of her horns. In many cases, females known to us or marked continued to lead herds for several (even 10+) years. The old cow "Babula" was aged 21 when stopped leading a group.

Fig. 15.10 Recognisable by her broken tail, the cow called "Babula" was long the leader of her herd. Her calmness of disposition had a beneficial influence on the behaviour of her whole group. *Photo* ZA Krasiński

She produced her last calf aged 23 years, spending the last 2 years of her life alone. She died in winter 1985, aged 24, and some 15 km from the home range she had occupied previously.

In today's population of "mountain bison" (a herd with an admixture of American bison blood), the leader in a mixed group is generally also an old and experienced cow (Nemtsev et al. 2003).

Observations of marked bison are not sufficient to support or dismiss the hypothesis that family ties influence the composition of groups of bison—as both Karcov (1903) and Wróblewski (1927) were convinced. However, it would seem that such ties play some role in the forming of winter aggregations. Only future genetic studies will offer a definitive insight into the role familial relationships play as groups of bison take shape.

Chapter 16
The Spatial Structure of Populations

16.1 Populations Area

The area occupied by the population in the Polish part of the Białowieża Forest increased as that population grew. By 1966, when numbers were once again up over 100, the population occupied a c. 70 km^2 area at the centre of the Polish part of the Białowieża Forest. Some bulls were by then extending over yet-wider areas (Krasiński 1967). Indeed, by 1963, several (2–6) bulls had started to stray into the Puszcza Ladzka area (north–west direction), while males had

M. Krasińska and Z. A. Krasiński, *European Bison*,
DOI: 10.1007/978-3-642-36555-3_16, © Springer-Verlag Berlin Heidelberg 2013

also made an appearance within Browsk forestry sub-district (north–east), having crossed the Narewka river. By the 1970s, the area penetrated intensively by bison had increased to 140 km^2 (or c. 25 % of the whole Forest area) (Fig. 16.1). Nevertheless, by their very nature mixed groups had a more aggregated spatial distribution, while the bulls tended to be dispersed across the whole area. It was still usual to meet with mixed groups most often in the central part of the area occupied by the population, while bulls were in both the centre and the south a part of the Forest as yet rarely frequented by any mixed group (Krasiński 1976, 1978a, b). Notwithstanding an increase in the population to more than 250 by the late 1980s, the area occupied by mixed herds and most of bulls had not in essence increased. Nevertheless, certain bulls were ranging yet farther around the Forest, even leaving it in spring and autumn in order to utilise the cultivated fields and meadows beyond (Krasińska et al. 1987).

It was only in the late 1980s and early 1990s that some cows, adolescents and calves began to head south–west into Starzyna forestry sub-district, as well as north–east into the Browsk sub-district (see Fig. 17.4). In such a way, some 70 % of the Forest was occupied by the end of the 1990s (Fig. 16.2). In the mid 2000s, the area occupied by the population was further augmented by fields situated to the north–east and south–west of the Forest (see colour insert). Bulls were still met with throughout the Forest, and were also observed on meadows and fields beyond its limits (Figs. 16.3, 16.4). Thus, as the numbers of bison spending certain

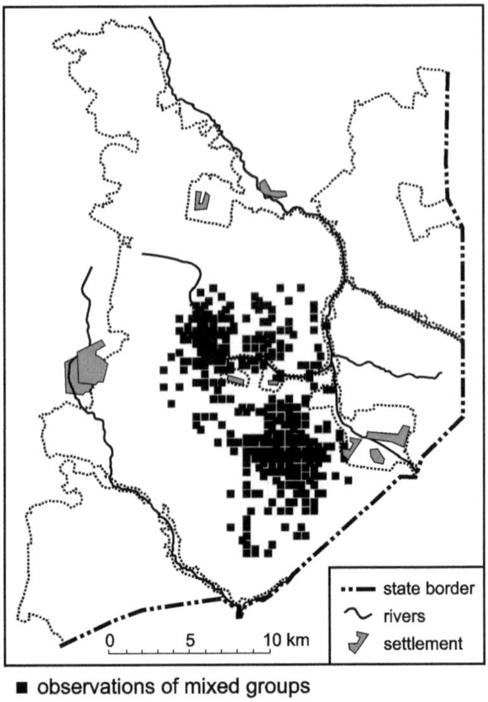

Fig. 16.1 The area occupied by the main part of the population (comprising mixed groups and most bulls) in the Białowieża Forest in the years 1969–1970. *Source* Krasiński (1978a), modified

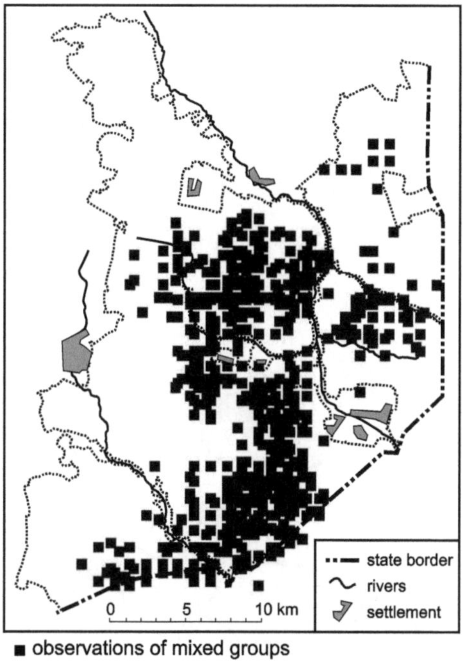

Fig. 16.2 Locations of bison frequenting mixed groups during the vegetative period of years 1989–1998 in the Polish part of the Forest. *Source* Krasiński et al. (1999)

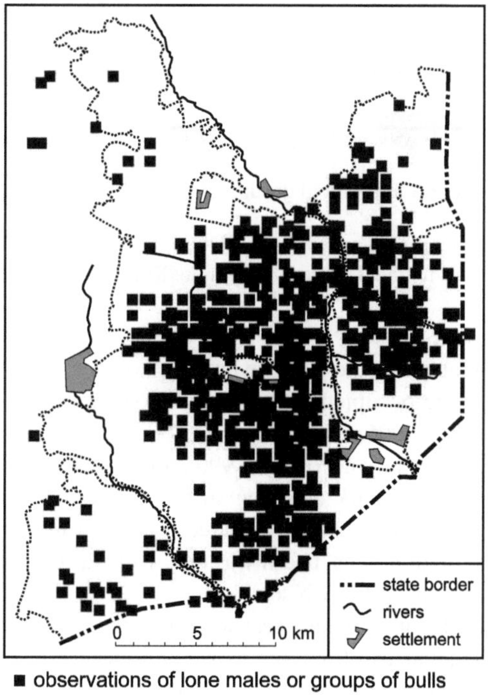

Fig. 16.3 Locations of single males and groups of bulls during the growing seasons of years 1989–1998 in the Polish part of the Forest. *Source* Krasiński et al. (1999)

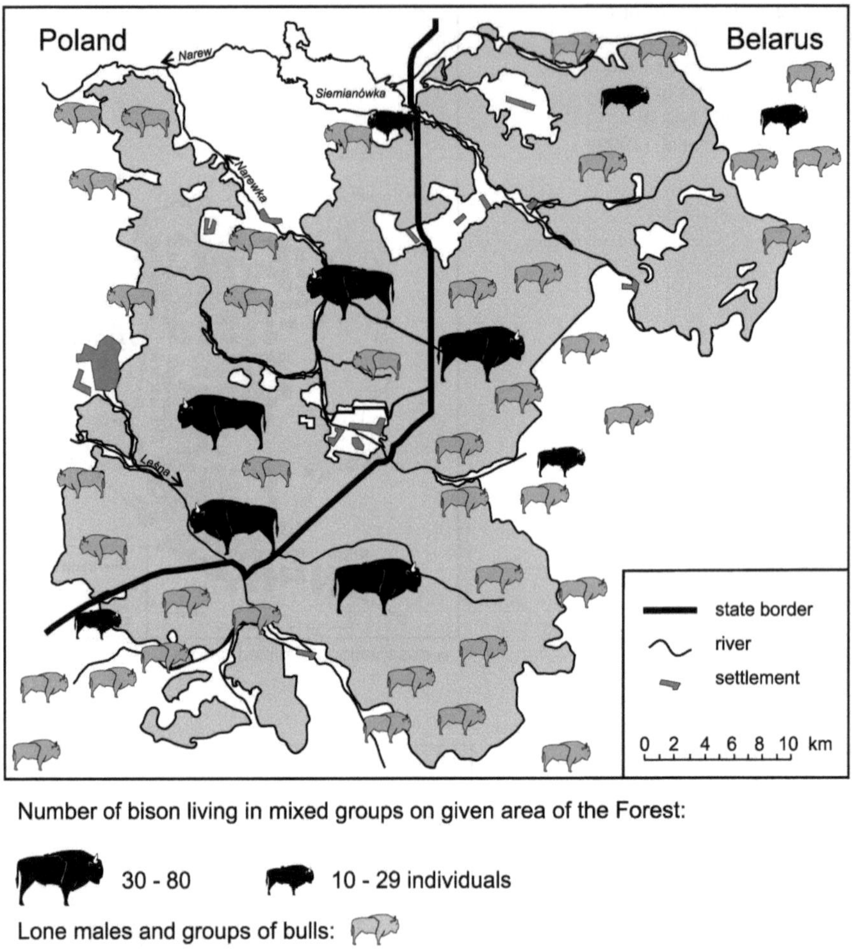

Fig. 16.4 The distribution of bison in the Polish and Belarussian parts of the Forest, 1998–2001. Data for the Belarussian part from: Krasiński et al. (1999) and Bunevich (2003)

seasons of the year beyond the Forest boundaries increased, so—steadily—did the range occupied by the population overall, to reach around 610 km^2 in the years 2005–2006, and 780 km^2 in 2009 (Kowalczyk 2010).

By the end of the 1990s, one mixed herd was making use of the part of the National Park enjoying strict (conservatorial) protection during the growing season. Penetration of this area was nevertheless most intensive in years with a good crop of acorns, and it was only from 2001 to 2002 that this area came to be utilised permanently by bison.

In turn, winter seasons up to the end of the 1960s saw all cows, adolescents and calves plus some bulls confining themselves to just some 5 km^2 or so of land around the places of supplementary feeding in the centre of the Forest (Krasiński

1978a). However, from the early 1970s on, the greater part of the population was occupying a wintering area of around 20 km^2. This territory comprised three separate parts (in the vicinity of forest compartments 391, 422 and 251), but all were located at the centre of the Polish part of the Białowieża Forest, in association with the supplementary feeding sites (see Fig. 15.9). From winter 1988/1989 onwards, the area occupied in winter ceased to be contiguous in character, the change reflecting the first return of cows with calves and adolescents to wintering grounds in the Starzyna (south–west direction) and Browsk areas (Fig. 16.5). By the 1990s, it was possible to talk about wintering territory comprising a central area encompassing places of supplementary feeding in compartments 391, 421/422, 448, 446 and 251 (with 60–70 % of the population); as well as a south-western area around compartment 697 and a north-eastern area around compartments 162 and 193, holding much of the rest of the population. The total area occupied by the wintering population was nevertheless greatly dependent on weather conditions. The mild and rather snow-free winters of 1976–1993 were associated with average occupation of some 52 km^2 (range 22–81). The average overwintering area in harsher and more snowy winters was just 22 km^2 (range 18–27) (Krasińska and Krasiński 1994).

In the late 1990s, the population present within Browsk forestry sub-district split into two groups, one continuing to gather around the feeding sites in compartments 162 and 193, while the other (of around 30 individuals) became dispersed, eschewing the supplementary feeding opportunities and heading off for the north-eastern margins of the Polish part of the Forest (see colour insert; Fig. 16.5). By 2011, around 112 individuals were already leaving the Forest proper to spend time to the north of it, finding shelter in small blocks of coniferous forest, but heading off on to the meadows to obtain food from under the snow and/or hay from the stores maintained by local farmers. Others periodically left for the meadows in search of haystacks around Babia Góra village. There was also an increase in the level of penetration of the BNP area of strict protection, this area supporting 10 bulls in winter 2010/2011, compared with just 1–3 individuals previously.

Research on marked bison confirms earlier observations to the effect that cows lead a sedentary life in winter, while bulls are much more mobile. Out of 31 marked males, 9 (or 20 %) headed out on to the cultivated fields surrounding settlements within the Forest or beyond its limits in the years 1976–1990. All were aged four or more. All eight of the bulls captured on such fields and transferred to the centre of the Forest would ultimately stray beyond its boundaries once again. Bull no. 14 was even captured and transferred in three consecutive years, only to stubbornly make his way back to fields beyond the Forest, if each time in a different place (Krasińska and Krasiński 1995). Bulls rapidly become habituated to taking advantage of the attractive food that winter crops represent.

The ever greater dispersal of bulls reflected growth in the population overall, periodic abandonment of supplementary feeding by bulls at feeding sites in compartments 448, 446 and 746, and foresters' complete cessation of the supplementary winter feeding of other ungulates (Fig. 16.6). In the 1960 and 1970s, single bulls had very willingly taken advantage of the small racks of hay then scattered throughout the Białowieża Forest.

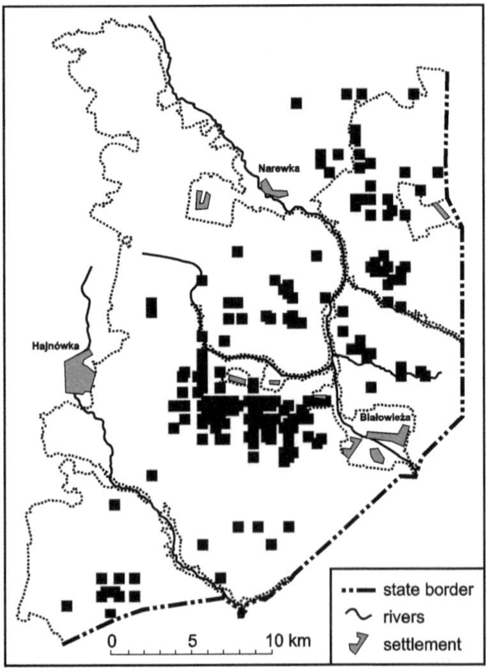

Fig. 16.5 The locations of mixed groups of bison during the winter periods of years 1989–1998, in the Polish part of the Białowieża Forest. *Source* Krasiński et al. (1999)

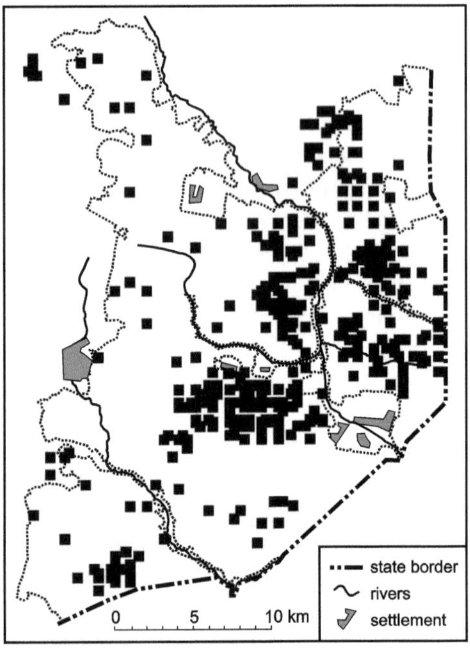

Fig. 16.6 The locations of single bulls and groups of bulls during the winter periods of years 1989–1998, in the Polish part of the Białowieża Forest. *Source* Krasiński et al. (1999)

In the Belarussian part of the Forest, the population of Lowland bison had already taken shape by the early 1970s, the 50–60 individuals, it comprised occupying a territory of as much as 90–100 km^2 (Korochkina 1973). Nevertheless, a kind of stabilisation then took hold throughout the 1980s, a 3-fold increase in overall population size not being associated with any major change in the area of land occupied by the core population (i.e., cows, adolescents, calves and also in part bulls), this remaining at 90–100 km^2 (or 1/6 of the Forest area) throughout (Korochkina and Kochko 1983; Bunevich and Kochko 1988). The density of the bison population here was high, and it co-existed with large numbers of other ungulates. The inevitable result was a depletion of food resources, and in consequence a lowering of the reproductive rate and an increase in mortality (Bunevich and Kochko 1988). With a view to bison densities being reduced somewhat, 37 were captured and transferred to central and north-eastern areas of the Forest in the years 1983–1986 inclusive. The mid-1990s thus saw the area occupied by the main part of the Belarussian population divided into four identifiable parts, i.e., the southern (130 km^2), the central (70 km^2), the northern (50 km^2) and the areas beyond the Forest near Nowosiółki (25 km^2). In contrast, bulls were already present over 60 % of the Forest area by that time (Bunevich 1994). From the late 1990s onwards, two small groups had taken up permanent residence beyond the limits of the Forest, while two others were inclined to cross the boundary from time to time (Fig. 16.4). By 2000, the overall area occupied by the Belarussian population as a whole had reached 600 km^2, the equivalent of 70 % of the entire eastern part of the Białowieża Forest area (Bunevich 2003c; Kozlo and Bunevich 2009).

In the Belarussian part of the Forest, the winters of the early 1970s were characterised by a concentration of animals in an area of just 3 km^2 around places of supplementary feeding (Korochkina 1972a, 1973). This wintering area did not increase much in subsequent years, even though the population continued expanding (Korochkina and Kochko 1983, Bunevich and Kochko 1988). Only at the end of the 1990s was there a marked expansion in the wake of the deliberate translocation of some bison into the Forest northern region. By that stage, bison were concentrating around 15 permanent places of supplementary feeding, seven of these being used by mixed aggregations, while eight were utilised by bulls only (Bunevich 1989b, 1991, 1994; Krasiński et al. 1999).

Another population of Lowland-line bison is that of the Borecka Forest (north–east Poland), which occupied a home range of 8.77 km^2 in the mid-1980s. As numbers increased, the area increased to 10.7 km^2, which represented around 60 % of the area of the Forest in the years 2000–2006 (Żoch 2007). The home range of the STADO ZACHODNIOPOMORSKIE (Western Pomeranian herd) in north-eastern Poland also grew with an increase in population, to reach as much as 108 km^2 in 2008 (Janicka et al. 2008). A similar spread was to be noted for the Lowland-line bison inhabiting the Nalibokskaya Pushcha (Belarus)—from 350 km^2 in 2004 to 400 km^2 in 2010 (Kozlo and Bunevich 2009; Shakun 2011).

The same kind of phenomenon has been observed in another population of Lowland-line bison inhabiting forest-field habitat of the Knyszyn Forest (north–east Poland). In the first years following the introduction there, the animals occupied some 67 km^2, but by 2005–2007 this had increased to 250 km^2, of which forest accounts for only 74 km^2 (6 % of the whole Knyszyn Forest) (A. Lickiewicz et al. unpublished data).

The home range of the Lowland-Caucasian bison in the Bieszczady Mountains has also increased. In the period 2001–2008, the change was from 144 to 300 km^2, in line with an increase in numbers from 140 to 271 (Perzanowski and Januszczak 2004). In turn, the small population of the same line living in the Vologda Region (northern European Russia) has a home range extending over 10 km^2 of the southern taiga sub-zone (Tyapougin and Gusarov 2004; Gusarov 2011).

16.2 Density

Over the years 1976–1989, the density of the bison population within the area of the Białowieża Forest actually occupied was of between 13 and 16 individuals/10 km^2 (Krasińska and Krasiński 1994). From 1989 onwards, the beginning of the increase in the amount of territory occupied by the population as a whole gave rise to a decline in average population density to just six individuals/10 km^2. This level remains applicable, although winter concentrations are many times greater. For example, in the 1970s, there were 200 wintering individuals/10 km^2 (Krasiński 1978a). In fact, winter densities of bison are also prone to variation in line with the influence weather conditions. Figures of between 39 and 140 individuals/10 km^2 occurred between 1976 and 1993.

As the growth of the Belarussian population in the years 1970–1984 was not associated with any overall increase in the territory occupied, the density at which individuals inhabited the area rose from 5.4 to 12.2 individuals/10 km^2 (Bunevich and Kochko 1988). The action to resettle bison across the Forest—taken in the years 1983–1986—resulted in a lowering of the population density (Bunevich 1994). Over the period 1998–2003 as a whole, bison were present in the Belarussian part of the Forest at a density of 6–7 individuals/10 km^2 (Kozlo and Bunevich 2009).

16.3 Individual Home Ranges of Bison and Areas Occupied by Mixed Groups

Detailed information on home-range sizes and locations was obtained in the years 1993–2002 through research on radio-collared bison living on either side of the national border. Some 34 specimens (24 males and 10 females) were marked, 25 (20, 5) being on the Polish part and 9 (4, 5) on the Belarussian (see colour insert). Later some 45 bison (26 males and 19 females) were radio-collared between 2005 and 2010, some of them also fitted with the satellite transmitters.

As has been mentioned, the overwintering bison gathering together around places of supplementary feeding occupy distinctly small areas (Fig. 16.7). The home ranges of bulls covered 11 km^2 on average, although examples ranged

Fig. 16.7 In winter, bison mainly spend time in small areas around the sites offering supplementary fodder. Their mobility is restricted by the deep snow cover, which is negotiated via trampled-down tracks. *Photo* ZA Krasiński

from 0.7 to 34 km^2. The marked bulls spent the winter in all forms of bison social organisation: alone, in male-only aggregations, or in mixed aggregations. Bulls moved in and out of the mixed or single-sex aggregations that were within 1–4 km from them. In turn, the home ranges of cows covered 8 km^2 on average (range 0.8–32). Equally, the areas occupied by marked cows reflected the home ranges of the whole mixed group which they formed a component of. The groups in question had 70–100 individuals on the Polish side, compared with 25–80 on the Belarussian. Each mixed herd, including marked cows, spent the entire winter in the vicinity of just one supplementary feeding station. Furthermore, the telemetry studies confirmed earlier observations regarding the confining effects of persistent snow cover and low temperatures (Krasińska et al. 2000; Rouys and Krasińska 2001). During the telemetric study, the smallest home ranges occupied in winter were of 1.5–2 km^2 during the snowy and frosty winter of 1995/1996, while the largest (32–34 km^2) coinciding with the mild winter of 1994/1995 (Fig. 16.8).

Work carried out in the mid-2000s offered confirmation of the fact that the winter home-ranges of bison are much smaller than the areas occupied in the growing season. It was at the same time noted that the intensity of supplementary winter feeding has an impact on the mobility of bison during the winter, and hence on the size of the range. Animals fed supplementarily to a significant degree

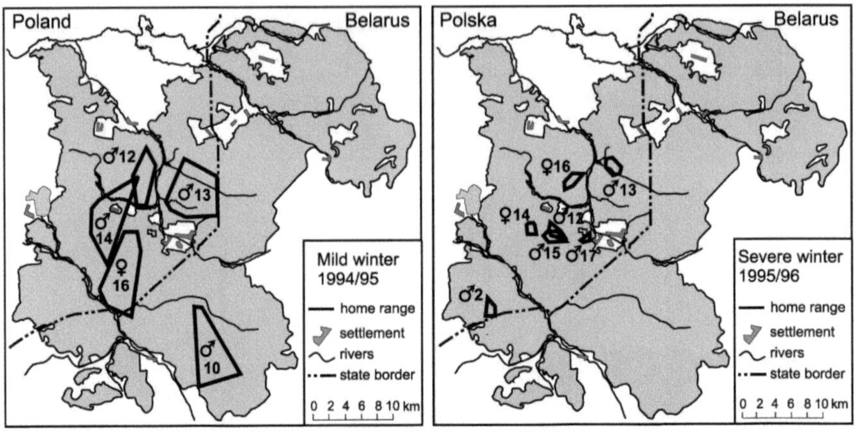

Fig. 16.8 Sizes of winter home ranges of Białowieża Forest bison in relation to weather conditions. *Source* Krasińska et al. (2000)

occupied a smaller area (of some 2.8 km^2), and covered shorter distances in the course of a 24 h period (just 0.5 km on average). In contrast, those experiencing a less-intensive feeding effort ranged over 16.2 km^2, and covered an average distance of 0.8 km. The two regimes were also compared with the situation where not extra fodder whatsoever was on offer, and in this case the respective values were (29.5 km^2 and 0.9 km respectively) (Kowalczyk et al. 2008).

Preliminary results of work on the spatial population structure of Lowland-Caucasian line bison living in the Western Bieszczady Mountains point to the animals' occupying home ranges that (at between 59 and 123 km^2) are far larger than those noted in lowland forest conditions. The differences may reflect the Bieszczady bisons' habit of making a late autumn migration down to areas of lower elevation than are occupied in summer, as well as the fact that the animals here only make limited use of winter supplementary feeding (Perzanowski and Januszczak 2004; Paszkiewicz 2004).

Between spring and autumn, bulls were found to be moving around a home range of average size 70 km^2. In general, the older the bull, the larger its home range. Thus, 4–6 year-old males occupy an area of 44 km^2 on average, while those aged six and over (which participate in breeding) penetrate areas covering 84 km^2 on average. Differences mainly resulted from the greater mobility of older bulls in the breeding season. The largest home ranges of all (136–152 km^2) are those patrolled by bulls spending the winter on the edges of the Forest, only to return to central areas that are inhabited by the females for the rut (Krasińska and Krasiński 2000; Krasińska et al. 1997b, 2000). Bulls present singly or in small groups were occupying home ranges similar in size to those of much larger mixed groups. This suggests that home ranges in bulls are largely influenced by sexual activity, rather than feeding. Bison living in a forest-field landscape in which coniferous forest prevails occupy a far larger area than their counterparts in the Białowieża Forest, on average 130 km^2, with

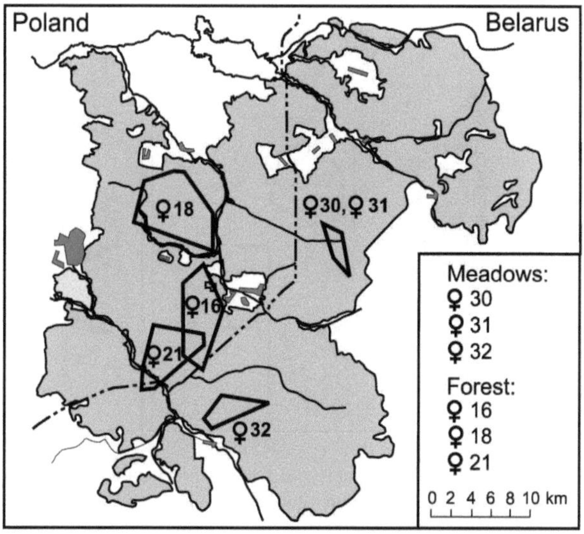

Fig. 16.9 Comparison of home-range sizes of cows frequenting the meadows of the Belarussian part of Białowieża Forest and forest areas of the Polish part, 15 August to 15 October 1997. *Source* Krasińska et al. (2000)

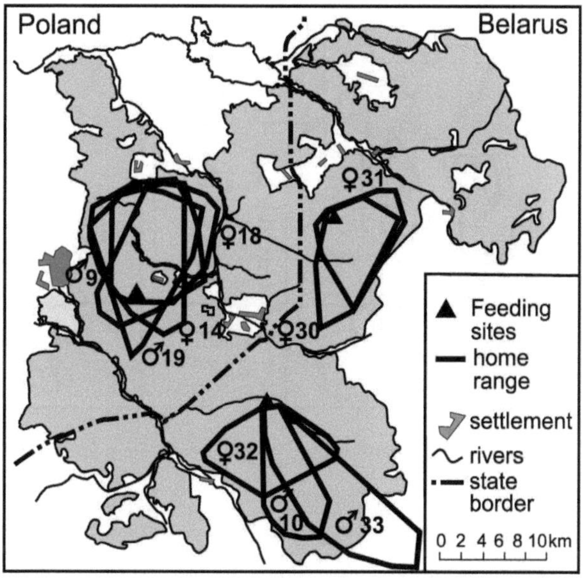

Fig. 16.10 The distribution of summer home ranges of bison from different winter aggregations in the Polish and Belarussian parts of the Forest. *Source* Krasińska et al. (2000)

a core area of 40 km^2 (or 20 % of the home range) (Kowalczyk 2010). Within the Bieszczady population, the growing-season home ranges of four radio-collared bison were found to cover 1.0 to 62.1 km^2, (average 26.8 km^2), these therefore being much smaller than the ones typical for bison in the Białowieża Forest. This may reflect the much greater food base present in the Bieszczady Mountains, as opposed to lowland forests, some 20 % of the western sub-population's area being taken up by meadow communities (Perzanowski and Januszczak 2004).

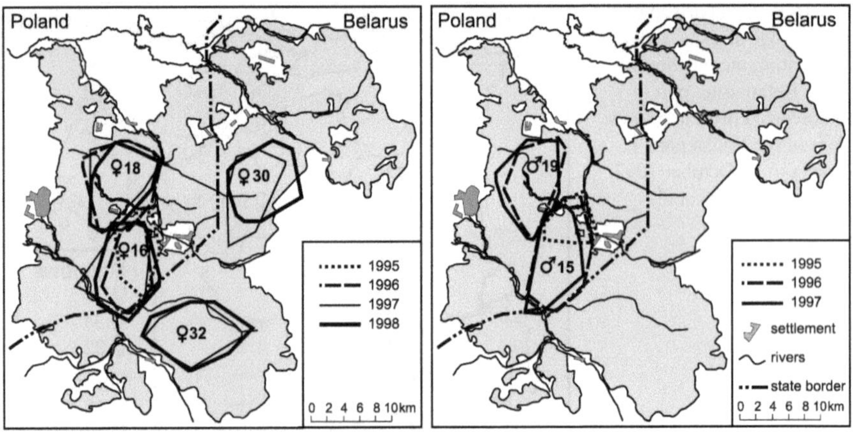

Fig. 16.11 Fidelity of males and females of bison to their spring–autumn home ranges (1 April-30 November) in subsequent years. *Source* Krasińska et al. (2000)

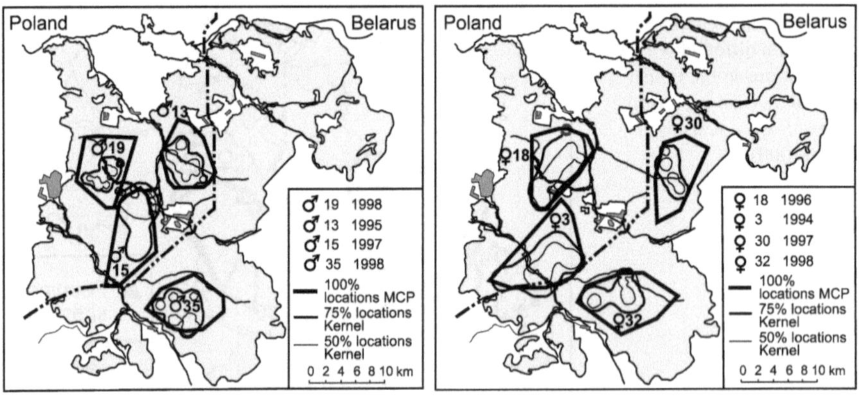

Fig. 16.12 Core home-ranges of radio-collared males and females. *Source* Krasińska and Krasiński (2007)

Home range sizes in mixed groups are dependent on the size and accessibility of the food base. In the spring–autumn period, the individual home ranges of cows averaged 69 km^2 in size (range 45–97 km^2). Ecosystems differing markedly in the richness of the food base (depending on the balance between meadows and forest occurring) were found to manifest significant differences in the sizes of mixed groups' home ranges (Fig. 16.9). Over on the Belarussian part, the August–October groups of 40–70 individuals feed on mown meadows and forest clearings. They thus occupied small home ranges (of 24 km^2 on average). At the same time, small groups (of 7–14 bison) were covering forested areas almost twice as large (45 km^2 on average; Krasińska and Krasiński 2000; Krasińska et al. 2000).

No significant differences between the sizes of individuals' home ranges in the Polish and Belarussian parts of the Forest have been noted, even though the latter are known to offer a poorer feeding base than the former. The paradox may be explicable in terms of the frequent use the eastern bison make of mown forest clearings and mown meadows.

Mixed groups arising in spring, as the different winter aggregations disperse, occupy areas distinct from one another in the growing season (Fig. 16.10). This suggests that the ties linking individuals in each mixed winter aggregation are strong enough to influence the choice of the common areas occupied in the growing season. This spatial fidelity is also indicated by the fact that home ranges of the same marked individual observed over several years are located in the same part of the Forest and show a high (87–91 %) degree of overlap (Fig. 16.11).

It is important to realise that the main existential needs of a bison are met on but a tiny fraction of the overall home range. The centre of activity is referred to as the core area and is usually seen to be situated by sources of water (Fig. 16.12), as well as in the vicinity of clearings or meadows whose specific vegetation ensures an abundance of food in different seasons of the year (Fig. 16.12) (Krasińska et al. 2000). This is confirmed by studies carried out on marked individuals whose collars were fitted with GPS. The determination of centres of activity using the MCP method revealed that a 50 % probability of an encounter with a bison coincides with occupation of just 0.5 % of the entire home range (Kowalczyk 2010).

16.4 Migrations and Dispersal

The European bison was once regarded as a sedentary animal, to the extent that Lowland bison were not recorded in the historical literature as participating in definable seasonal migrations. Bojanus (1827) wrote that bison passed all their time in the same places, not seeking any alternative refuges. What was more, animals that were for some reason driven out would faithfully return to their original spot when they could, he said. However, according to Sztolcman (1926), a pair of bison set off for the Siedlce (centre of Poland) area in 1846, spent a couple of months there, and then returned to the Forest.

Indeed, to this day the environment in which they live has not required of them long-distance movements in search of either food or sources of water. Where a migration does take place, the cause most likely relates to depletion of a food resource due to a rise in population necessitating the search for new habitat.

Bulls have nevertheless been shown to cover great distances in Poland, Belarus and Lithuania. The best known case involving Poland concerns the wanderings of the 4-year-old bull PULPIT. He was brought to the Bieszczady Mountains from the Niepołomice Reserve, but began moving off in a north-westerly direction in 1966. He then proceeded to cover some 300 km in 28 days, being recaptured at Żabno near Tarnów. In 1981, another adult bull appeared not far from Roztocze National

Park, some 350 km from the Bieszczady. A search for the bull made on 27 October 1981 gave no result, but that November a lone bison turned up within Mińsk Mazowiecki Forest District. This bull was shot on 7 December 1981 near Cegłowo, not far from Mińsk, and was aged at 7 years. Since there was no other information about migrating bison at that time, this animal is thought to be the same as the one from the Bieszczady population. He had wandered for some 700 km (Fig. 16.13).

Departures of bulls from the Borecka Forest forest complex are noted relatively often and invariably involve movement north-eastward over approximately 30 km to the Romincka Forest (on the border with Kaliningrad), eastward for 40 km to Olecko or 80 km to the Suwałki area, or southward 40 km in the direction of Orzysz or as much as 100 km in the direction of Łomża (Fig. 16.13). In 1993, a small mixed group comprising a bull, a cow, a heifer and a calf covered a distance of 50 km in the Gołdap area (Krasiński and Krasińska 1992, 1994b).

Some roaming bulls are found to have been injured by snares, while others manifest pathological changes. On 3 June 1981, it proved necessary to shoot a 5-year-old bison near Bakałarzewo, 40 km east of the Borecka Forest. The male in question had complete atresia of the prepuce, only being able to urinate via a fistula. This poor animal also had extensive purulent pneumonia.

In the Forest's western (Polish) population, the first movements beyond the limits of the forest took place in 1956 (Fig. 16.13). In turn, in 1969, an adult bull made

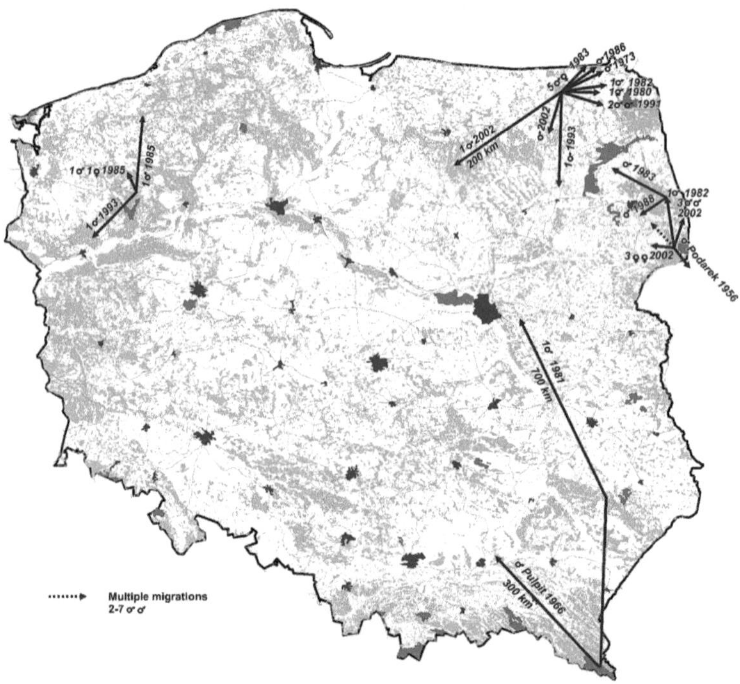

Fig. 16.13 Long-distance migrations of bison in the years 1956–2002. Figure by K. Stachura

a move to the Knyszyn Forest—a distance of 30 km (Krasiński et al. 1994b). In autumn 2002, three cows appeared in small woods close to Zbucz (some 10 km beyond the Forest). They proceeded to spend the winter there, giving birth to calves in spring and passing the entire summer in that area. In that same year, three bulls wandered around the Forest District of Żednia (some 15 km from the Forest).

In contrast, it is not reasonable to regard the periodic departures for cultivated fields as migration, or even the wandering beyond the limits of the forest that take place from time to time (Fig. 16.13). Males may sometimes arrive in orchards or vegetable gardens in this way—mostly but not only in the villages and towns close to the forest edge (Fig. 16.14). Bulls have made their way into the very centre of Białowieża village (Fig. 16.15), while three other males got into Hajnówka cemetery (Fig. 16.16).

Thus far, the only more-detailed information on migratory movements in European bison has been that concerning the Polish population in the Białowieża Forest. Over the 46-year period 1964–2010, it was on average possible to observe nine males a year beyond the Forest boundaries. Agricultural land was most often frequented by lone males (these accounting for 50 % of observations), while only more rarely (in 25 % of observed cases) were 2–3 bulls involved. The phenomenon has been most likely to involve males aged between 5 and 18 years. On average, bulls put a distance of 1.8 ± 0.2 km (range: 0.1–33.9 km) between themselves and the forest edge. The numbers of bulls wandering was dependent on the size of the population and snow depth (Kowalczyk et al. 2013).

Fig. 16.14 In March 1995, a disorientated bison found its way into the yard of a house in Hajnówka. *Photo* ZA Krasiński

Fig. 16.15 An adult (c. 7-year-old) bull earned the name "POBOŻNY" (meaning "pious") for himself, following his wanderings around the Białowieża Orthodox Church churchyard on 4 January 1975. He had previously frequented fields in the village, and so was captured and transferred to the Borecka Forest on 21 January 1975. *Photo* ZA Krasiński

Fig. 16.16 Bison pass through Hajnówka Cemetery in 1987. *Photo* ZA Krasiński

Living in far larger groups that bulls, cows show a different kind of spatial behaviour related to the abundance and availability of nutritional resources. Overall, the numbers of cows straying beyond the boundaries of the Forest rose steadily from the time the population exceeded 260, and show a significant correlation with population size (Kowalczyk et al. 2013). This observation lends support for the contention that—in autumn and winter at least—the food base of Poland's Białowieża Forest is not adequate to support a population of bison of close on 500 individuals (Krasińska et al. 2010). Cows come out of the Forest up to an average distance from its boundaries of 2.1 km (range: 0.1–17.2 km), which did not differ significantly from males. In the years 1994–2010, those boundaries were crossed by an average of some 44 cows (26–106) heading for arable land. Most observations (94 % in bulls and 93 % of cows) were up to 5 km from the forest edge. The ranging of bison in the vicinity of the Białowieża Forest was strictly seasonal. Most observations (78 % in males and 88 % in females) were recorded from November to April. Number of males moving out of the forest was fluctuating during the study period, while number of females was steadily increasing. Increasing utilisation of areas beyond the Forest habitats may be driven by different factors, but most probably is related to range expansion and preference of bison for open habitats. The strong seasonal pattern of bison movements indicate that the partial seasonal migration were initiated in Białowieża population (Kowalczyk et al. 2013).

Typical seasonal migrations are to be noted in the population that inhabits the field-forest landscape of the Knyszyn Forest areas. There, the growing season is passed in the forest, while autumn sees a move out on to farmland, where the animals remain until spring (Krasińska and Krasiński 2007; Hofman-Kamińska and Kowalczyk 2010, 2012).

In contrast, the Belarussian part of the Białowieża Forest bore witness to migratory movements at an earlier stage and with greater intensity than on the Polish side. The migrations made by males would begin there as summer ended and the rut began, the cases mostly involving movements to the north and east. In 1965, 3 young males wandered 10 km beyond the Forest, while in the next year they put as much as 30 km between themselves and it. August 1968 saw a 5-year-old bull make it to the Brest area (70 km from "home"). In turn, in the years 1968–1971, single males moved some 70–180 km away, one individual covering 350 km in total and hence arriving in Lithuania (Korochkina 1972a, 1973). In the next few years, any bison that participated in migrations tended to fall foul of poachers (Bunevich 1999c). For example, in 1998, radio-collared bull no. 33 left the Forest and travelled 20–30 km from it, where it was poached.

The bison in the Caucasus Mountains do make seasonal migrations, however, moving down into the river valleys in winter, on account of the thick snow and attendant shortage of food that characterises the higher altitudes (Filatov after Kulagin 1919). Likewise, the bison present in the Bieszczady Mountains commence with their seasonal movements in late autumn, heading northwards towards areas that are lower down than those occupied in summer (Perzanowski and Januszczak 2004; Paszkiewicz 2004).

Chapter 17
Population Management

17.1 The Carrying Capacity of the Environment

If the biodiversity of the Białowieża Forest is to be retained, all large herbivorous mammals within it should be present at numbers below the ecological carrying capacity of the environment. The needs of the European bison should nevertheless be assigned priority status, since there is no other forest complex in Poland in which such a large population of bison could function.

M. Krasińska and Z. A. Krasiński, *European Bison*,
DOI: 10.1007/978-3-642-36555-3_17, © Springer-Verlag Berlin Heidelberg 2013

The bison population in the Białowieża Forest should function in numbers ensuring preservation of the genetic variability, appropriate level of reduction, good physical and health condition. Establishing a wide margin between minimal and maximal size of the population would allow to control its number in accordance with the current environment and population status.

17.2 Principles of Regulating Population Sizes

Up to now, selective-reduction culling of bison within Białowieża F National Park has been based on forecasts produced using the *Vortex* population viability simulation programme. This makes use of data on the size, structure, reproduction and mortality of the bison population within the Polish part of the Forest (Pucek et al. 1995).

Analysis of changes in population structure makes it clear that the most rapid reduction in numbers can be achieved by limiting numbers of cows aged ≥4. In fact, however, actions of that kind are only rarely undertaken, and even then confined to sick animals that are in a poor condition and/or very old, and choose to spend their time away from groups. The thinking behind this is of an ethical nature on the one hand (the elimination of pregnant females seems distasteful), but is also based on population considerations, since a high rate of reduction among cows might give rise to an unexpectedly abrupt decline in numbers. It is thus much safer for the population if the removal of adolescents and calves at a sex ratio of 1:1 is practised, albeit at a level not exceeding 50 % of the numbers thereof (Pucek et al. 1995). The drawback here is that it is necessary to wait much longer to achieve a reduction in numbers.

Culls should give rise to a stabilisation of population size at a level determined to be optimal, while at the same time bringing about corrections in the age and sex structure of the population with a view to the obtainment of the recommended structure whereby 15 % of animals are calves, 25 % adolescents aged 2–3, 35 % cows and 25 % bulls. Such a structure ensures the maintenance of a good level of reproduction.

The size of the bison population of the Białowieża Forest has been influenced by rates of reproduction, natural mortality and sizes of culls. Within the group of bison removed in the years 1971–2002, the male:female ratio in all ages classes did not depart significantly from 1:1. Individuals singled out for culling were all those displaying symptoms of illness, above all males suspected of having posthitis/balanoposthitis.

17.3 Supplementary Feeding and Winter Refuges

from Karcov 1903

The European bison is an animal of the temperate zone. Its natural environment comprised deciduous and mixed/deciduous forests, these being quite capable of assuring individuals a diversity of food. In turn, the winter period and the shortages of nutritional components associated with it were key factors limiting populations of the species. Regular supplementary feeding to circumvent this problem has led to the formation of large aggregations in winter (Fig. 17.1).As early as in the 16th and 17th centuries, stacks of hay began to be put out on meadows in the Forest to provide food for ungulates during winter (Hedemann 1939). This supplementary feeding activity entailed mowing the Forest's meadows and storing the hay in stacks or sheds. Those under an obligation to pursue this activity were the so-called "osocznicy" (see Sect. 7.1). Up until 1884, there were only a few permanent places at which supplementary fodder was on offer. Specifically, there were 4 barn-feeding racks, 8 feeding sizes and 10 small racks. From 1895 onwards, work began on the construction of barns in which to keep hay, with racks located close by into which hay could be put out on a daily basis. Alongside the places of supplementary feeding were salt-licks supplied with clay mixed with a salt solution. The number of feeding sites increased year after year, the aim being to prevent excessive concentrations of game animals in given places, and to ensure access to hay for all—younger and weaker animals included. By 1902, the number of barn-stores had risen to 64, while there were 27 feeding sizes and 190 small racks. Considerable thought was put into the positioning of the barns, which were built in glades or areas of sparse forest, close to thickets of young spruces, and close to drinking places. Karcov (1903) wrote that, when

Fig. 17.1 Waiting for the supply of hay. *Photo* ZA Krasiński

supplementary feeding was first commenced with, only some bison gathered around the feeding sites, and even these frequently departed from them in search of natural food. A group coming to a feeding site in autumn did not remain in its vicinity, but wandered from one to the next, across an area of 5 or 6 square "wiorsts" (one wiorst = 1066 m). After satisfying their hunger, bison usually lay down in the vicinity. They were capable of eating 16 kg of good hay per day. When temperatures rose, the bison dispersed across the Forest, even when food was still being offered. An important augmentation of the diet was the aspen bark chewed by animals. To meet this demand, around 7,000 aspen trees were cut each year (Karcov 1903).

Observations of the first bison released from the Reserves into the Forest in the years 1952–1953 showed that the winter period would be the most difficult in their lives. Having become accustomed to regular supplementary feeding in Reserves, bison responded to any more major snow falls by entering settlements in search of hay. It was due to this behaviour that the practice of putting out hay on feeding sites in forest compartment no. 419 was initiated in 1954. Two more feeding sites were built in 1958 specifically with bison in mind, one of these being in compartment 390C, west of the Breeding Reserves, the other in 422B, by the eastern boundary of the Reserves in a clearing that had also served in the supplementary feeding of bison during the 19th century (Fig. 15.9). The feeding site in compartment 391C was built in 1960. In the first 10 years of free breeding, bison made use of hay over a period of 150–160 days. The hay was supplied once a day. Even in this earliest period, adult bulls had already taken to gathering at separate feeding sites (Fig. 17.2).

Fig. 17.2 Bison wait impatiently for hay brought in by horse and cart. Today it is the familiar sound of the tractor that encourages the animals to approach. *Photo* ZA Krasiński

The fidelity to traditional winter refuges is great, such that—even if bison spend a certain amount of time in other places—they will usually join the large aggregations at feeding sites eventually (Fig. 17.3). The emergence of new and permanent winter refuges requires a longer period. The regular supply of hay is important. In 1963, a group of 2–6 bulls began to overwinter in the southern part of the Białowieża Forest. These bison made use of the hay put out in hunters' feeding racks, or stored in stacks on mown meadows on the Perebel river valley. In 1989, some mixed groups dispersed south-west into the Starzyna forestry sub-district. In 1992, new feeding site was established in the forest compartment 697 (Fig. 17.4).

Since some bison wandered off into Browsk forestry sub-district (north-east) in 1989, and decided to stay there, it was essential for new places of supplementary feeding to be set up for them. In 1992, a feeding site was established in forest compartment 162C, with a second in the neighbouring forest compartment 193C introduced in 1997 (see Figs. 15.9, 17.4; Table 17.1).

The remaining feeding sites in forest compartment are used from time to time, primarily by bulls that happen along whenever hay is put out. In turn, across the exploited part of the Białowieża Forest, several haystacks or feeding sites have been installed but the bison chose not to utilise them, so they were taken away again (Table 17.1).

Two protective zones established in 1978 assumed the name of "Bison Winter Refuge" coinciding as they did with the places the animals tended to concentrate in winter, within forest compartments 391 and 422. These protective zones are subject to restrictions on utilisation in forestry, hunting and tourism. The "391"

Fig. 17.3 In 1971, bison lingered by the hay-rick at the "Sośninnik" clearing within forest compartment 247 of Hajnówka Forest District. In subsequent years, they again joined up with the large aggregation in compartment 391. *Photo* ZA Krasiński

Table 17.1 The main feeding sites of the European bison in the years 1958–2011 in the Polish part of th Białowieża Forest. Comp.—No of forest compartment

Feeding sites	Year of creation	Number of bison in 2011
391	1958	49
422	1958	50
Starzyna Comp. 697	1992	33
Hwoźna Comp. 162, 193	1992	51
Postołowo Comp. 211	1987	15
"Nieznany Bór" Comp. 441	2010	18
Siemianówka	2007–2009	112
Meadows and fields outside of BF		
Vicinity of villages Olchówka, Zabrody, Siemianówka, Babia Góra		
Altogether bison in mixed aggregations at feeding sites in 2011		**328 (68 % of population)**
Others feeding sites:		
Comp. 251	1987	8
Comp. 334B	1967	4
Comp. 446A	1988	Used periodically by bulls
Comp. 475AB	2007	6

Source Krasińska and Krasiński (2007), modified

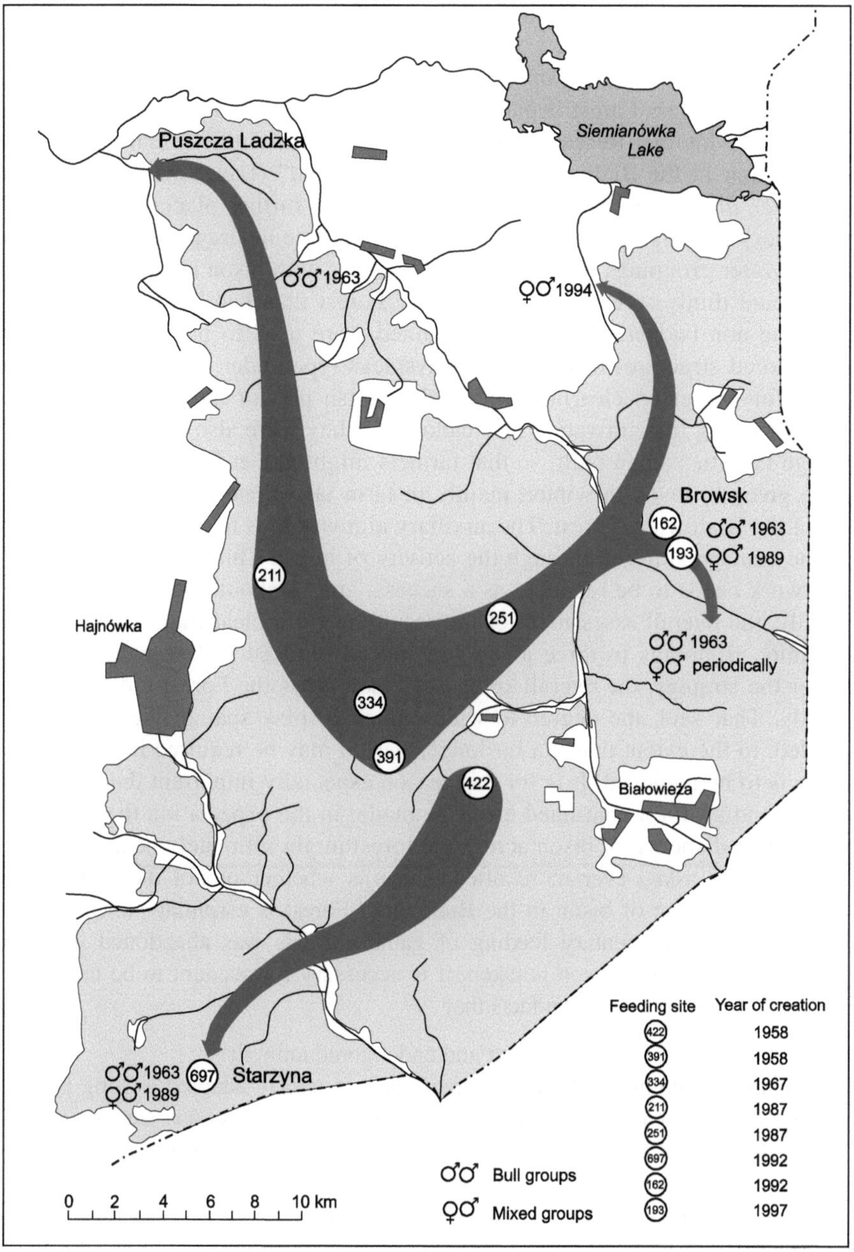

Fig. 17.4 Main direction of bison dispersal in the Polish part of Białowieża Forest. Years of first observations of bull (♂♂) and mixed groups (♂♀) in different parts of the Forest are given. Figure by R. Kowalczyk

zone covers some 173.7 ha. Winter refuge "422" covers 150 ha and occurs in compartment 422 and partly 394.

In the mid 2000s, work on a larger scale to improve the food base of the (Polish) Białowieża Forest began. This took place within the framework of an EU-funded Mammal Research Institute LIFE Project entitled "European bison conservation in the Białowieża Forest—Bison Land" (Ławreszuk et al. 2010). This *inter alia* encompassed the development of further places offering supplementary feeding opportunities, the aim being to encourage a breakup of the larger winter groupings that had become typical, with bison ideally being more widely and thinly spread across the entire Forest than had been the case hitherto. The non-budgetary resources obtained were used to build 19 permanent open roofed structures of the roofed haystacks type under which hay might be stored for winter on clearings within the Polish part of the Forest. Contracts commissioning the cultivating of meadows for hay were also entered into, with storage facilities again built, so that farmers might gather hay with a view to it being given to bison in winter, mainly in farm land areas beyond the boundaries of the Białowieża Forest. The auxiliary aim was thus to compensate farmers for the damage suffered through the activity of bison. This commissioned contract work needs to be regarded as a success, and so should be continued with. Equally, the overall assessment of the results made it clear that, while the size of winter groupings in three areas was indeed reduced in line with the goals set for the strategy, the overall spread of bison cross the Forest only increased slightly. That said, the degree to which bison display spatial fidelity must be recalled, to the extent that it a far longer period may be required if a change of habits is to be achieved. It is for this reason especially important that efforts to secure funding for a continued effort be made, in the expectation that a greater winter-time dispersal of bison across the forest might ultimately be achieved.

A question asked ever more often concerns whether or not the winter supplementary feeding of bison in the Białowieża Forest is essential—all the more, so since the supplementary feeding of game animals was abandoned in 1996. However, as such a decision is taken, it is necessary for account to be taken of a range of factors, including the facts that

- the European bison is a protected and endangered animal;
- the bison requirement of food is great, as it is poorly adapted to long periods without food, or to the extraction of energy from endogenous sources (a prolonged period of hunger may disrupt the flora of bacteria and protists needed if the digestive process is to proceed smoothly);
- a shortfall in the amount of food may give rise to increased mortality;
- it is necessary to count on increased conflicts with farmers living in villages or other settlements in or on the outskirts of the Forest as bison become more and more accustomed to leaving and heading out over agricultural land bison driven off or disquieted by people may also become aggressive;
- the abandonment or curtailment of supplementary feeding would have to be associated with an abrupt decline in numbers—something that would be unfavourable from both the population and genetic points of view (the unavoidable

increase in animals' winter mortality would also be taken awry by both public opinion and the media).

- the Białowieża Forest cannot offer enough natural food, e.g. on meadows where there may be food for animals under the snow.

A mild winter can see limits set on the supplementary feeding of bison. There are now winters with limited snow cover, while after the autumn is passed, there may be a masting of oak trees ensuring that bison hardly take hay at all.

The present system of supplementary winter feeding for bison has its pros and cons. The concentration of bison in their winter refuges makes servicing the population a great deal easier, this being important in difficult weather conditions, such as when snow limits access to many areas. The presence of small enclosures for catching bison at places where animals concentrate also allows for their capture as necessary. Inventorying, review and selection are also facilitated. The impact of bison on forest (i.e. damage) is likewise limited to relatively small areas. The disadvantages of the system in turn include the fact of gathering in just a small area for some 4–5 months in winter, providing for the very effective contamination of the environment with different kinds of pathogens, the ease with which disease can be spread from one animals to another, and—in spite of efforts to prevent it—the hindered access to hay in the case of weaker individuals. As has already been noted, recent telemetric research showed an influence of the intensity of supplementary winter feeding on the composition of the diet in bison (Kowalczyk et al. 2011).

17.4 Captures and Transport

In the Middle Ages, many of the rich and powerful kept wild animals on their estates. European bison were among the animals finding their way into their menageries. These animals had first to be captured somewhere in the forests where they were present naturally. In his poem, Mikołaj Hussowski (1523) recalled the way in which bison might be caught. "Two walls are placed widely apart, but narrowing down as they lead forward into a kind of tight wedge". From this short description, it is clear that the principles of capture have not changed so much since that time. Captures of bison in the Białowieża Forest can be assumed to have involved pens built to hold animals prior to the hunt.

Today's captures of bison in the Białowieża Forest also involve herding into a narrow-ended enclosure into which a transporting cage is incorporated. Each Reserve at Białowieża has such a setup. Captures from the free-ranging herd are also achieved using structures built at the winter refuges in 1973 in forest compartment 391C and in 1976 at forest compartment 422B. In the years 1971–1992, some 235 (112, 123) bison of differing ages were captured and transferred to other centres. Single individuals, especially adult males, were captured using live traps (Fig. 17.5).

Fig. 17.5 An effective means of capturing bison is to guide them into special live traps. The openwork cage built of strong poles can very readily have food introduced into it—and this is taken willingly by the animals. Once they entered the cage, the latch is lowered and the bison are effectively trapped. Notwithstanding the fact that one bull is caged, a second does not get away, but waits peacefully for his associate. Equally, it has sometimes happened that one bull has freed its compatriot by turning over a cage. *Photo* ZA Krasiński

Table 17.2 Measurements of cages used to transport of European bison

Sex and age of animals	Internal measurements of cage in cm		
	Height	Breadth	Length
Adult bull	180–190	78–80	260–300
Adult cow	170–175	70–75	250–280
Young 2–3 year old	160–165	65	200
Calves up to 1 year old	145–150	60	185

Source Krasińska and Krasiński (2007)

The transport of bison makes use of wooden cages of dimensions adjusted to those of the animals themselves (Table 17.2). Cages for adult bulls are reinforced with metal squares. On longer journeys, animals are fed and watered through an opening in the front wall of the cage. This reflects the fact that any lack of regular feeding and watering in ruminants may lead to disruptions of the digestive process due to changes in the composition of the bacterial and protozoan flora in the rumen.

The transport of bison over short distances may be achieved without a cage if pharmacological sedatives are used. Thus far, full immobilisation of bison prior to transport has not been applied, on account of the subsequent difficulties that would be experienced with putting animals into a cage.

17.5 Immobilisation and Marking

The immobilisation of animals can be achieved with pharmacological agents (Fig. 17.6). As the mechanical restraint of such a large and strong ruminant as the bison is dangerous for both animals and people, the pharmacological alternative is now applied routinely as research is being carried out (e.g. with radio-collars being fitted) or veterinary investigations and treatments pursued. The animals are drugged with the aid of a special dart-gun operating over short distances (30–40 m) and powered by compressed air. Any greater distances will require a weapon with a powder charge, or else one powered by compressed carbon dioxide (CO_2). The maximum range of such a weapon is an estimated 70–90 m, although in practice it is not effective over more than 50 m (Krasiński 1971).

Where the sedation of a bison is required, it is important that the appropriate agent be selected for the purpose, and the optimal dosage determined. The agent selected should meet the following criteria: (1) it should not be toxic systemically, (2) it should leave the animal powerless for a certain period of time, (3) it may be applied by way of intramuscular injection, (4) it should be absorbed rapidly and produce a rapid systemic reaction, (5) it must be characterised by a wide margin of safety between the dose that leaves an animal incapacitated and the lethal dose, and (6) it should be effective in small amounts, bearing in mind the limited capacity of the dart-gun method. It is also important that there be an antidote to the agent.

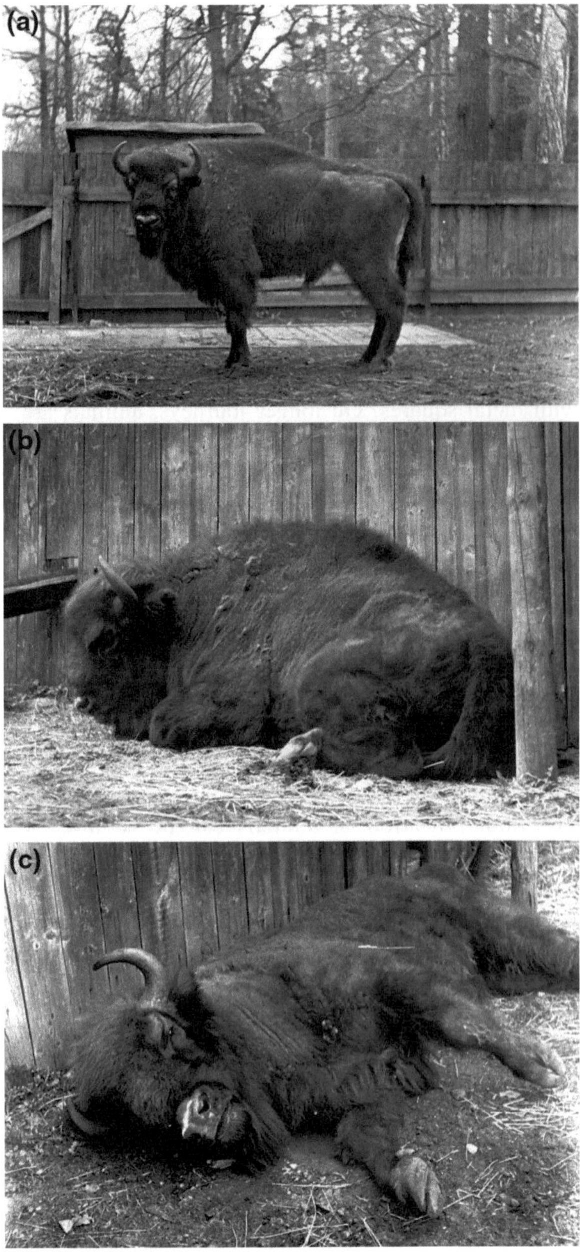

Fig. 17.6 The first attempts at immobilising (sedating) bison were made in the Reserves. *Photo* ZA Krasiński. **a** A bison that has received pharmacological agent M99, now called *Immobilon*. **b** The most desirable situation involves the animal lying in a physiological position on its sternum. **c** The sleep induced may be a deep one, the animal lying on its side. In such situations, it is necessary to make sure that no choking on the animal's gut contents is taking place (which happens readily in ruminants)

Some pharmacological preparations of nothing more than historical significance now include succinyl choline, or nicotine alkaloids. Widespread use is now made of opium alkaloids, like etorphine hydrochloride (Kania et al. 1973) and the thiazine derivative xylazine (Krasiński 1971). The first preparation goes under the commercial name LARGE ANIMAL IMMOBILON, (formerly known as M99). The standard packet of LARGE ANIMAL IMMOBILON contains 10.5 ml of solution (2.45 mg of etorphine hydrochloride and 10 mg of acepromazine), as well as the antidote LARGE ANIMAL REVIVON as 10.5 ml of solution containing 3.26 mg of diprenorphine hydrochloride.

All applications of etorphine should be made in strict compliance with the procedures indicated, since the preparation is highly toxic to people. In any event, the antidote for people—known as NARCAN—should be on hand at all times, although Revivon may also be used as a last resort.

IMMOBILON is extremely effective on bison, even in small doses. In practice, the author (ZAK) makes use of between 0.5 and 0.7 ml of IMMOBILON and the same amount of REVIVON per 100 kg of bison body mass (Table 17.3) (Krasiński et al. 1982). Larger doses of Immobilon were used to immobilise 20 bison at Barcelona Zoo (Peinado et al. 1999). Adult males are particularly sensitive to the action of etorphine, so doses of 3 ml/individual have proved effective. In the case of adult females, this dose is not always sufficient to bring animals down. Etorphine has a powerful painkilling effect, allowing it to be used as veterinary procedures are carried out.

Xylazine is made by Bayer, under the name ROMPUN. It has a sedative effect, induces sleep, produces numbing and leaves the skeletal muscles flaccid. Most often present on the market is a 2 % solution (Rompun 2 %), or else dry powder containing 500 mg of xylazine in a packet. Maximum concentrations of up to 20 % are permitted (Krasiński 1971).

Table 17.3 Immobilisation of the European bison using Immobilon/Revivon in 1992–2001

Age years	n	Approximate body masses			Immobilon ml/indiv.	Immobilon ml/100 kg body mass	Time appearance of full immobilisation (min.)			Revivon ml/indiv.	Revivon ml/100 kg of body mass	Time recovery to full activity (min.)		
		Mean	min	Max			Mean	min	Max			Mean	min	Max
Females														
5–9	20	424	340	540	3.2 (3–3.5)	0.7	9	5	12	3	0.7	10	5	15
Males														
3	2	303	240	390	2	0.7	6	6	6	2	0.7	12	9	15
4	3	366	260	430	2.3 (1.5–3)	0.6	8.5	5	12	2.2 (1–3)	0.6	24	6	60
5	7	467	345	660	2.6 (2–3)	0.6	6.1	4	8	2.7 (2–3)	0.6	16.7	8	45
≥6	25	634	436	840	2.8 (2–3)	0.5	7.2	5	12	3.3 (3–5)	0.5	12.5	5	40

Data concerning body masses of bison from: Krasińska and Krasiński (2002)
Indiv individual. *Source* Krasińska and Krasiński (2007)

For the purposes of ecological research, it is essential that selected bison be marked or identifiable in some way. Since the 1960s, bison have been branded with numbers using liquid nitrogen applied to the skin of the hindquarters. Skin is first exposed, before the brand that has been immersed in liquid nitrogen for 1–2 min is applied. After 2 months have passed, the place of the branding is indicated by the number (8 cm in height) applied being visible in white hair on the coat (Krasiński et al. 1982).

The study of free-ranging bison has also been very much facilitated by radiotelemetric research. The first attempts to put collars with radio transmitters on to bison were made in 1985 and 1986, although several years would in fact pass before telemetric equipment really suitable for tracking the species arrived, this continuing to operate without problems in the forest environment over a longer period of time (Krasiński and Krasińska 1993). From 1993 onwards, the method was applied over a 17-year period in the study of the Białowieża Forest bison population. The battery-powered transmitters usually kept working for 3 years (maximum 7 years), while their range in the forest environment was of between 2 and 5 km. GPS tracking has also been applied, and does offer much more abundant research material. However, the downside is the short period over which the transmitters work—in the range several months up to 2 years.

17.6 The Assessment of Trophies

In Poland, the Minister of the Environment may consent to the culling by shooting of bison. In the Borecka Forest (and previously also in the Bieszczady Mountains), the sale of hunting licences was permitted, although this only allowed for the hunting of particular bison identified as individuals. Limited culling of identified individuals is also practised in Belarus, even in the eastern part of Forest. In Ukraine, the free-ranging population of bison falls mainly under the control of hunting reserves anyway, so licences to take bison are on sale. Likewise photographs in Russian publications dealing with hunting make it clear that bison are sometimes hunted there.

The method of an international assessment of bison trophies (CIC), i.e. the skull with horns, was devised by Ignacy Stachowiak (1979). Three dimensions to the horns are taken account of (Fig. 17.7): (1) the maximum distance between horn curves, (2) the circumference of the horn at its base as measured on the right and left sides, and (3) the length of the horn along its external curve. These data are assigned the relevant numbers of points, the total serving as the overall assessment criterion. Points intervals for the determination of medals were laid down in Bucharest in 1978 by an international commission, a gold medal applying to 170.00 points and over, a silver for 150.00–169.99 points and a bronze for 130.00–149.99 points.

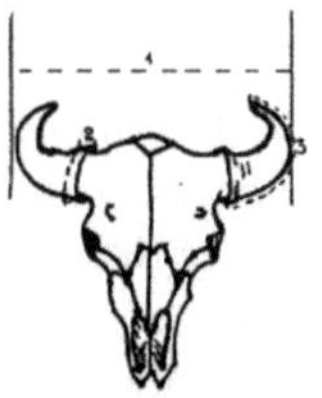

1. Maximum distance between horn curves
2. Circumference of the horn at its base
3. Length of horn

	Component of valuation		Measurement	Sum	Average	Factor	Score
I Measurements	1. Maximum distance of horn in cm			X	X	0.5	
	2. Circumference of horn in cm	Left				3.0	
		Right					
	3. Length of horn in cm	Left				1.0	
		Right					
	Total sum of measuremnt points						
II	Addition for beauty				0 - 3 pkt		
III	Deduction for defects				0 - 3 pkt		
	Total valuation						

Fig. 17.7 An assessment sheet by which to evaluate a bison trophy. *Source* Krasińska and Krasiński (2007)

The only hunter's licence to be issued for a European bison of the Białowieża population came out of the Ministry of Agriculture and Forestry in January 1986. Hunters from Canada took a 6-year-old bull that was out on a cultivated field near the village of Chytra. The trophy achieved a points total of 170.2, qualifying it for a gold medal in accordance with the above criteria. The highest ever assessment characterised a Białowieża skull (from the collection of BNP) which achieved no fewer than 207.05 points. This was put on display at the 1978 Bucharest Hunting Exhibition.

In the Białowieża Forest, the culling of identified bison is carried out by National Park employees. Material for research is taken from each specimen killed, the skulls passing into the scientific collections of the Mammal Research Institute, Polish Academy of Sciences based at Białowieża. There are now more than 740 bison skulls there.

17.7 Determining the Age

The age of a bison may be determined by reference to outline, size and body mass (see Sects. 3.2 and 3.3). As captive-bred bison are all subject to breeding documentation, their ages are known precisely. It is on the basis of these data that models have been devised allowing ages of bison in the free-ranging population to be determined.

17.7.1 Size and Horn Shape

The basic measurements applying to horns concern length and basal circumference, the distance between horn bases, the maximum distance between the curves of the horns and the distance between the horn-tips. Detailed measurements of horns and descriptions of how these are obtained are supplied in the work by Krasińska and Krasiński (2002).

Calves are born with horn buds that begin to develop rapidly. Two-month-old animals already have horns around 2 cm long. Sex-related differences are visible much earlier for the shapes and sizes of horns than for other bodily dimensions. Even among 9 month-olds, the horns in a male are already longer and fatter than those of a heifer. The first 12 months of life see horns increase in length by around 5 cm in each quarter of the year, the circumference of the horn base extending by some 2–3 cm in the same period. Two-year-old males already have markedly greater horn dimensions than females of the same age, and such size disparities are maintained throughout life.

The horns of adult bulls are longer, thicker and set more widely apart than those of cows. The maximum dimensions achieved by bulls' horns are: length 65 cm, distance between horn ends 75 cm, maximum distance 79 cm and basal circumference at base 36 cm. The corresponding figures for adult cows are: 52, 55, 70 and 28 cm.

Changes in the shape and dimensions of bison horns with age are presented in photographs alongside schemes supplying average values for the five main parameters, i.e. length of horn along its external curve (A–B), circumference of the horn at its base (F), breadth between horns (F–B), distance between horn ends (C–A) and maximum distance between horn curves (D–E) (Figs. 17.8, 17.9, 17.10, 17.11, 17.12, 17.13, 17.14, 17.15, 17.16, 17.17). Among bison aged 12 months or less, there can be no separation of the sexes on the basis of minimal differences in horn dimensions. At the age of 6 months, bison still have straight horns, a slight inward curvature having been established by the age of 12 months, however (Figs. 17.8, 17.9). The age-class of young bison is even nicknamed the spike (Shackleton 1968). By the time they are 2 years old, bison have markedly inward-curving horns, the curvature being greater in females (Fig. 17.10). Differences in shape between males and females become distinct by the time bison are 3 years old (Fig. 17.11). The sizes and shapes of horns are similar in 1–3 year-old European and American bison (Fuller 1959). There is strong horn growth among bison in the first 3 years of life (Fig. 17.11), but this is less considerable in subsequent years (Fig. 17.12).

The surface of the horns of young bison is matt, but that of older specimens becomes glossy. It is among 7–8-years-old adults that the horns attain full

Fig. 17.8 Determining a bison's age by its horns. *L to R* horns of an animal around half a year old, 2 and 3 years old. *Photo* ZA Krasiński, modified

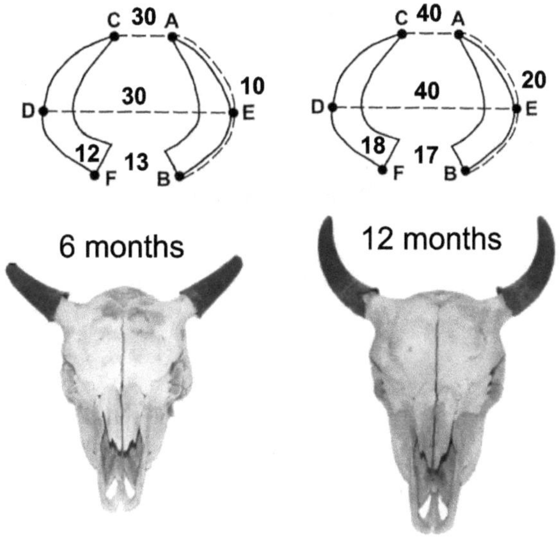

Fig. 17.9 Bison aged 6 months have straight horns, but they are beginning to turn slightly inwards by the time animals have reached 12 months. At this age, the shape and dimensions to the horns of males and females are very similar. Dimensions to horns are given in cm. For the designations regarding horn measurements, see Sect. 17.7.1

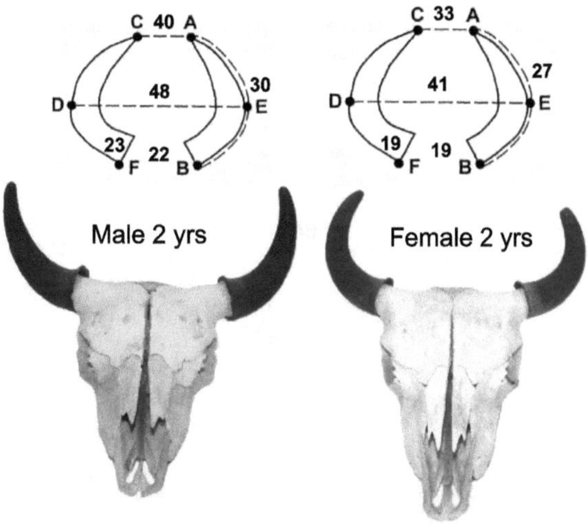

Fig. 17.10 By age 2, a male has longer and fatter horns than a female, while the latter has horns more fully turning inwards. For designations, see Fig. 17.9

development. Those of males are much thicker and wider than those of females (Fig. 17.13). Males over 10 years old frequently have worn down and rounded ends of horns due to rubbing against trees. Horns are also shorter for this reason (Fig. 17.14). A comparison of the shape and appearance of horns in a 9 and 14-years-old male is made in Figs. 17.15, 17.16. In contrast, even 20-years-old females do not have worn-down horns (Fig. 17.17).

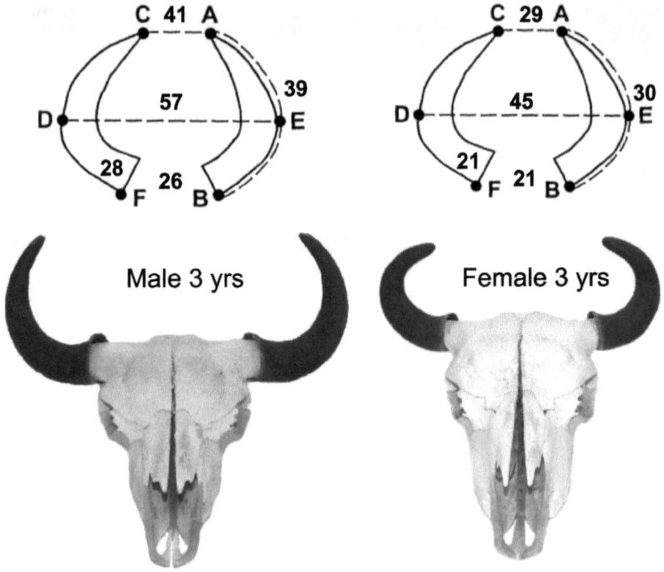

Fig. 17.11 By the age of 3, bulls and females have horns of markedly different dimensions and shape. The horns of males are larger in all their dimensions. However, 3 is the age at which the intensive growth of horns ceases. For designations, see Fig. 17.9

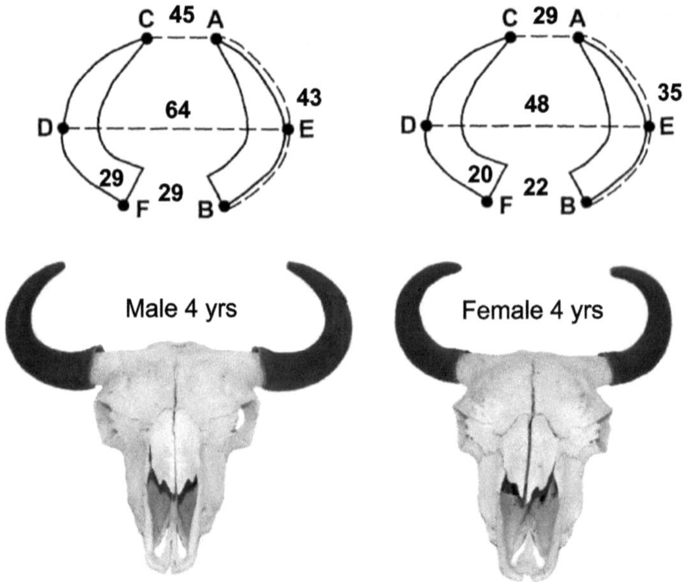

Fig. 17.12 Four-year-old males and females have horns that have fully assumed their final shape, there being a degree of curvature typical for each sex. For designations, see Fig. 17.9

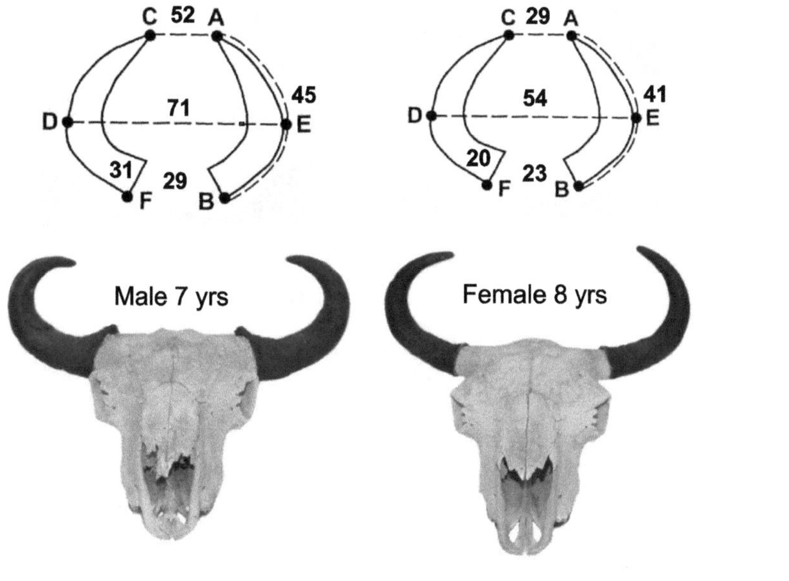

Fig. 17.13 When they reach 7–8 years of age, bison horns attain their maximum dimensions for the given individual. Dimensions are given in cm. For designations, see Fig. 17.9

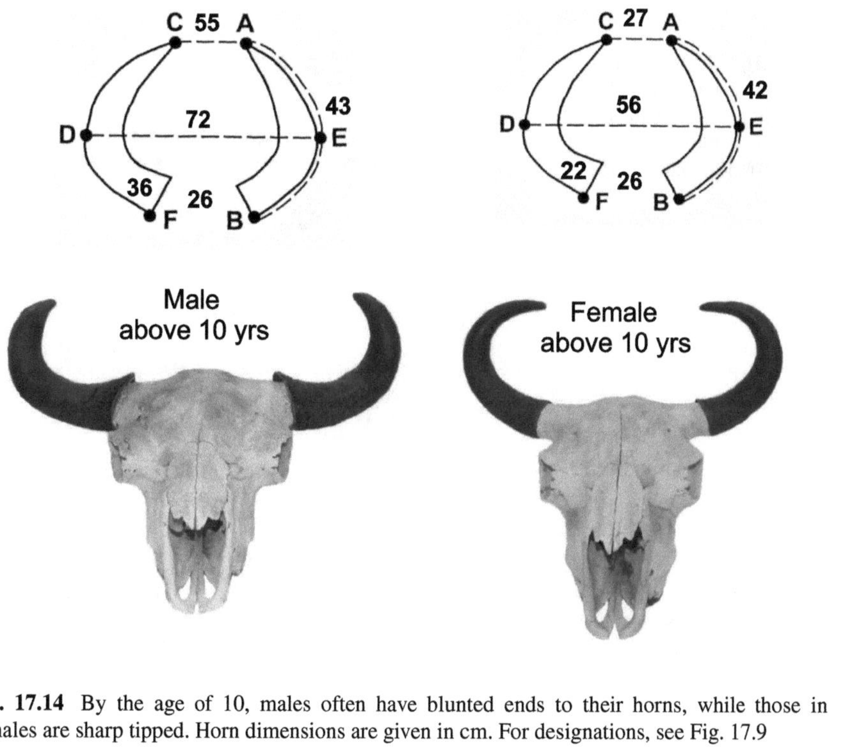

Fig. 17.14 By the age of 10, males often have blunted ends to their horns, while those in females are sharp tipped. Horn dimensions are given in cm. For designations, see Fig. 17.9

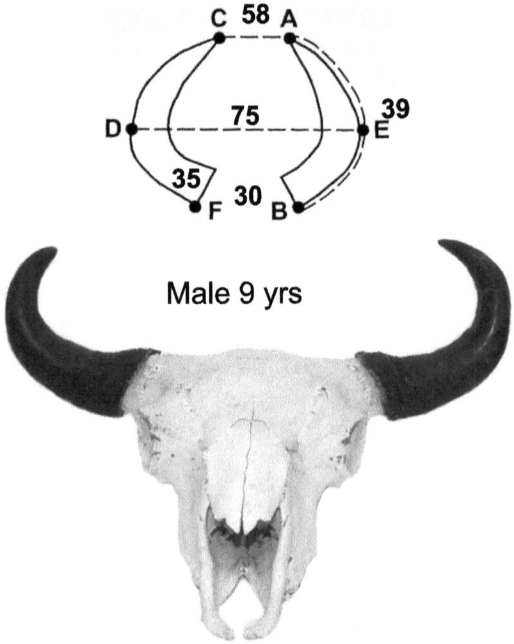

Fig. 17.15 The typical horns of an adult male. For designations, see Fig. 17.9

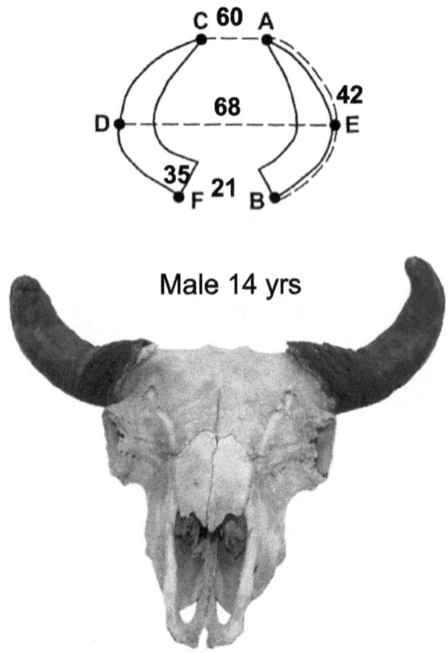

Fig. 17.16 The horns of an old male with markedly damaged points. Irregular rings are visible at the bases. For designations, see Fig. 17.9

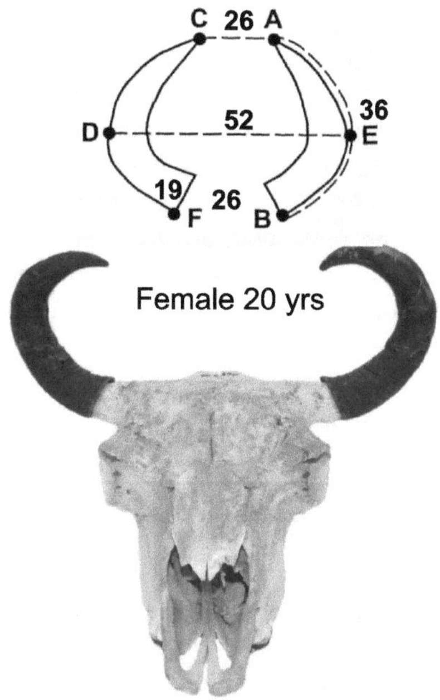

Fig. 17.17 The corneous sheaths of a female are distinctly upwards pointing, while the ends of the horns are sharp. For designations, see Fig. 17.9

A further comparison involved the measurements applying to two museum bison dating back to 1830, an old male and a female (collections of the National Park Museum at Białowieża). The basal circumference of the male's horns was of 45 cm, and was thus much greater than is noted among the European bison alive today (a maximum of 36 cm). The maximum distance of the horns was of 78 cm, compared with the maximum of 79 cm recorded at present. Other dimensions to the horns of the museum male and all those relating to the female are close to the average values noted for today's bison.

17.7.2 Determining Age by Reference to the Teeth

Bison are members of the ruminant sub-order Ruminantia of the Order Cetartiodactyla. The premolar and molar teeth are covered in enamel folds and half-moon shaped ridges allowing plant food to be ground up. The full dentition of the species includes 32 teeth (as in domestic cattle), the course of evolution having brought about the disappearance of the upper incisors and canines. The dental formula (number of molars, premolars, canines and incisors in each side of the mouth) in bison is thus described as follows:

Centre of mouth

Maxilla	3 3 0 0	0 0 3 3	12 teeth
Mandible	3 3 1 3	3 1 33	20 teeth

The deciduous teeth breakthrough in the first 3-month period, although the full formation of their crowns takes until the 6th month of life (Table 17.4). The ages at which the different molar teeth are cut are described in Table 17.5. The overall growth of the crown of each molar lasts 5–13 months (Węgrzyn and Serwatka 1984). The age at which the deciduous incisors are replaced by permanent ones is given in Table 17.6. These data agree with those presented by Flerov (1979). From the moment the milk teeth are lost, the incisors and canines completely shape their crowns within some 3–6 months. The replacement of premolars and incisors occurs 2–5 months earlier in the female than the male.

Table 17.4 Age at which deciduous teeth erupt in European bison

Mandible		Maxilla	
Name of tooth	Age in months	Name of tooth	Age in months
Id_1	Before birth	Lack Id^1	–
Id_2	Before or after birth	Lack Id^2	–
Id_3	Before birth	Lack Id^3	–
Cd_1	Before or after birth	Lack Cd^1	–
Pd_2	1–1.5	Pd^2	Before birth
Pd_3	Before birth	Pd^3	Before birth
Pd_4	Before birth	Pd^4	Before birth
Cranial lobe	Before birth	Cranial lobe	2–3
Medial lobe	1–1.5	Caudal lobe	
Caudal lobe			
Deciduous premolars are almost fully formed in Europeans bison 4 months old			

Data from: Węgrzyn and Serwatka (1984, modified). Explanations: Id_1 Id_2 Id_3 incisors; Cd_1 canine. Pd_2 Pd_3 Pd_4 premolars in the mandible; Pd^2 Pd^3 Pd^4 premolars in the maxilla. *Source* Krasińska and Krasiński (2007)

Table 17.5 Age at which molars erupt in European bison

Mandible		Maxilla	
Name of permanent molar	Mean age in years	Name of permanent molar	Mean age in years
M_1	0.5	M^1	1
Cranial lobe		Cranial lobe	
Caudal lobe		Caudal lobe	
M_2	1.5	M^2	1.5–2
Cranial lobe		Cranial lobe	2–2.5
Caudal lobe		Caudal lobe	
M_3	2.5	M^3	2.5–3
Cranial lobe	3	Cranial lobe	2.5–3
Medial lobe	3.5	Caudal lobe	
Caudal lobe			
In the European bison molars are permanent dentition			

Data from: Węgrzyn and Serwatka (1984, modified). Explanations: M_1 M_2 M_3 molars in the mandible; M^1 M^2 M^3 molars in the maxilla. *Source* Krasińska and Krasiński (2007)

Table 17.6 Age at which deciduous teeth are replaced by permanent teeth in European bison

Mandible		Maxilla	
Name of tooth	Mean age in years	Name of tooth	Mean age in years
I_1	2	Lack I^1	–
I_2	2.5	Lack I^2	–
I_3	3	Lack I^3	–
C_1	3.5	Lack C^1	–
P_2	2	P^2	2.5
P_3	2.5	P^3	2.5
P_4	3	Pd^4	3

Data from: Węgrzyn and Serwatka (1984, modified). Explanations as in Table 17.4, C permanent canine. *Source* Krasińska and Krasiński (2007)

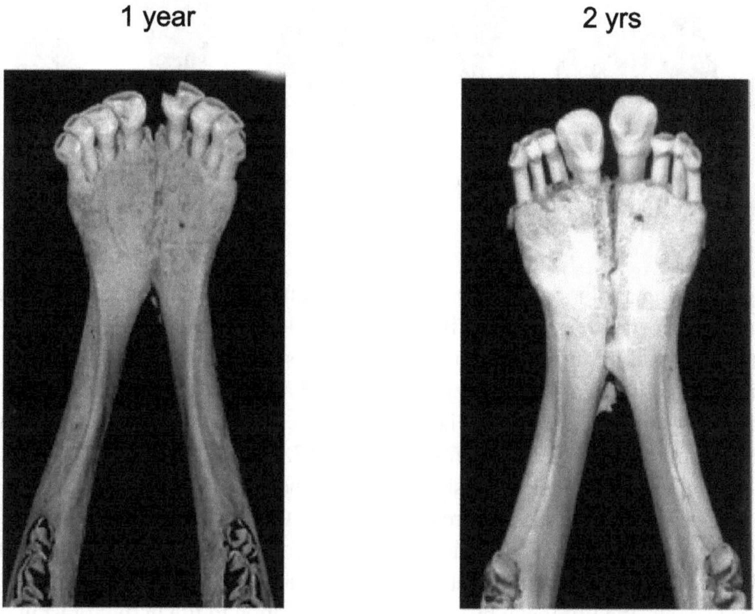

Fig. 17.18 One-year-old bison have fully developed crowns of their deciduous teeth. From the age of around two the process of replacing teeth begins, the first to be lost being the central incisors. *Source* Krasińska and Krasiński (2007)

Bison aged 12 months have a set of 8 deciduous teeth in the mandible (Fig. 17.18). At around 2 years of age, the process by which milk teeth are replaced by permanent ones gets underway. The replacement begins with the central incisors and then takes in further incisors. By the age of around three and a half, the replacement of the set of deciduous teeth is complete, and at the age of 4, bison have a full set of permanent teeth in the mandible (Fig. 17.19). Molars emerge at the same time in both male and female bison. The cutting of

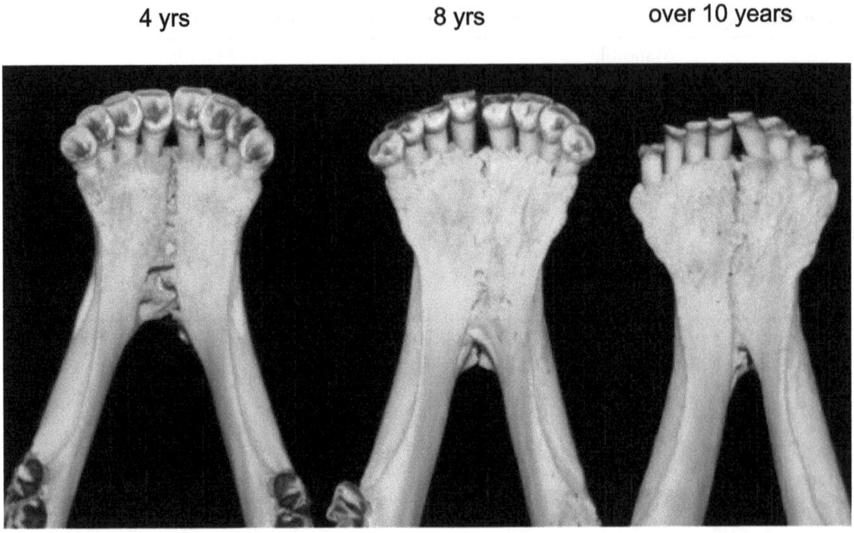

Fig. 17.19 The mandible of a 4-year-old bison has a full set of permanent incisors. By the age of 8, animals show clearly worn-down crowns, the process advancing with age. *Source* Krasińska and Krasiński (2007)

the mandibular molars and exchange of the premolars takes place earlier than the replacement of teeth in the maxilla. In general terms, the majority of teeth (other than the canines) erupt and are replaced in the same way among bison as in late-maturing breeds of cattle (Węgrzyn and Serwatka 1984).

In old bison, the crowns of the incisors have become severely worn—at times barely protruding beyond the gum line (Fig. 17.19). This is particularly visible in free-ranging bison, and is obviously linked with the chewing of tree bark, at times full debarking. Wróblewski (1927) drew attention to this, writing that the process began in middle age and intensified gradually. Bulls as young as 15 can even experience difficulties with the chewing of beets. As the ancestors of bison evolved in steppe and tundra habitats, the European bison's level of adaptation to life in the forest in general, and bark removal in particular, is not yet far-reaching. Furthermore, bulls are more intensive barking than cows, hence the greater degree of wear on their incisors.

17.7.3 Cement Layers

This method entails the microscopic counting of annual rings of cement laid down in decalcified and microtome-sliced sections of tooth. Attempts to determine age on this basis were made by Klevezal and Pucek (1987). However, the cement rings in bison teeth were not found to be very clear cut, the authors noting that a correct assessment would require selection of an appropriate part of a tooth, as well as the acquisition of a certain amount of skill where the counting process was concerned.

Chapter 18
The Significance of the Białowieża Bison for the Preservation and Recognition of the Species *Bison bonasus*

A leading role in saving the European bison from extinction was played by Poland in general and the Białowieża Forest in particular. The work to reinstate the species at Białowieża began on the Breeding Reserve. It was from here that the first bison went free after many years in captivity. It is also the Forest here that supports more free-ranging individuals than anywhere else on Earth. Furthermore, it is by way of bison originating at Białowieża that new populations and breeding centres in Poland and across Europe have been established.

In Poland, all of the closed breeding centres for bison—and all the free-ranging populations except that in the Bieszczady Mountains—include Białowieża bison. In the years 1945–2011, some 530 (220, 310) bison were shipped out from the area, 205 (96, 109) of these being from the Breeding Reserves and 325 (124, 201) individuals from the free-ranging population. The destinations of bison leaving Białowieża were Austria,

M. Krasińska and Z. A. Krasiński, *European Bison*,
DOI: 10.1007/978-3-642-36555-3_18, © Springer-Verlag Berlin Heidelberg 2013

Belgium, the former Czechosłovakia and later Czech Republic, France, Germany, Hungary, Moldova, the Netherlands, North Korea, Romania, Spain, Sweden and the former USSR. It was in turn thanks to bison from the Belarussian part of the Forest that new breeding centres elsewhere in Belarus and in Ukraine and Russia were set up.

The largest breeding centre for European bison in the western part of the continent is in Margeride, France, which again has Białowieża bison. As of 31 December, 2010, Margeride had 37 (19, 18) bison. The first 9 (4, 5) individuals from the Białowieża Breeding Reserves were sent there at the end of May 1991 (Fig. 18.1). A second set of 14 (3, 11) were transferred in early April 1992. The

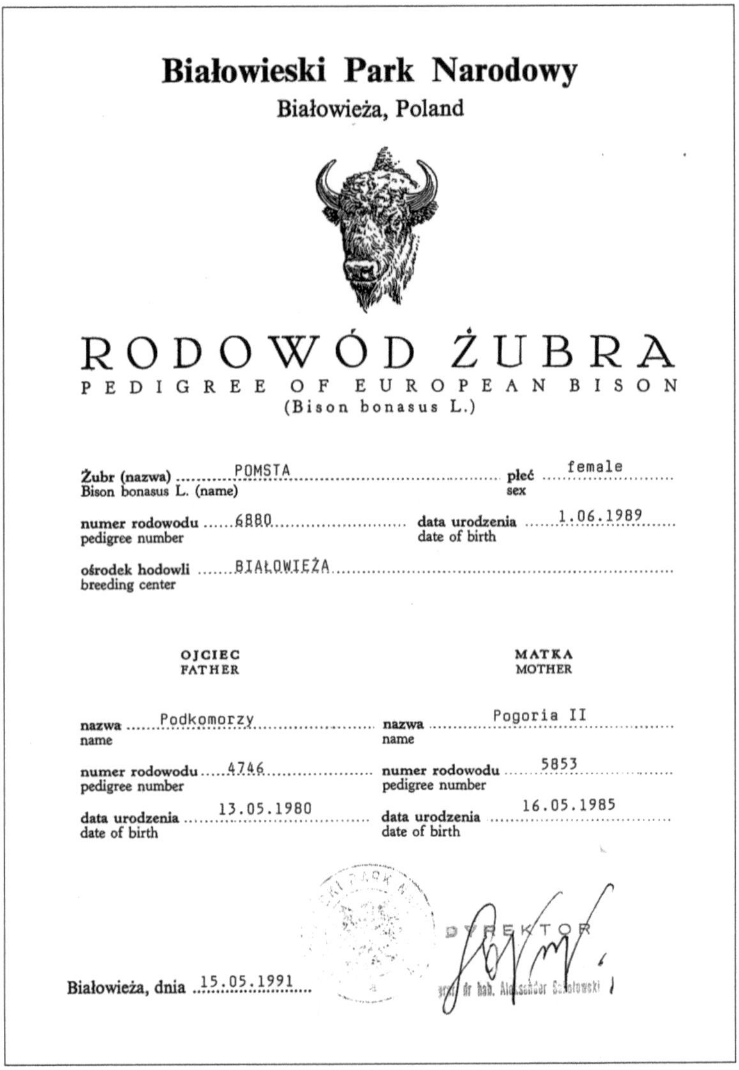

Białowieski Park Narodowy
Białowieża, Poland

RODOWÓD ŻUBRA
PEDIGREE OF EUROPEAN BISON
(Bison bonasus L.)

Żubr (nazwa) / Bison bonasus L. (name): POMSTA
płeć / sex: female

numer rodowodu / pedigree number: 6880
data urodzenia / date of birth: 1.06.1989

ośrodek hodowli / breeding center: BIAŁOWIEŻA

OJCIEC / FATHER	MATKA / MOTHER
nazwa / name: Podkomorzy	nazwa / name: Pogoria II
numer rodowodu / pedigree number: 4746	numer rodowodu / pedigree number: 5853
data urodzenia / date of birth: 13.05.1980	data urodzenia / date of birth: 16.05.1985

Białowieża, dnia 15.05.1991

DYREKTOR

Fig. 18.1 The pedigree of cow F 6880 POMSTA, which was sent to Margeride, France, in 1991. *Source* Krasińska and Krasiński (2007)

latter animals were from the free-ranging population, and so only received their names and pedigree numbers after they had been captured and placed on the Reserve. In line with the established practice, the names of these bison begin with KA, and include KARESA, KASANDRA and KABERNET. The bison at Margeride are present in a large enclosure. This part of the Reserve may only be visited with a guide, who goes in with a horse and trap. It is always possible to see a few bison that are held in the small paddock (see colour insert).

A second large Western centre for the breeding of Lowland line European bison is at Lelystad in the Netherlands. The fenced off 30 ha reserve here is divided into two enclosures. As of 31 December, 2010, there were 27 (8, 19) bison here. The Reserve has been registered in the European Bison Pedigree Book since 1993. The first bison to be transferred here came from the wild population in the Białowieża Forest (Krasiński 2000).

A third such centre is Hardehause II (Germany), which in 2004 took delivery of four Lowland bison from Białowieża and Międzyzdroje, these being placed in a newly-constructed enclosure covering 80 ha. As of 2010, there were 9 (1, 8) Lowland-line European bison here. The same overall location has also been involved since 1959 in the breeding of Lowland-Caucasian bison, albeit in the context of the so-called Hardehausen I centre. As of 2010, Hardehausen I had 21 (9, 12) animals. As of late 2010, the large closed breeding centres at Damerower Werder (Mecklenburg, Germany) had 34 bison (11, 23), while Avesta (in Sweden) had 31 (14, 17)—each centre possessing Białowieża/Caucasian bison. The dispatching of European bison to different countries and breeding centres has increased the level of security the species enjoys.

In 1949, Prof. Kazimierz Krysiak, Director of the Department of Animal Anatomy at the Veterinary Faculty of Warsaw Agricultural University, began to organise a Centre for Research into the Anatomy of the European Bison based in Warsaw. The goal was for the Centre to collect and study anatomical material deriving from the European bison, before going on to compile a monograph on the species' anatomy (Krysiak and Świeżyński 1967; Kobryń 2006). Wide use of the material was also made in the course of research carried out by other scientific institutions of the aforementioned Faculty.

Among the Centre's publications, the main items were works on the skeletal system (Pilarski 1956; Pilarski and Roskosz 1957; Empel 1962; Roskosz 1962; Empel and Roskosz 1963; Roskosz and Empel 1961, 1963, 1965; Kobryńczuk and Roskosz 1980; Kobryńczuk 1985). Recent years have seen a furtherance of this kind of work, with several new studies concerning the postnatal development of the skull, as well as cranial features associated with sexual dimorphism. This work was based around a large amount of material present in the collections of the Mammal Research Institute PAS based in Białowieża (Kobryńczuk et al. 2008a, b; Krasińska et al. 2008).

A familiarity with the musculature of the species was also gained, however, along with its tendons, fasciae and joints (Krysiak 1951a; Świeżyński and Pilarski 1956; Węgrzyn and Serwatka 1961; Świeżyński 1962; Kobryńczuk 1976). A number of studies were devoted to the anatomy of the GI tract (Pytel 1969; Pytel and Węgrzyn 1976; Gill 1968; Pytel and Wojnowska 1980), the structure of the skin (Pilarski 1962), the vascular system (Węgrzyn 1968; Węgrzyn and Kupczyńska 1986; Pytel et al. 1988) and the male reproductive system (Świeżyński 1968). Principles by which bison could be aged from their teeth were also worked out (Węgrzyn and Serwatka 1984).

In later years, work on measurement was continued with, as well as research into the non-metric features of the bison skulls deriving from the large scientific collection of the Mammal Research Institute PAS at Białowieża (Kobryńczuk and Krasińska 1987, 1991; Baranov et al. 1997). Changes in weight and body size during a bison's postnatal development were studied, as were changes in the weight and sizes of testicles (Krasińska and Krasiński 2002; Krasińska et al. 2009). The post-natal development of spermatogenesis was also researched (Czykier et al. 1999; Czykier and Krasińska 2004; Sysa and Matuszewska 2006).

Białowieża bison have been the objects of a number of physiological studies, including some of a long-term nature (Gill 1999, 2006) (Fig. 18.2). Among other things, this work has focused on red and white-blood-cell indices, the level of certain biochemical indicators and enzymes in blood serum (Wołk 1983, 1990; Wołk and Józefczak 1984, 1988; Gill 1990, 1992a, b, 1995; Peinado et al. 1999; Wołk and Krasińska 2004) and the seral contents of micro- and macroelements (Dmowski 1998; Kośla 1993; Kośla et al. 2004). Work has also been done on the reproductive biology of Białowieża bison (Jaczewski 1958; Krasiński and Raczyński 1967; Daleszczyk 2002), as well as their behaviour (Caboń-Raczynska et al. 1983, 1987; Daleszczyk and Krasiński 2001; Daleszczyk 2002, 2004, 2005).

The free-ranging population of European bison in the Białowieża Forest has been the subject of monitoring, since it was first established. There is a wealth of documentation from the 1950s onwards relating to changes in population size, levels of natural mortality, age, social and spatial structure of the population. Analyses of these data may provide a basis upon which to forecast the functioning,

Fig. 18.2 Professor Janusz Gill (*on the right*) was the scientific worker engaged in study of the Białowieża Forest for the longest. Here, he is seen in conversation with the first guard of the free-living herd, Grzegorz Sawicki. *Photo* the authors' archive

and plan the management of the Białowieża population itself, as well as other newly-established ones elsewhere.

Back when the free-ranging population of Białowieża bison was only just beginning to develop, Pucek (1967) drew attention to the need for a series of ecological studies on the animals. Research of this kind was commenced in the late 1960s and continues to this day. Thanks to it, we now have an intimate knowledge of the composition of the diet among bison in different environments (Korochkina 1969bc, 1971a, 1972b; Borowski and Kossak 1972; Gębczyńska et al. 1991), the population's social and spatial structure, and the strategies pursued by animals as they utilise different environments in both parts of the Forest (Krasiński 1967, 1976, 1978a; Korochkina 1969a, 1971b, 1972a, 1974; Korochkina and Bunevich 1982, 1983; Krasińska et al. 1987; Bunevich and Kochko 1988; Bunevich 1994, 1999a; Krasiński et al. 1994a; Krasińska and Krasiński 1995; Daleszczyk et al. 2007). In the years 1970–1980, work was done on home-range sizes and group size and structure, by way of the observation of individuals marked using liquid-nitrogen branding of their hindquarters (Krasiński et al. 1982; Krasiński and Krasińska 1993).

Ecological studies took a great step forward with the onset of radio-telemetric research from 1993 onwards (Fig. 18.3, see also colour insert). Through this, the home ranges of individual bison became known, as did their spatial distributions across the two parts of the Forest (Krasiński et al. 1999; Krasińska et al. 2000). Also better known now are the use bison make of their habitat, and the types of damage they do on farmland (Daleszczyk et al. 2007; Hofman-Kamińska and Kowalczyk 2010, 2012; Kamiński et al. 2010; Kowalczyk 2010). The influence the intensity of supplementary feeding in winter is able to exert on the animals' diet

Fig. 18.3 A bison should only be approached if it has first been sedated. This bull has been radio-collared allowing his movements to be traced. *Photo* ZA Krasiński

has also been studied (Kowalczyk et al. 2011), as have the matters of the sizes of home ranges in winter and movements made in the course of 24 h (Kowalczyk et al. 2008). There has also been a continuation of work on the ecology of the bison population, and its conservation (Mysterund et al. 2007). 2012 also brought the publication of musings as to whether the European bison ought to be regarded as a "refugee" species (inasmuch as that it may be living in suboptimal forest habitat), and what consequences for the protection of the species that might entail (Kearley et al. 2012).

A series of studies were also devoted to the history of the species and its protection. An analysis of the situation *Bison bonasus* found itself in globally was carried out, and recommendations for the species' further conservation devised (Pucek et al. 1995, 2004; Kowalczyk et al. 2010; Tokarska 2010). A strategy for the protection of the European bison in Poland (Olech and Perzanowski 2007), Russia and Belarus (Flint et al. 2002; Kozlo 2001) was also devised as well as a protection plan for the Białowieża Forest (Krasińska and Krasiński 2002a; Krasiński et al. 2010), and for the BIESZCZADY population (Perzanowski and Olech 2004) and the Western Pomerania-based STADO ZACHODNIOPOMORSKIE population (2005).

Very importantly, levels of genetic diversity in the Białowieża population were determined, by reference to values for the inbreeding coefficient (Olech 1987, 1989, 1998, 2003, 2006), as well as proteins, mtDNA and kappa-casein genes (Gębczyński and Matuszewska-Guszkiewicz 1987; Hartl and Pucek 1994; Burzyńska and Topczewski 1995; Kamiński and Zabulewicz 1997; Burzyńska et al. 1999, Wójcik et al. 2009), MHC genes (Radwan et al. 2007; Babik et al. 2012); microsatellites (Gralak et al. 2004; Luenser et al. 2005; Tokarska et al. 2009a, b; Tokarska et al. 2009b) and SNP microchips (Pertoldi et al. 2009ab; Pertoldi et al. 2010).

A slightly different kind of ecological research carried out in the Forest in the years 1997–2001 sought to determine the fate of bison carcasses as these were consumed by mammals and birds of various species, in harsh winter conditions in particular. The most frequent scavengers were ravens and foxes. The carrion was used over long periods—of 106 days on average—the ultimate effect being the consumption of 80 % of the original live weight (Selva et al. 2003).

As has already been noted elsewhere, work on the state of health of the Białowieża bison population has been in progress for many years now (Fig. 18.4), especially as regards the elucidation of causes of disease posthitis/balanoposthitis among males (Kita et al. 1990, 2003; Piusiński et al. 1997; Jakob et al. 2000; Wolf et al. 2000; Borchers et al. 2002; Thiede et al. 2002; Kita and Anusz 2006a, b; Lehnen et al. 2006; Speck et al. 2007; Krasińska and Krasiński 2010; Bunevich 2010). Also the subject of research is the influence of the worsening condition of animals in the Białowieża Forest in the years 2000–2003 on biochemical values for serum and enzyme activity in blood (Wołk and Krasińska 2004).

Work to better understand the fauna of endoparasites affecting bison was carried out from the moment the free-ranging population of the Białowieża Forest was established (Drożdż 1967). Later it was continued with through the

Fig. 18.4 This ultrasound scan of an adult male is being carried out by Dr O. Wolf and A. Lange of the Institut für Zoo—und Wildtierforschung and Free University of Berlin. *Photo* ZA Krasiński

examination of culled animals, the work revealing a steady increase in intensities of infection by many different parasites, as well as increase in the number of different parasites to be noted in the bison (Dróżdż et al. 1989, 1990, 1994, 1998a, b, 1999, 2002a, b, 2003; Demiaszkiewicz 1988, 1995; Demiaszkiewicz et al. 2008, 2009; Demiaszkiewicz and Pyziel 2010; Cabaj et al. 2005, 2006, 2008, 2012; Osińska et al. 2010). Bison emerged as suitable hosts for a great many parasitic species also present on other ungulates (Table 19.2).

A start to work on external parasites was in turn made at the end of the 1990s (Izdebska 1999, 2000, 2001a, b, c, 2003, 2005, 2006, 2011; Kadulski and Izdebska 1996; Izdebska and Cydzik (2010).

Chapter 19
Threats to the Species *Bison bonasus*

The fate of the European bison offers a good example of how very quickly a species can be brought to the edge of extinction, and how hugely difficult it can subsequently be to make it safe once more. Furthermore, while all the effort made to reinstate this species has been an undoubted success, there remains a need to ask if we can even now rest easy knowing that the European bison's future is assured. For, while the situation in the early 21st century is incomparably better than that faced as the restitution efforts were just beginning, there are new threats on the horizon. These are primarily of a global nature, connected with the "bottleneck" the species was forced to pass through and hence common to all European bison, wherever in the world (and in whatever circumstances) they may find themselves. Beyond this there is a second group of threats linked to the local conditions in which given wild populations live (Krasiński and Krasińska 2006).

The consequence of the "bottleneck" is the limited genetic variability characterising all the European bison now alive. In turn, a result of the drastic decline in

M. Krasińska and Z. A. Krasiński, *European Bison*,
DOI: 10.1007/978-3-642-36555-3_19, © Springer-Verlag Berlin Heidelberg 2013

numbers may be inbreeding depression (Olech 1987; Pucek 1991, 1994; Pucek et al. 1995, 2004), which may lead to the appearance ofgenetic defects, less-flexibity and tolerance to unfavourable environmental changes, or a lowering of resistance and consequent vulnerability to disease. Among the problems that may be genetically derived are such testicular anomalies as cryptorchidism-noted in 10.6 % of the males over 1 year old subjected to post-mortem examinations in the years 1968–2005; as well as hypoplasia or atrophy (reported from 13.6 % of examined males) (Krasińska et al. 2010). Also perhaps worthy of being added to this category is the high incidence of testicular cysts—reported in 60.7 % of the males culled from the free-ranging Białowieża population in the years 1995–2000 (Matuszewska and Sysa 2001, 2002; Matuszewska et al. 2004; Gill 2002, 2006; Sysa and Matuszewska 2006).

19.1 Global and Local Threats

Historical factors conspired to ensure the very limited gene pool that today's bison possess. Slatis (1960) determined that just 7 Lowland bison were founders of the population, i.e. M 45 PLEBEJER, F 42 PLANTA, M 87 BILL, F 89 BILMA, M 15 BEGRUNDER, F 16 PLAVIA and M 147 BISMARCK (see Table 2.1). However, the role played in the Lowland bison gene pool by most of these founders is growing ever smaller (Pucek et al. 2004). The role of the bison pair M 45 PLEBEJER and F 42 PLANTA is prevalent (more than 80 % contribution) (Olech 2003). The three founders M 15 BEGRUNDER, M 87 BILL and M 147 BISMARCK left only one descendent, the female F 524 BESTE, born in Berlin Zoo and arriving in Białowieża on May 13th 1947 via the zoos in Wrocław and Łódź. These data make clear the flimsy foundations upon which the further existence and fate of the whole species is based.

A fundamental limitation imposed on free-ranging populations of animals as large as the European bison relates simply to the size of an area in which they can go on existing. The usual situation applying to most of the world's populations is that any further spread is effectively impossible. The only real chance of increasing the area over which the species exists lies with the vast spaces existing in Russia. Equally, there are small areas that inevitably support just small populations. Where the populations resident in a forest-field environment are concerned, a threat is posed by growing conflicts with farmers (as in the cases of the populations PUSZCZA KNYSZYŃSKA (the Knyszyn Forest) and STADO ZACHODNIOPOMORSKIE (the Western Pomeranian herd). Poaching is a further source of very real concern where the bison living in the Caucasus Mountains are concerned.

In turn, within the Białowieża Forest, it is possible to identify the following threats to the bison population:

- limitations on food resources resulting from:
  - the overgrowth of riparian meadows,
  - the fencing-off of young plantations;
- periodic water shortage due to droughts and snow-free winters;
- competition with other ungulates;
- the concentration of the population due to winter supplementary feeding;

- conflicts with forestry management (reflecting damage done in young plantations and mature stands);
- conflicts with agriculture (periodic feeding on arable fields);
- lack of natural selection;
- poaching;
- lack of opportunities for migration between isolated populations.

19.2 Threats to the Health of Bison

Until recently, there were no diseases specific to European bison. Unfortunately, recently a disease of the urogential systems in males inhabiting the Białowieża Forest, as well as other free-living populations and closed breeding centres, has arisen. The appearance of this disease of so-far-unknown etiology may represent a major threat to the future of the species. The book entitled "Health threats for the European bison particularly in free-roaming populations in Poland", which is under the editorship of Kita and Anusz (2006a), presents the results of research on different bison diseases. Among other things, the book summarises the infectious diseases afflicting *Bison bonasus* (L.) (Kita and Anusz 2006b).

The most important of these diseases—capable of posing a threat to animals' health—include those dealt with in the following sections.

19.2.1 Posthitis/Balanoposthitis

The disease of bison involving inflammation of the prepuce is known as posthitis. In more advanced cases it becomes balanoposthitis, inasmuch as that the penis becomes entirely inflamed. Thus far the disease has no Polish name, hence the resort to the Latin/English form posthitis, although now it is suggested that the disease be termed necrotic inflammation of the prepuce (Krasiński and Krasińska 2003, 2006). Cases of the swelling of the male sexual organs with a purulent discharge giving rise to infertility were observed at the Pszczyna animal collection in the years 1907–1908 (Czudek 1930). Then, in 1965, symptoms of disease of the sexual organs were observed in MOCHNACZ at the Mordovskij Reserve in the Kyrgyz SSR (Tereshkin 1966; after Shabailo and Pererva 1989). However, there is now no way of determining if these were the first cases of posthitis, or were simply caused by some other factor. The late 1970s brought descriptions of the first two cases of posthitis in the eastern part of the Forest (Korochkina and Kochko 1982), while the free-ranging population in the Białowieża (Polish) Forest had its first two cases of posthitis reported in adult bulls in 1980 (Krasiński 1994b; Krasiński et al. 1994a). In 1982, the disease was noted in two bulls at the Białowieża Breeding Reserves, the animals involved being the 7-year-old PORT (M 3521) and the 6-year-old PONCZ (M 3752). In that same year (1982), pathological changes in the vicinity of the sexual organs were to observed in one 3-year-old and one 11-year-old male forming part of the Borecka Forest free-ranging population.

Insert IV. 1, 2 Bison in the far west of their range in Margeride, Massif Central, France. Neither ground strewn with boulders nor the very hot summers have stood in the way of the development of what is Western Europe's largest breeding centre for the species. *Photo* ZA Krasiński

Insert IV. 3, 4 Bison photographed in the Cherga Reserve in the Altai in 1990. While this is far to the east of the species' natural range, the animals have adjusted perfectly well to the severe winter conditions and mountainous terrain (above). The central part of the Caucasus is inhabited by so-called "mountain bison". They have an admixture of American bison blood, and for this reason could not be considered as pure-bred European bison (below). *Photo* ZA Krasiński

Insert IV. 5, 6 An adult bull became trapped by poacher in a steel snare attached to a tree. It had tightened around his horns. The animal was freed, having first been sedated. *Photo* ZA Krasiński

Insert IV. 7, 8 Several specimens of the *żubron*—a hybrid between domestic cattle and European bison identifiable by its colouration and build, as related to the breed of cow used in the cross. *Photo* ZA Krasiński and the MRI PAS Archive

Single cases were also reported from there in subsequent years. On 4 November 2000, a 7-year-old bull was culled and found to be displaying signs of fibrinoid and purulent inflammation of the prepuce. Similar symptoms were observed in another adult male in October 2001. The bison in the Borecka Forest derive from the Białowieża population. In 1984, posthitis was recorded in a male

Table 19.1 Incidence of posthitis disease among males of the free-ranging European bison population in the Polish part of the Białowieża Forest, 1980–2010

Year	Number of males in the population	Number of diseased males			
		Culled	Dead	Total	Number of diseased animals as a percentage of all males in the population
1980	115	1	1	2	1.7
1981	117	2	1	3	2.6
1982	119	7	2	9	7.6
1983	123	3	0	3	2.4
1984	118	4	0	4	3.4
1985	110	10	1	11	10.0
1986	99	9	1	10	10.1
1987	101	10	2	12	11.9
1988	98	8	3	11	11.2
1989	107	5	0	5	4.7
1990	121	0	3	3	2.5
1991	119	2	0	2	1.7
1992	114	13	0	13	11.4
1993	105	15	1	16	15.2
1994	114	3	1	4	3.5
1995	111	5	0	5	4.5
1996	101	13	2	15	14.8
1997	117	4	1	5	4.3
1998	117	9	0	9	7.7
1999	123	11	0	11	8.9
2000	130	1	1	2	1.5
2001	141	8	0	8	5.7
2002	146	4	0	4	2.7
2003	151	10	1	11	6.6
2004	158	7	0	7	4.4
2005	167	8	3	11	6.6
2006	167	13	1	14	8.3
2007	187	5	2	7	3.7
2008	201	11	1	12	6.0
2009	197	14	1	15	7.4
2010	199	14	4	18	9.0
Total		**230**	**34**	**264**	**6,6 ± 3,8**

Data for 1980–2009 after Krasińska and Krasiński (2010). *Source* Krasińska and Krasiński (2007), updated

at the Show Enclosure in Gołuchów (Gill 2002, 2006). In 1989, 3 bulls were culled in the NADVIRNJANSKA free-ranging population in Ukraine, because these also displayed symptoms of disease of the urogenital organs (Shabailo and Pererva 1989). Disease was one of the reasons for the complete elimination of this population in 2002 (Kryzhanovskij, personal communication). Wolf et al. (2000) then described 4 cases of posthitis that had occurred in captivity in the Bayerischer Wald (Germany).

In the free-ranging population of the Białowieża Forest, posthitis was recorded in a total of 264 males between 1980 and 2010 inclusive. Of these, 230 were culled by shooting and 34 died (Table 19.1). These data take account of all the males dying in which symptoms of posthitis were noted, even though the immediate cause of death may, for example, have been collision with a train. In the years 1980–2010, there were between 2 and 16 males with symptoms of the disease each year. The proportions of all males in the population showing symptoms were ranged from 1.5 to 15.2, the average for the period being 6.6 % (Table 19.1) (Krasińska and Krasiński 2010). If any disease is present in more than 10 % of all males, the incidence can be thought of as a high one. The highest ever proportion of males showing symptoms of posthitis was reported in 1993 (Table 19.1 and Fig. 19.1). Until now, it has been impossible to eradicate the disease, even though a policy of culling sick animals was pursued. Posthitis has been found in males of all age groups, although 37.7 % of cases relate to those aged 2–3 and 14.4 % of calves contract it before they have reached the age of 1 (Fig. 19.2). As symptoms of the disease are found in calves and 2–3 year-old adolescents, i.e. animals in no way participating in reproduction, it is clear that the posthitis is not a sexually transmitted disease. In fact there is a presumption founded upon observation that most of the males found displaying early-stage posthitis during the winter—and thus scheduled for culling—were infected by the disease in different months of the growing season immediately preceding that same winter (Krasińska and Krasiński

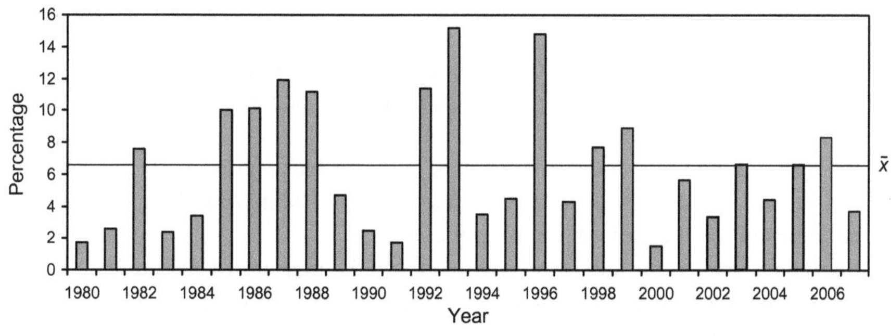

Fig. 19.1 The percentages of males culled or dying showing symptoms of posthistis/balanoposhitis disease—recalculated as the proportion of the whole males in population in the Polish part of the Forest. *Source* Krasińska and Krasiński (2007), updated

Fig. 19.2 The occurrence of posthitis/balanoposthitis diasease in European bison males of the Polish population in relation to age, 1980–2002. *Source* Krasińska and Krasiński (2007)

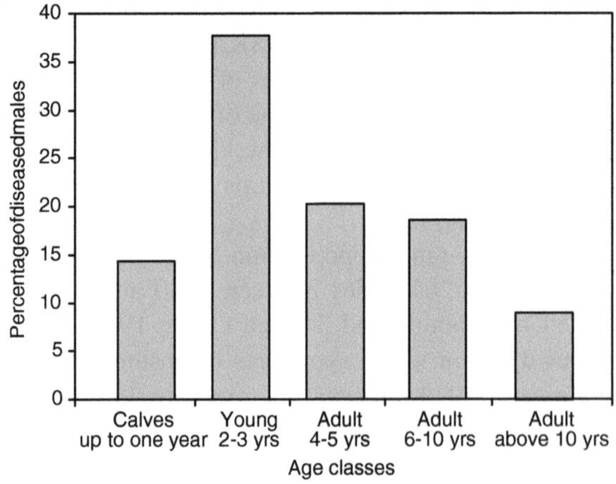

2010). Cases are also observed (if more rarely) among males at the Białowieża National Park's Breeding Reserve.

In the Belarussian free-ranging population of the Białowieża Forest, posthitis was recorded in a total of 151 males between 1980 and 2008. Of these, 127 were culled by shooting and 24 died. In the years 1980–2008, there were between 1 and 16 males with symptoms of the disease each year. The proportions of all males in the population showing symptoms ranged from 0 to 14.8 %, the average for the period being 5.4 %. Posthitis has been found in males of all age groups, although 30 % of cases relate to those aged 2–3, while 0.5 % of calves contract it before they have even reached the age of one (Bunevich 1999b, 2010). Overall, the work by Bunevich makes it clear that posthitis was observed much sooner in the Belarussian population much earlier than on the Polish side—most probably even from 1966 onwards, albeit as a rarity until 1980 (Bunevich 2010).

The threat to the two separate populations inhabiting two parts of the Białowieża Forest is a joint one, hence the research being done in parallel on both sides of the national border (Shashenko et al. 1994a, b; Krasochko et al. 1997; Bunevich 2010).

We have divided the degrees of advancement of the disease into separate stages on the basis of anatomical and pathological changes accompanying the posthitis. Thus:

P+ denotes a single focus of necrosis and purulence present on the margins of the prepuce (although even at this stage a foetid slimy/purulent discharge is already present on the hairs of the preputial tuft);

P++ denotes several foci of inflammation on the margins of the prepuce;

P+++ indicates that the disease process takes in the entire preputial orifice (*ostium preputiale*);

P++++ indicates that purulent and necrotic change has now taken in the prepuce and penis, while foetid discharge is present in the preputial cavity (*cavum preputiale*);

P+++—refers to the state following prolonged pathological change, which often entails the self-amputation of the penis, with atresia of the preputial orifice (*ostium preputiale*) and the creation of a fistula, or more rarely the retention of urine and consequent uremia.

Pathologically affected tissues are shown by bacteriological study to support bacteria of the genera *Corynebacterium* and *Bacillus*, as well as *Staphylococcus koagulazo* negative, *Pseudomonas aeruginosa* and *E. coli* (Kita et al. 1990, 2003). The anaerobic bacterium *Fusobacterium necrophorum* has also been noted (Jakob et al. 2000). Nevertheless, the research done to date has identified neither the source of any infection, nor any single pathogenic agent that can be linked unambiguously with the disease. Studies initially centred around the diseases inducing similar symptoms in domestic cattle, among them infectious bovine rhinotracheitis-infectious pustular vulvovaginitis (IBR-IPV), which is viral. The sexual form of IBR-IPV induces the presence on the sexual organs of both males and females of foci of inflammatory change. However, the results of serological studies of bison blood seeking the presence of antibodies against the IBR-IPV virus were negative in every case (Kita et al. 1990). A negative result was also obtained when 3 bison from the Breeding Reserve were tested in 2002 at the State Veterinary Institute in Puławy. In the light of this, certain authors are of the view that a large role in the tendency to become ill with posthitis may be being played by genetic factors associated with the limited genetic diversity of the bison population resulted from increased inbreeding level (Burzyńska et al. 1999; Gill 1999, 2002, 2006). Other studies showed no significant association between DRB3 alleles or genotypes and susceptibility to posthitis in male bison (Radwan et al. 2007) and inbreeding level and susceptibility to that males disease (Matuszewska et al. 2004). Since the disease is mainly confined to Białowieża Forest (the Polish and Belarussian parts), it may reasonably be assumed that it is this area that needs to be comber for the initial local factor or agent (or vector) that transferred the disease to male bison.

19.2.2 Q Fever and Lyme Disease

Work has been done on Q fever, which is caused by the bacterium *Coxiella burnetii*, and on Lyme disease, which is induced by the spirochaete *Borrelia burgdorferi*. Both diseases are tick-borne. In 1985, a focus of Q fever was noted in the Białowieża Forest area, among bison, domestic cattle and sheep in the local villages (Anusz et al. 1986). Antibodies specific to *Borrelia burgdorferi* were noted in 16 % of bison studied. Ticks of the species *Ixodes ricinus* are abundant in the Białowieża Forest, so it is quite possible that bison may represent an important element in the epidemiology of Lyme disease (Siński et al. 1996). Q fever, and especially Lyme disease, can affect people (Wachnik 1983; Flisiak et al. 2000). No link between these diseases and the posthitis noted in bison has been found.

19.2.3 Trypanosomes (Trypanosoma)

Trypanosomes are blood parasites reported in bison living in the Białowieża Forest in the 19th century. According to Wróblewski (1927), they caused swellings in sub-epidermal tissue. Contemporary parasitological studies point to the presence of trypanosomes in 42.4 % of the bison tested (Kingston et al. 1987, 1992), as well as in 20 % of the domestic cows in the Białowieża area (Demiaszkiewicz and Lachowicz 1991). There has, however, been no success with linking the presence of trypanosomes with the disease posthitis in bison.

19.2.4 Mycoplasmosis

Mycoplasmosis is induced by bacteria that attack all animal species and human beings. They can induce pathologies in different organs, although they are vulnerable to environmental factors and disinfecting agents. Certain less-virulent strains have adapted to life in the external environment or are present as commensals. Illness may arise where an animals' immunity is impaired, but a full account of the role played by *Mycoplasma* in bison will require further study. *Mycoplasma* spp. cause various diseases including genital disorder. Sera of 50 bison with balanoposthitis, moreover 48 healthy males and 49 female bison from free-ranging population in the Białowieża Forest were investigated for antibodies against *Mycoplasma bovis* and *M. bovigenitalium.* Sera of three diseased and one healthy bull had antibodies against *M. bovis*, but all tested females were negative. Serological tests for the antibodies to *Mycoplasma bovigenitalium* revealed its presence in 33 (66 %) out of 50 diseased males bison (Thiede et al. 2002).

19.2.5 Foot-and-Mouth Disease (Aphtae Epizooticae)

The name foot-and-mouth makes plain the main foci of this disease around animals' bodies—serosal blisters are present in the mucous membrane of victims' muzzles, leading to dribbling, while inflammed areas between the hoofs are sufficient to induce disablement. This viral disease is extremely contagious. Bison are very vulnerable to infection by FMV. In the years 1953–1954 it cost the lives of some 35 animals on Reserves in southern Poland (Podgurniak 1967). The threat of the disease arises periodically. In late 1961/early 1962, a focus of foot-and-mouth disease was noted in Pogorzelce, a few kilometres from the Bison Reserves. Fortunately effective prophylaxis resulted in the safeguarding of the bison against the disease. A further threat arose in the years 1965–1966, thanks to the appearance of foot-and-mouth in Belarus, close to the boundary of the Belarussian part of the Białowieża Forest. In turn, in 1972, the disease gave rise to tangible economic losses in the UK. The last threat was noted at the beginning of 2001, in which there were outbreaks in the UK initially, and then also in other parts of Europe. From 1990 on, protective vaccinations of cattle against the disease were dispensed with, leaving the most important preventative measures as the earliest possible introduction of prophylactic action once a threat has appeared.

19.2.6 Tuberculosis

Tuberculosis in bison has never been noted in the Białowieża Forest, and post-mortem examinations have never revealed the presence of the typical symptoms. Poland has been regarded as free of the disease since 1975, following effective action to combat bovine TB. However, this does not denote the total elimination of tuberculosis, only compliance with the relevant provisions in international law (Żórawski and Lipiec 1998). In the Bieszczady Mountains of SE Poland (within Brzegi Dolne Forest District), a 3-year-old female that died in 1996 was found to have been afflicted by a generalised process of tuberculosis giving rise to the extensive pathological anatomical changes *Mycobacterium bovis* is capable of inducing (Żórawski and Lipiec 1997). The group of around 25 bison present in that area formed part of the Bieszczady population. The source of infection of bison might have been domestic cattle, since several foci of bovine TB were reported in the years 1992–1996 in areas in which bison were also present (Żórawski and Lipiec 1997; Welz et al. 2005). Subsequently, cases of tuberculosis were noted in a cow that died in 2010, and in a bull culled in 2011 (Brewczyński and Welz 2011). Epizootic analyses done subsequently in domestic cattle and selected species of wild mammal of the Bieszczady Mountains suggest that the pathogen causing the disease can pass from the former to the latter (Salwa et al. 2011). As is well-known, the spores capable of producing TB infections are exceptionally resistant, and certainly capable of remaining viable in the environment for several years. It is for this reason that the disease is so difficult to combat. Tuberculosis and brucellosis represent a major problem for several populations of American bison in the USA and Canada (Williams et al. 1993; Joly et al. 1997; Lisle et al. 2001; Rhayan et al. 2001; Mackintosh et al. 2002), and that species is obviously a close relative of the European one.

- **Blue Tongue Disease**

Blue tongue disease poses a new threat to the European bison. This is a viral disease passed on by sick animals that carry it, though with small midges of genus *Culiciodes* acting as vectors. Blue tongue originated in Africa, but was transferred to Europe with imported animals. In 2007, it emerged *en masse* in Western European sheep, producing a very high 80 % mortality rate among animals infected. Domestic cattle also proved vulnerable, and 2007 also brought the deaths of no fewer than 10 European bison at the Hardehausen Breeding Centre in the German Land of North Rhine-Westphalia. *Post mortem* examinations of the victims oedema and blue discolouration of the tongue, as well as erosions and ulcerations of the buccal cavity, hence the Polish synonym for the disease of "false foot and mouth" (Glunz 2008).

19.2.7 Parasites of the European Bison

The European bison offer exceptional opportunities for parasitological study, having been eradicated in the natural state, only to be reintroduced into the wild after a gap of some years, albeit with an intermediate stage involving captive breeding in semi-natural conditions. All of this signifies the opportunity to compare the parasitic fauna of the bison present in the Białowieża Forest in the late 19th and early 20th centuries (as described by Kulagin (1919), Wróblewski (1927), in the conditions of the Breeding Reserves (Dróżdż 1961), and in today's free-ranging population (Dróżdż 1967; Dróżdż et al. 1989, 1994, 1999).

Thus far, some 38 helminth (flatworm and roundworm) species have been reported from the bison present in the Forest (Table 19.2). To be added to these are 14 species of ectoparasite and 3 protist species (*Trypanosoma wrublewski*, *Sarcocystis sp.* and *Eimeria sp.*). A decided majority of these species would also be typical for domesticated ruminants. Only the trypanosome *T. wrublewski*, the biting-louse *Bisonicola sedecimdecembrii* and the mite *Demodex bisonianus* can be regarded as specific to or typical of the European bison.

Two endoparasites widespread in bison are the liver fluke *Fasciola hepatica* (present in 44 % of animals studied) and the lungworm *Dictyocaulus viviparus* (nematode) recorded in 58 % of studied animals (Demiaszkiewicz et al. 1999a). More major infestations of both of these produce clear clinical symptoms in bison, among them weight loss, poor development and diarrhoea. On occasion, they may even lead to an animal's death. Both of the parasites in question are also typical for livestock, such that infections in both directions are possible where the same pastures are used.

Nematodes of the stomach and gastrointestinal tract are very widespread, with 100 % of animals potentially being afflicted (Dróżdż et al. 1999, 2002). The helminthofauna of free-ranging bison has been augmented by 15 species initially characteristic of members of the deer (red and roe deer and moose). The three of these that have seemingly now adapted to bison in a more permanent way are *Ostertagia leptospicularis*, *Spiculopteragia boehmi* and *Nematodirus roscidus*, all being both widespread in the population and capable of generating severe infestations. *Ostertagia leptospicularis* seems to be the most dominant of the three (Dróżdż et al. 2002).

Comparisons of the parasitic faunas of free-ranging bison in the Białowieża and Borecka Forests, as well as the Bieszczady Mountains, have also been made. The Bieszczady population brings forth only single cases of the presence of liver fluke, while lungworms are not found. Equally, it was for the first time anywhere in Poland that 4 bison in the Bieszczady Mountains were parasitised

Table 19.2 The helminth fauna of free-ranging European bison in the Białowieża Forest, Borecka Forest and Bieszczady Mountains

		Percent of infected bison		
No.	Species of parasite	Białowieża Forest	Borecka Forest	Bieszczady
1.	*Fasciola hepatica*	84	20	12
2.	***Parafasciolopsis fasciolaemorpha***	**5**	**–**	**–**
3.	*Dicrocoelium dendricum*	16	–	–
4.	*Paramphistomum cervi*	+	x	x
5.	*Moniezia benedeni*	+	–	–
6.	*Chabertia ovina*	33	x	x
7.	*Oesophagostomum radiatum*	14	x	x
8.	*Oe. Venulosum*	71	x	x
9.	*Trichostrongulus axei*	25	60	46
10.	***T. capricola***	**12**	**40**	**15**
11.	*T. vitrinus*	–	–	8
12.	*Ostertagia ostertagi*	100	100	69
13.	*O.lyrata*	87	80	8
14.	***O. leptospicularis***	**87**	**80**	**100**
15.	***O. kolchida***	**87**	**80**	**61**
16.	***O. antipini***	**12**	**–**	**–**
17.	***Spiculopteragia boehmi***	**87**	**60**	**53**
18.	***S. mathevossiani***	**19**	**–**	**–**
19.	***S. asymetrica***	**2**	**–**	**–**
20.	*Cooperia oncophora*	100	–	–
21.	*C. surnabada*	62	–	8
22.	***C. pectinata***	**12**	**–**	**46**
23.	*C. punctata*	19	20	–
24.	*Haemonchus placei*	37	–	–
25.	***Ashworthius sidemi***	**100**	**–**	**100**
26.	*Nematodirus helvetianus*	87	20	–
27.	***N. roscidus***	**100**	**80**	**–**
28.	***N. europeaus***	**25**	**60**	**–**
29.	***Namatodirella alcidis***	**2**	–	–
30.	*Dictyocaulus viviparous*	94	80	–
31.	*Thelasia gulosa*	5	–	–
32.	*Onchocerca lienalis*	70	x	75
33.	*O. gutturosa*	47	x	x
34.	*Setaria cervi*	3	–	–
35.	*Trichiuris ovis*	43	x	x
36.	*Aonchotheca bilobata*	75	–	100
37.	***Mazamastrongylus dagestanicus***	**12**	–	–
38.	***Bunostomum trigonocephalum***	**22**	–	–
	Number of species	**38**	**13**	**11**

"*x*" not tested, – no determined presence of a parasite, "+" confirmed presence of a parasite, typical parasites of ungulates are in bold. *Data from* Dróżdż et al. (1989, 1999, 2002a, b), Demiaszkiewicz et al. (2008, 2009, 2010). *Source* Krasiński and Krasiński (2007), updated. Worked out and modified by A. Demiaszkiewicz

by the bloodsucking nematode *Ashworthius sidemi* in February 1997 (Dróżdż et al. 1998a). Originally characteristic of Asian deer species (mainly the sika), the latter's widespread introductions brought the parasite with it into neighbour countries of Poland like Ukraine, Slovakia and Belarus, ultimately allowing for the penetration of our country (Dróżdż et al. 2000, 2002a, b). Further study has made it clear that the Bieszczady Mountains now constitute an extensive focus for Ashworthiosis: 100 % of red deer, roe deer and bison now seem to be infected with this dangerous parasite. The species was first detected in the Belarussian part of Forest in 1999 (Dróżdż et al. 2002a) and—inevitably—specimens were also obtained from bison on the Polish side just 2 years later. Shockingly, work done in 2003 made it clear that as many as 87.5 % of bison had already become infected by this nematode, and at intensities of infection reaching 1,411 worms per host (Dróżdż et al. 2003; Demiaszkiewicz and Pyziel 2010). Clearly, there is a high dynamic to the process of invasion and infestation involving this introduced species. Furthermore, most bison in the Belarussian population sustaining infestations of Ashworthius sp. were found to be suffering reduced infestations of other species of nematode (Kochko 2003). There is now a real danger that Ashworthiosis will intensify throughout the Białowieża Forest ruminants, in just the same way as it has in the Bieszczady Mountains (Dróżdż et al. 2002b). Indeed, in recent years the parasite has been found in every single culled bison (Table 19.2), while the winter density of bison at supplementary feeding sites has also been shown to be associated positively with intensity of infection with this species (Radwan et al. 2010).

Białowieża bison in turn play host to 4 species of nematodes of the large intestine, i.e. *Chabertia ovina, Oesophagostomum radiatum, O. venulosum* and *Trichocephalus ovis* (Dróżdż et al. 1990). The ends of the wintering periods passed by dense aggregations of bison are associated with both the extensification and intensification of parasitism by *Oe. radiatum* and *T. ovis* (i.e. more animals becoming infected with greater numbers of worms per individual). This is doubtless associated with the accumulation over a small area of large amounts of bison faeces containing the invasive forms of the parasites, particularly at the winter supplementary feeding sites. The nematodes in question are also abundant in livestock, ensuring that hay bought in from farms in the vicinity is also capable of harbouring the invasive forms.

Bison onchocercosis requires separate consideration, on account of the suspicion that the parasites involved may in some way be behind the appearance of posthitis in bison (Demiaszkiewicz et al. 1999b). Cases of bovine onchocercosis in Europe are attributable to the two nematodes species *Onchocerca gutturosa* (which infects the nuchal ligament) and *O. lienalis* (which focuses on the gastrolienal ligament, as well as the area beneath the capsule of the spleen). In each case, European bison are seen to be a new host organism (Demiaszkiewicz 1988, 1995). The parasite's developmental cycle proceeds via intermediate hosts in the shape of blackflies of the family *Simuliidae*, as well as ticks of genus *Ixodes*. The mature female nematodes, having located in the relevant place of parasitisation in the final host (e.g. bison), produce large numbers of larvae that make their way to the host's skin. The first reports of *O. lienalis* larvae both involved bison displaying symptoms of posthitis. Furthermore, areas of the body in which the parasite larvae concentrate are characterised by small areas of necrosis (Demiaszkiewicz et al. 1988, 1999b). The spring-summer period sees parasite larvae concentrate just beneath the epidermis, in order that contact with the intermediate hosts might be facilitated. *Onchocerca lienalis* has been reported from some 65.3 %, while *O. gutturosa* is present in 36.5 % of studied bison (Demiaszkiewicz 1995).

Recent research is also yielding information on *Neospora caninum*—an intracellular protozoan parasite whose final hosts are the dog, the coyote, and probably also other true dogs of subfamily *Caninae*. The bison is just one of a number of intermediate hosts, this status first being noted (by way of the presence of antibodies) in a female culled in 1988. Further work on the serum of 320 Lowland and Lowland-Caucasian line bison culled in the years 1984–2003 from the free-ranging Białowieża, Borecka Forest and Bieszczady Mountain populations brought "positives" in 7.3 % of the animals studied (Cabaj et al. 2005). However, in the period 2004–2011, positive antibody presence was noted in 10.9 % of tested animals from several bison populations (Cabaj et al. 2005, 2006, 2009, 2010, 2012). Fortunately, the presence of *Neospora caninum* in bison has not yet been associated with any tangible negative influences on health.

Mention should also be made of the occurrence of *Sarcocystis sp.* in the heart muscle of bison. The protist responsible was first reported from European bison by Szwejkowski (1954), being present in 85.4 % of the animals studied by Osińska and Piusiński (1997).

A recent finding from Pyziel et al. (2011) is the presence in bison faeces of oocytes of no fewer than three new species of *Eimeria*, to be added to the four species recorded previously.

European bison play host to 14 species of ectoparasite (parasitic arthropods), 13 of these known to be present in the Białowieża population (Izdebska 1999, 2000, 2001a, b, 2006a, b, 2011; Izdebska and Cydzik 2010) (Table 19.3). The two species specific to bison are the aforesaid biting-louse *Bisonicola sedecimdecembrii* and the mange-mite *Demodex bisonianus* (Kadulski and Izdebska 1996, Izdebska 2000, 2003). Each has been noted in around half of all bison tested. Clear signs of louse infection have been reported in two young animals, suggesting that young bison are more vulnerable to invasions by this parasite. *Demodex biosonianus* mainly locates in the skin of a bison's eyelids (both upper and lower), the elongate body shape and small size allowing the mite to take up residence in the Meibomian glands or ducts (Izdebska 2006). Two species reported more or less accidentally are *Ixodes hexagonus*—a tick of hedgehogs only apparently transferring to bison in the Niepołomice Reserve, and *Demodex bovis*, a mite specific to domestic cattle that is seemingly confined to bison of the Białowieża Reserve. The *Demodex* mange-mites are small arachnids parasitising both domestic and wild animals. While the different species vary little in appearance, they are characterised by distinct host specificity (Stefański 1968).

Bison also sustain external parasites typical for other animal species, e.g. *Lipoptema cervi*—for members of deer. The winged deerfly emerges from its pupa in the autumn, and the species occurs *en masse* in the Białowieża Forest, proving a

Table 19.3 Ectoparasites of the European bison, updated

No	Species	Place of occurrence: Puszcza Białowieska free-ranging and captive population	Bieszczady moutains	Other regions of Poland (Niepołomice, Smardzewice, Zoos)	Remarks
1.	*Bisonicola sedecimdecembrii* (Insecta, Phthiraptera, Trichodectidae)	+	+	+	Biting louse, specific parasite of European bison (prevalence above 50 %); mallophagosis is manifested only rarely
2.	*Lipoptena cervi* (Insecta, Diptera, Hippoboscidae)	+	–	+	Deer ked—a typical parasite of Cervids though noted rather sporadically on bison (prevalence about 8 %)
3.	*Melophagus ovinus* (Insecta, Diptera, Hippoboscidae)	+	–	–	Sheep ked is a sheep-specific parasite (it was only found on few bison in captivity)
4.	*Chrysozona* (*Haematopota*) *pluvialis* (Insecta, Diptera, Tabanidae)	+	–	–	Common horse fly was found only in the Białowieża Forest in the summer; however, this is a common species in Poland
5.	*Dermacentor reticulatus* (Acari, Ixodida, Ixodidae)	+	–	+	Marsh tick—in winter, ticks were noted on above 90 % of European bison from the Białowieża forest (free-ranging population)
6.	*Ixodes ricinus* (Acari, Ixodida, Ixodidae)	+	+	+	Common tick—frequent and numerous in Poland; it occurs commonly on European bison (in winter—prevalence almost 40 %)
7.	*Ixodes persulcatus* (Acari, Ixodida, Ixodidae)	+	–	–	Taiga tick—rare in Poland (only in the Białowieża forest); occasionally on bison
8.	*Ixodes hexagonus* (Acari, Ixodida, Ixodidae)	–	–	+	Hedgehog tick—only one individual found on the European bison (incidental parasite)
9.	*Chorioptes bovis* (Acari, Astigmata, Psoroptidae)	+	+	+	Chorioptic mite—prevalence above 50 %; occasional manifestations of chorioptosis
10.	*Psoroptes equi* (Acari, Astigmata, Psoroptidae)	+	–	–	Psoroptic mite—noted occasionally on bison
11.	*Sarcoptes scabiei* (Acari, Astigmata, Sarcoptidae)	+	–	–	Sarcoptic mite (itch mite)—occasionally on bison

(continued)

Table 19.3 (continued)

No	Species	Place of occurrence			Remarks
		Puszcza Białowieska free-ranging and captive population	Bieszczady moutains	Other regions of Poland (Niepołomice, Smardzewice, Zoos)	
12.	*Demodex bisonianus* (Acari, Prostigmata, Demodecidae)	+	+	+	Demodecid mite—specific to European bison (prevalence about 50 %)
13.	*Demodex bovis* (Acari, Prostigmata, Demodecidae)	+	–	–	Demodecid mite—specific parasite of cattle, noted incidentally on European bison
14.	*Demodex* sp. (Acari, Prostigmata, Demodecidae)	+	–	–	Unknown species of demodecid mites—found only on few individuals of European bison from: Białowieża Forest

+ confirmed presence of a parasite, – no evidence of presence of a parasite. *Data from* Izdebska (2001a, b, c, 2005, 2006, 2011); Izdebska and Cydzik (2010). *Source* Krasińska and Krasiński (2007), updated. Worked out and modified by J. N. Izdebska

nuisance even for people. Another arthropod found on bison, although characteristic of sheep, is the sheep ked *Melophagus ovinus* (Izdebska 1999, 2001a, b). Then there are the three further species of mite that have been reported: the chorioptic mange-mite of cattle *Chorioptes bovis*, the sheep scab mite *Psoroptes ovis*, and the itch or scabies mite *Sarcoptes scabiei* (Izdebska 1999, 2000). Demiaszkiewicz (1991) described a case of scabies induced by *Ch. bovis* in an 8-month-old male eliminated from the free-ranging Białowieża population in February 1988. The animal's skin had extensive hair loss over the thorax, abdomen and hind limbs. The skin in those areas was thickened, folded and scaly in texture. The generalised pathological changes in the skin were caused by a severe infestation of the scab mite.

The Białowieża bison also play host to three tick species: the dominant castor-bean tick *Ixodes ricinus* and ornate cow tick *Dermacentor reticulatus*, and the rare taiga tick *Ixodes persulcatus*. *IIxodes ricinus* most often locates around the inguinal areas, abdomen, neck and sides of bison, while *D. reticulatus* favours the ears in winter—at times one ear can support several tens of the parasites.

In general, parasitic arthropods influence the health of bison by damaging their skin, to the point where inflammation, loss of hair or even necrosis ensue. Damaged skin obviously also offers an entry route for various other infections, perhaps also facilitating the transmission of pathogens (this latter impact remaining a largely unknown quantity at this stage).

Bison on the Breeding Reserves are treated systematically against parasites twice a year. A more complex question concerns the justification for applying the same measures to the free-ranging populations. Not surprisingly, opinions are divided. The most rational approach, in harmony with nature, would be to take such prophylactic steps as would prevent bison from concentrating excessively, obstruct contacts with domestic cattle, seek to enhance animals' overall condition, ensure access to a diversity of natural food and encourage a cautious approach to any transfers of animals. Only in certain specific circumstances can the use of anti-parasitic agents really be contemplated.

Chapter 20
Bison as Protected Animals

In the Middle Ages, there were certain European countries (Poland and Germany among them) in which the right to hunt bison was a *regalium*, i.e. one reserved for the King. In the Lithuanian Statutes (a compilation of laws in force in the Grand Duchy of Lithuania dating back to 1529), reference is made to the protection of bison. Severe punishments for the killing of animals without permission are also laid down in the *Ustawa na włoki* of 1557, which stated that "… no animal shall be killed on Our land, and especially in and beyond Our (i.e. Royal) forests, guns shall be stored away and no animal killed on pain of death" (Hedemann 1939). The killing of bison was thus the subject of special royal permits, e.g. that issued by Stefan Batory while at Malbork in 1577, allowing Crown Under-Chancellor Jan Zamojski to kill 4 bison and 4 moose in the Białowieża Forest (Hedemann 1939). In 1802, Tsar Alexander of All the Russias issued a decree that brought bison under protection, prompting locals to dub the bison *carskaya skotinka* ("the Tsar's cattle").

20.1 The Legal Status of the European Bison in Poland

The first contemporary act of Polish law—bringing the bison under total protection—was a Regulation issued by the national authorities in 1938.

As of today (2006), the European bison in Poland enjoy protection by virtue of the Nature Conservation Act of 16 April, 2004 (*Dziennik Ustaw* Official Journal of Laws no. 92, item 880). It was on the basis of this Act that the Minister of the Environment introduced a list of wild animal species subject to protection (*Dz.U.* 220, item 2237 of 2004). The list in question included European bison *Bison bonasus* (L.). In exceptional circumstances, the Minister of the Environment enjoys the right to allow listed plants and animals to be harvested or made subject to other activities otherwise proscribed or restricted. The Treasury in turn assumes responsibility for damage done to crops or forestry by bison.

M. Krasińska and Z. A. Krasiński, *European Bison*,
DOI: 10.1007/978-3-642-36555-3_20,

20.2 The European Bison in the International Law

European bison were first listed as endangered species in the so-called Red Data Book compiled in 1966 by the International Union for the Conservation of Nature and Natural Resources (IUCN). The Red List is updated regularly—alas in the main through the addition of species and adjustment (mostly raising) of the level of endangerment species face. In this company, the European bison is one of rather few animals whose level of endangerment has gone down. Thanks to the protective measures pursued, the world population of the species has risen to a level capable of ensuring its relative safety.

In 2000, the European bison was categorised as an "EN" species (i.e. ENDANGERED) in line with the old criteria applied in the 1994 Red List from the IUCN. However, since 2001 new criteria have been in place, whereby:

CR	CRITICALLY ENDANGERED
EN	ENDANGERED
VU	VULNERABLE

On the basis of these criteria, the European bison is now "merely" VU for VULNERABLE. The categorisation first and foremost reflects the size of the population in existence, as well as trends to be noted for its numbers. A Polish Red Data Book of Animals came out in 1992 (Głowaciński 1992), and assigned the European bison *Bison bonasus* to category 0 ("out of danger"). At the beginning of the reintroduction effort the species was naturally deemed endangered, and then later vulnerable. Bison are also included in the Red Books of the Russian Federation, Ukraine, Belarus and Lithuania. Poland also has a Red List of Endangered and Threatened Species by Głowaciński (1992, 2002). While both publications have the same aim, the Lists are to be verified and renewed more regularly than the Books.

European bison are also subjects of the international conventions and agreements detailed below.

The Washington Convention, CITES (i.e. the 1973 Convention on the International Trade in Endangered Species of Wild Flora and Fauna held in Washington on 3 March, 1973). The aim of this agreement is to protect plant and animal species by limiting international trade in live specimens, parts and derived products. Poland ratified the Convention on 3 November, 1989. The list constituting an annex to a Regulation of the Minister of the Environment of 27 February, 2001 **does not mention the European bison** in any of the three groups determining degrees of regulation.

The Berne Convention (i.e. the 1979 Convention on the Conservation of European Wildlife and Natural Habitats which was held at Berne on 17 September, 1979). This instrument puts particular emphasis on the protection of endangered species and their habitats, notably where action serving this objective requires cooperation between several states. In Annex III to the Convention, reference is made to members of the family Bovidae, i.e. to the European bison and the chamois (Rupicapra rupicapra) in the Polish case.

The Habitats Directive (i.e. Council Directive 92/43/EEC on the conservation of natural habitats and of wild fauna and flora) enacted on 21 May, 1992 and changed by directive 97/62 EEC. This directive is a piece of European Community Law requiring EU Member States to safeguard their biodiversity through the preservation of the natural habitats for wild fauna and flora, while taking into consideration economic, social and cultural requirements, as well as regional and local specifics. The main measure through which habitats are protected is the *NATURA 2000* network of protected areas. Annex II to the Directive mentions species of plant and animal whose protection requires the designation of so-called "Special Areas of Conservation". In the case of family Bovidae, the species listed include the European bison *Bison bonasus*, with an indication that this is a "priority species".

Chapter 21
Co-operation Over the Restitution of the European Bison in the Białowieża Forest

What had originally been the Białowieża Forest after World War II found itself bisected by a state boundary between the Soviet Union (Belarussian SSR) and Poland. Being in the western part, the European Bison Breeding Reserves all remained in Poland. On 18 July 1946, five of the animals present there (3, 2) were transferred to a newly established Reserve in the eastern part of the Forest, across the border. This initiated the breeding effort there, and gave a start to the bilateral cooperation over the breeding and restitution of the European bison that was developed by way of a series of conferences.

The first such Conference was held on 12–15 April 1961 at Białowieża, and at Kamieniuki on the Belarussian side. Participants from the then Soviet Union included Prof. G.P. Dementiev, head of the Nature Conservation Committee of the Soviet Academy of Sciences, Mikhail A. Zablocky, long-time Director of the Prioksko-Terrasnyjj Bison Reserve, W. S. Romanov, Director of the "Belovezhskaya Pushcha" Hunting Reserve Holding, and Ludmila N. Korochkina, who was involved with bison over many years, from both the administrative and scientific points of view. The Polish delegation in turn comprised Prof. Tadeusz Szczęsny, Director of the Nature Protection Board at the Ministry of Forestry and the Wood Industry, Dr Jan Żabiński, the eminent naturalist and editor of the European Bison Pedigree Book, Prof. August Dehnel as founder of the Białowieża-based Mammal Research Institute, Polish Academy of Sciences and Member of the State Nature Conservation Council, Prof. Stanisław Piwowarczyk, specialist in contagious diseases of animals, Engineer Andrzej Jaroński—an employee of the Ministry of Forestry involved with bison in the 1950s and 1960s, Engineer Jerzy Szymczak—then Director of Białowieża National Park, Engineer Julian Ścibor—chief of the Bison Breeding Centre at Białowieża and Veterinarian Wiaczesław Demiaszkiewicz, who was in charge of bison health. A key result of the meeting was the decision that the only European bison that would be present in either part of the Forest were animals of the Białowieża or Lowland line.

The second conference, convened on 13–15 November 1963, took place in both halves of the Forest: at Białowieża (Poland) and Kamieniuki (Belarussian

M. Krasińska and Z. A. Krasiński, *European Bison*,
DOI: 10.1007/978-3-642-36555-3_21, © Springer-Verlag Berlin Heidelberg 2013

SSR). The head of the Soviet delegation was Prof. G. P. Dementiev, while other participants were S. G. Kalugin, a scientific employee of the Caucasus Reserve. On the Polish side, the participants at the first Conference mentioned above were joined by Prof. Kazimierz Krysiak, Head of the Animal Protection Commission of the State Nature Conservation Council, Prof. Zdzisław Pucek of the Mammal Research Institute PAS, and Włodzimierz Pirożnikow, Head of the Bison Breeding Centre at Białowieża (Fig. 21.1). The Conference agreed upon a timetable for the exchange of Białowieża-Caucasian line bison for Lowland bison on the Belarussian side. Poland undertook to supply the necessary Lowland bison. The materials from the First and Second Conferences were gathered together in a 1965 publication of the Ministry of Forestry and the Wood Industry entitled "The Breeding of European Bison in the Białowieża Primeval Forest".

The greater part of the third conference sessions (Białowieża–Kamieniuki, 18–21 April 1967) were given over to discussing ways in which multifaceted scientific research on the free-ranging bison populations might be pursued. Studies

Fig. 21.1 A photograph taken on 15 November 1963, at Kamieniuki in the Belarussian part of the Białowieża Forest. It shows participants at the Second Polish-Soviet Conference on "Progress with the restitution of the European bison" in front of the Museum buildings. In the *bottom row* are, from *left* Prof. S. Piwowarczyk, Prof. K. Krysiak, Prof. T. Szczęsny, Prof. Romanov, a member of the Soviet delegation and Engineer A. Jaroński. The *upper row* in turn comprises, from *left* Dr. W. Demiaszkiewicz, W. Pirożnikow, an Officer of the Border Protection Service, Dr. M. Zablockij, Engineer J. Szymczak, Dr. J. Żabiński, two members of the Soviet delegation, *below* Prof. Dementiev and above Engineer R. Filipowicz, L.N. Korochkina, S. T. Kalugin, Prof. Z. Pucek and a member of the Soviet delegation. *Photo* the authors' archives

at this stage were to establish the scientific foundations for further work seeking to reintroduce the species. A noteworthy aspect of this conference was the participation of Prof. Vladimir G. Heptner, principal author of the massive work entitled "Mammals of the Soviet Union" (1966). Also taking part was Evgeniya G. Kisieleva, director of the Okskijj Bison Reserve in the Russian SSR. Prof. Szczęsny continued to head the Polish delegation, while further participants included the scientific workers studying the bison on both sides of the state border. Materials from the third conference entitled "Progress with the restitution of the European Bison" was published in 1969 by the Ministry of Forestry and the Wood Industry's Nature Protection Board.

The fourth conference, held on 23–26 November 1971 at Białowieża, featured a large number of scientific papers and reports presenting the status of and prospects for bison breeding in both countries, against the background of the situation in the world as a whole. The first fuller reference was by then being made to the results of ecological studies on bison in natural conditions. On the basis of the evidence assembled, it was determined that the Forest as a whole should ideally support some 330–350 bison, of which 200–230 would live on the Polish side and 130–150 on the Belarussian. Resolution IV of the conference was signed by Prof. Z. Pucek and later President of the Belarussian Academy of Sciences Prof. L. M. Sushcheniya. It is thankful to what was established at the conferences that the two free-ranging populations of the Białowieża Forest are homogeneous from the genotypic and phenotypic points of view, and have been recognised as the priority species for the Białowieża Forest.

Subsequent years brought scientific meetings at either Białowieża or Kamieniuki, during which scientific staff presented the results of their work on bison, including joint studies pursued through the 1990s throughout the Forest area.

Chapter 22
The Restitution of Bison in Captive Breeding in Poland

22.1 The First Attempts at Restitution of Bison in Poland

By the time the horrors of World War I had come to an end, there were only 54 European bison left anywhere in the world (Olech 2009). All that remained on what is today Polish territory were three bison in the private animal collection of the von Hochberg family in Pszczyna. The old animal collection of the Tsar in the Białowieża Forest had undergone total destruction, while Polish zoos did not possess a single surviving European bison. In consequence, first steps were then taken to procure bison from zoos in Germany and Sweden by Polish members of the International Society for the Protection of the European Bison, in partnership with Poznań Zoo (see Chap. 2, 3). It proved possible to buy a Białowieża-Caucasian bison in Germany (M 101 HAGEN), as well as pure Lowland bison cow F 96 GATSCHINA (GATCZYNA), which was born in 1911 in the Białowieża Forest. These were brought to Poznań Zoo on 11 May 1924. Both were then 13-year old. In Poznań, GATCZYNA gave birth to two calves, in 1924 the male M 114 SCHARLATAN and in 1929 the female F 171 SCHATULLE. Unfortunately, neither of these survived (Grochmalnicki 1933). HAGEN and GATCZYNA—both bought to Białowieża on 27 June 1931—were 20-year old at the time and could no longer produce offspring. In subsequent years, the restitution effort in Poland was based around specially established breeding centres. Alongside those at Białowieża and Pszczyna, further Reserves at Smardzewice and Niepołomice were also set up (in 1934 and 1938 respectively) (Fig. 22.1). The main objective in the first period of restitution was to increase numbers of bison, since each individual was of exceptional importance in saving the species. At present, the goal of these centres is to breed and raise bison of known pedigree, and to encourage such matings of individuals as will maintain genetic diversity, within the framework of the programme taking in Poland and Europe

M. Krasińska and Z. A. Krasiński, *European Bison*,
DOI: 10.1007/978-3-642-36555-3_22, © Springer-Verlag Berlin Heidelberg 2013

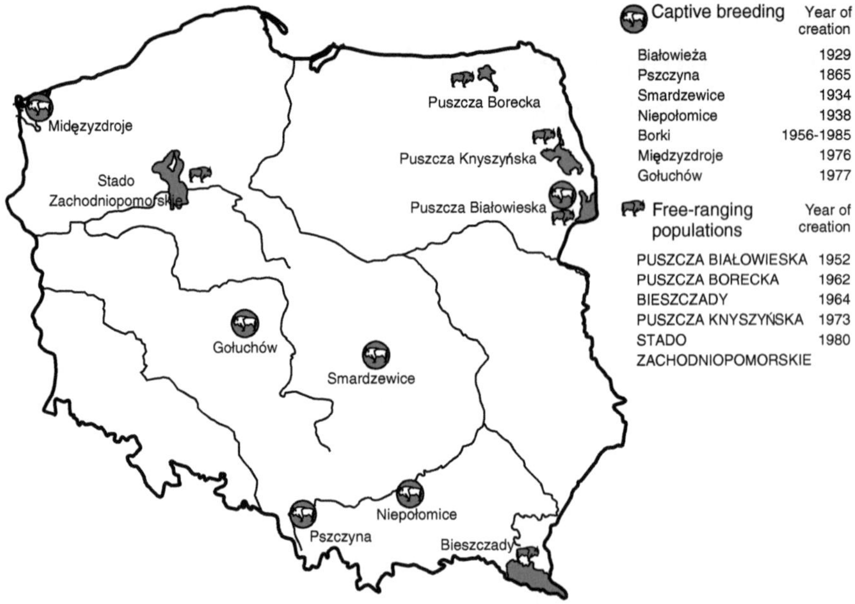

Fig. 22.1 Breeding centres for enclosed and free-living populations of bison in Poland. *Source* Krasiński (1993), modified

(Olech 1998). Reserve breeding also allows for the isolating and protecting of bison should any epizootic threat arise.

22.2 Bison Reserves

The restitution of the bison at the **Białowieża** Bison Breeding Centre has been discussed in depth in Chap. 8.

22.2.1 *Pszczyna (Pless)*

While on a trip to Russia, the Duke of Pless Hans Heinrich XI Hochberg received a precious gift from the Tsar in the shape of four European bison (one male and three females). In exchange for that, the Duke supplied 20 red deer for Białowieża (Karcov 1903; Czudek 1930). The bison in question came across to the woodlands at Pless (Pszczyna) in 1865 (Fig. 22.2), and they would go on to play a major role in the restitution of the Lowland subspecies of bison worldwide. This would have seemed an unlikely outcome back then, however, since the Białowieża herd at that time numbered no fewer than 724 animals, leaving nobody with any reason to anticipate extinction in the wild state just a relatively short time later. Further bison (five females) came to Pszczyna (Pless) in 1893, while two males were also transferred in from Berlin Zoo—one in 1880 and the other in 1900. Beyond that, a single male was obtained from Dresden in 1909. The young bull SCHARBOW imported from Germany in 1921 fell victim to a poacher's bullet (Czudek 1930), leaving 12 (4, 8) animals to be considered founders of the Pszczyna line (Krasiński 1996a). This name was assigned to the group from the moment the European Bison Pedigree Book appeared. The animals representing it obtained names beginning with the letters PL, since Pless was the name given to the town known as Pszczyna in Polish. The numbers of bison here rose steadily, to a 1918 maximum of 74 (19, 55) animals.

The symbolic hunts occasionally organised at Pszczyna (Fig. 22.3) did not cause any serious reduction in numbers, but the onset of World War I did (dramatically). Precisely three bison survived the War, the two males M 45 PLEBEJER (Wodan) and M 47 PLATON (Thor), as well as the female F 42 PLANTA (Fricka). The cow PLANTA, notwithstanding the fact that she was by then 19-year old,

Fig. 22.2 Bison at the menagerie of the von Hochberg family in Pszczyna, Upper Silesia (southern Poland). With its poor ground cover, pine forest has little natural food to offer, so bison in such habitat require intensive supplementary feeding. *Source* Karcov (1903)

managed to produce the young bull-calf named M 48 PLUNDER in 1923. Alas the latter succumbed in November 1924, but PLANTA soldiered on, giving birth to the young female F 49 PLAKETTE on 25 October 1924 (the father being PLEBEJER). The last calf born to PLANTA came into the world on 14 November 1926—M 158 PLATEN (Tyr). It was PLATEN and PLAKETTE (Fig. 22.4) that parented the male M 229 PLISCH (Rüdiger), the latter playing such a huge role in reviving the species at Białowieża (to which he was transferred on 17 April 1936) (Sztolcman 1926, Groeben 1932, Grochmalnicki 1933, Krasiński 1994a, 1996a). Over the period 1922–1936, all bison born at Pszczyna were sired by PLEBEJER, while subsequent ones were fathered by PLATEN and PLESS, their mothers being the sisters PLAKETTE and PLATANE. Museum exhibits of PLANTA and PLEBEJER can be seen in the Upper Silesian Museum in Bytom (Parusel 1996).

Fig. 22.3 The display of trophies following Kaiser Wilhelm II's hunt at Pszczyna in 1901. *Source* Karcov (1903)

All the Lowland-line European bison males alive today carry genes from PLEBEJER. An old and frail PLEBEJER was shot aged 20 at Pszczyna on 15 September 1937 (European Bison Pedigree Book 1932/37). The Director of the Pszczyna animal collection marked the death of this worthy bison by playing the following hunter's call (Geysztor 1938):

The King was dead!
The Forest King!
While the forest had shaken, mere moments before,
To the crash of his hoofs and his deafening roar.
Now here he lay, by the bullet's blow smitten.
Our bison was fallen, our bison was fallen.

The last bison at Pszczyna succumbed to an outbreak of foot-and-mouth disease in late 1953 and early 1954. For nearly a year and a half thereafter there were no members of the species at Pszczyna, but then—in 1955—bison of the Lowland-Caucasian line were brought in, such that there were 10 (2,8) present by the end of the year. Until 1964, there were only bison of this line at Pszczyna, but they were removed steadily after that, the last representative being transferred out of the breeding centre in 1974.

In turn, as noted above, November 1964 saw the arrival at Pszczyna—from Białowieża—of 8 (2, 6) young bison in two groups. From that moment on, Pszczyna resumed its work on the raising of pure Lowland bison. However, analysis of the pedigrees of Pszczyna bison makes it clear that the animals here are closely related to one another, and are thus characterised by higher inbreeding

Fig. 22.4 Bison at Pszczyna. From *left* F 49 PLAKETTE (1924–1944), F 42 PLANTA (1904–1931), M158 PLATEN (1926–1933) and M 45 PLEBEJER (1917–1937). *Source* G. Von der Groeben 1932. The cow PLANTA made a valiant contribution to saving her species. She not only lived to the ripe old age of 27 (longer than any other specimen in the Pedigree Book), but she also gave birth to the female PLAKETTE aged 20, and to the male-calf PLATEN when she was 22 (on November 14, 1926). It was thanks to the (in truth incestuous) mating of PLAKETTE and PLATEN in 1933 that bison M 229 PLISCH was born. Transferred to Białowieża in 1936, he was the starting point for almost all of the Lowland Line bison. This example makes it clear to what extent the European bison was truly saved at the last possible moment. *Source* Krasińska and Krasiński (2007)

coefficients than animals at any other Polish breeding centre for bison. The mean figure is 0.3310 (Olech 1987). Over the years 1998–2010, there were between 25 and 42 bison at Pszczna. As of December 2010, the PSZCZYNA centre had 42 (18, 24) individuals.

The European Bison Breeding Centre at Pszczyna is in fact located within the Kobiór-Pszczyna forest complex covering 140 km^2. The Reserve itself is a 7.4 km^2 area entirely fenced off from its surroundings. Within this zone, there are three enclosures—also fenced—currently covering some 70 ha. The central part of the Reserve has feeding sites, outbuildings, a capture pen and drinking troughs (Fig. 22.5). The forest here is mixed/coniferous and moist coniferous, as well as mixed deciduous forest. Pastures and meadows covering 19.3 ha are managed and mown to provide hay for the bison (Krasiński 1996a).

The European Bison Breeding Centre at Pszczyna has a more than 135-year tradition and has made a huge contribution to the restitution of the species. In consequence, it should be given priority treatment where forestry management tasks are concerned. Furthermore, the location of the Reserve, in a highly urbanised area not far from a major urban agglomeration, should justify its inclusion within the biomonitoring system.

Fig. 22.5 Bison by a feeding site at the reserve in Pszczyna. Living in the closed breeding centre, they receive supplementary fodder year round: rolled oats with barley (sometimes also other additions), as well as hay and fodder beets in the winter period. *Photo* ZA Krasiński

22.2.2 *Smardzewice*

This Reserve was set up in 1934, when the Polish diaspora in Canada chose to celebrate the founding of the World Union of Poles Abroad by offering four American bison to President of Poland Ignacy Mościcki. The 31.02 ha enclosure

built to house these animals was in the old pine forest. Specifically, the Reserve here is located on the Central Polish Lowland and in the Pilicka Forest, just 10 km from Tomaszów Mazowiecki and around 65 km from the large city of Łódź. The original name was Spała, deriving from the recreation area in which President Mościcki had a holiday residence. An observation platform from that era is still present on the Reserve today. In 1935, the Reserve was divided into two separate enclosures, one for American bison, and one for European/American bison hybrids removed from Białowieża. The first true European bison came here from Białowieża on 22 January 1936. This was the bull M 189 PUHACZ, which was followed by M 258 PURYTANIN. While the centre was unfortunately afflicted by the foot-and-mouth outbreak just before the War, the fact that all the animals present on the Reserve contracted the disease did not stand in the way of all returning to health after some 8–10 days.

The Reserve was closed down during the War, its bison being shipped off to the Reich. Post-War, the European Bison Pedigree Book began to refer to the Centre by its new name of the Smardzewice Reserve, this deriving from a small locality not far away. European bison began to be moved in here from 1949 onwards, at first males only. The first female to arrive—in 1954—came from the Niepołomice Reserve. She gave birth to the Centre's first calf in that same year, while the period 1958–1970 brought the designation of this Reserve as the place to hold surplus bulls. There were thus between 6 and 18 males here throughout that time (Krasiński and Krasińska 1997).

Fig. 22.6 Bison feeding in a clearing at the European Bison Breeding Centre in Smardzewice. An important role in reserve breeding is played by natural food to be found both in the forest and on clearings and meadows within enclosures. The numbers of bison maintained in each reserve need to be adjusted to the natural food resources. *Photo* ZA Krasiński

The breeding function at the Smardzewice Reserve resumed in 1971, when two bulls were brought in from Pszczyna and four young female Lowland bison of pure blood were translocated from Białowieża. Calves were born from 1973 onwards. Remaining bison of the Lowland-Caucasian line were moved out of here to the Bieszczady Mountains in 1975–1976, so from the end of the latter year this was a centre breeding Lowland-line bison only. Throughout the years 1998–2010 there were between 15 and 25 bison present. However, as of December 31, 2010, the figure for the Smardzewice Centre was 17 (8, 9) despite the recommended capacity for the Centre being set at 20 individuals.

The Reserve currently covers 56 ha, and is divided into five large enclosures and one small one. The whole area is fenced off. The Reserve area is 72 % coniferous forests, and 28 % deciduous forests. There are 2 ha of cultivated meadows within the Reserve (Fig. 22.6). Since 1976, the Smardzewice Reserve has come under the administrative management of Kampinoski National Park near Warsaw (Krasiński and Krasińska 1997). It therefore functions, not only as a breeding centre, but also—since 2003—as an educational tourist attraction.

22.2.3 Niepołomice

This centre was founded in 1938 within the Niepołomicka Forest close to the city of Kraków. It is now administered by Niepołomice Forest District, the Reserve having been taken over from Babia Góra National Park in 1998. There is a solid fence around the 65 ha Reserve area (Fig. 22.7), which has held between 20 and 32 bison in the years 1998–2010. As of 31 December 2010, the Niepołomice Centre had 26 (11, 15) individuals.

Fig. 22.7 The European Bison Breeding Centre in Niepołomice is surrounded by a solid concrete wall. The advantage of such an enclosure is its durability. *Photo* ZA Krasiński

22.3 Show Enclosures

22.3.1 Międzyzdroje

This centre on Wolin Island was founded in 1976 and is under the administration of the Woliński National Park. The Reserve covers 16.85 ha and is divided into two enclosures. The first bison to come here—F 3282 POMPEA and F 3524 PODWYZKA—arrived on 6 July 1976 from Borki in the Mazury region. The next transfer—on 7 April 1977—was of the male M 3040 POLARNY from Białowieża Reserve, as well as the cow F 3886 KAJUTA from the free-ranging Białowieża Forest population. The first calf was POMORZANKA, born at the Centre on 23 October 1976. The Reserve is deemed to have a capacity of 6–10 bison, and it supported between 4 and 10 throughout the period 1998–2010. As of 31 December 2010, the Międzyzdroje Centre had 5(1, 4) European bison.

22.3.2 Gołuchów

The Show Enclosure was created in 1977, on the land of the Taczanów Forest District, close to Kalisz. The 27. 97 ha enclosure is divided into two parts. Four bison moved in here in April 1977, three of these (one male and two females)

arriving from Pszczyna and a single female from Białowieża. There were between five and ten bison present between 1998 and 2010, while on 31 December 2010 there were 7(3, 4).

22.4 Zoological Gardens

As of the end of 2010, there were six Polish zoos (at Gdańsk–Oliwa, Łódź, Poznań, Warsaw and Wrocław) holding a total of 21(9, 12) European bison. The different zoos had between one and eight specimens at that time. Simultaneously, other animal collections (in Białystok, Bydgoszcz, Człuchów, Karolew, Kiermusy, Leszno, Pszczyna Park, Toruń and Ustroń) held a total of 30 (16, 14).

Chapter 23
Restitution of Free-Ranging Populations Around the World

23.1 Introduction

European bison display an impressive capacity to adapt to new conditions. Even the great grandparents (and mostly the yet-earlier ancestors) of the founders of the new free-ranging populations had not known what that felt like, yet this did not prevent what were ostensibly irrevocably pampered animals from readjusting (in fact without more major difficulties) to the harsher realities of life beyond a fence. In seeking to account for this state of affairs, we need to recall that European bison could never really be domesticated in the true sense of the word. Even in captivity, the animals retained (and still retain) a certain—and fully justified—mistrust of people, forever maintaining that reserve and distance in relation to them. By and

M. Krasińska and Z. A. Krasiński, *European Bison*,
DOI: 10.1007/978-3-642-36555-3_23, © Springer-Verlag Berlin Heidelberg 2013

large, other game species born and raised in captivity are simply unable to make the change to living out in the wild, either because they have lost their natural fear of people, or because they lack what it takes to find food in natural circumstances. Indeed, this comment is true of both carnivores (wolves and lynxes, for example), and ungulates (red and roe deer or moose). The attempts that have been made to reintroduce captive-bred mammals like lynxes to the wild are seen to require long-lasting and very intensive measures if animals are to readapt (Okarma 2000). Yet, the European bison—a species completely eliminated in the wild state—was capable of returning to nature.

While the increase in numbers in captivity left complete extinction of the European bison a remote possibility, the final reinstatement of the species could only really be said to have been achieved when free-ranging populations were again in place. Only free-ranging bison could be restored to the form their wild ancestors represented, shaped by the hard knocks nature had to offer, as well as the pressure natural selection was able to impose.

Furthermore, free-ranging populations would do most to raise overall numbers of animals—and herd sizes—worldwide. This was a task beyond the scope of captive breeding centres, in which space is always limited, along with budgets, and where the need to offer supplementary feeding year-round imposes logistical and staffing requirements where any larger numbers of specimens are involved.

23.2 The Establishment and Development of Free-Ranging Populations[1]

The first European bison to regain their freedom did so in the Białowieża Forest. The events took place in 1952 in the case of the Polish part and 1953 in the case of the Belarussian part, this single forest complex having been divided after the War by the presence of the state border. Free-ranging populations were established subsequently in other forest complexes, beginning with Poland's Borecka Forest in 1962. Two years later, work began to set up a free-ranging population in the Polish Bieszczady. Inturn, 1965 brought the founding of the first free-ranging group of Lowland-Caucasian bison in the Caucasus Mountains—at the CEJJSKIJJ RESERVE, as well as in Ukraine (in regard to the TSUMANSKA population). Thus, by 1965, there were five free-ranging populations in the world, albeit with a total population of just 218 individuals (itself representing 28 % of all the European bison on the Earth) (Table 23.1).

As further free-ranging populations were established, the numbers of individuals present in the wild grew. As of 1983, 50 % of the 2,600 bison listed by the European Bison Pedigree Book were in captivity, while the other 50 % were

[1] Names of European bison captive breeding and free-ranging populations after the European Bison Pedigree Book 2004.

Table 23.1 The world population of European bison, 1965–2010

Years	Captive breeding		Free-ranging populations		Total 100 %	Remarks
	Number	%	Number	%		
1965	562	72.0	218	28.0	780	
1966	605	70.4	254	29.6	859	
1967	593	63.0	349	37.0	942	
1968	628	61.6	391	38.4	**1,019**	**Number of bison exceeded 1,000 Individuals**
1969	711	62.1	434	37.9	1,145	
1970	700	58.3	500	41.7	1,200	
1971	737	57.2	551	42.8	1,288	
1972	833	57.0	629	43.0	1,462	
1973	901	58.0	652	42.0	1,553	
1974	984	59.1	680	40.9	1,664	
1975	979	56.7	748	43.3	1,727	
1976	1,075	57.7	789	42.3	1,864	
1977	1,139	57.1	855	42.9	1,994	
1978	1,192	56.5	919	43.5	**2,111**	**Number of bison exceeded 2,000 Individuals**
1979	1,211	55.1	987	44.9	2,198	
1980	1,247	56.6	**1,079**	46.4	2,326	**Number of bison in free-ranging populations exceeded 1,000 individuals**
1981	1,242	51.7	1,161	48.3	2,403	
1982	1,273	51.7	1,190	48.3	2,463	
1983	**1,300**	50.0	**1,300**	50.0	2,600	**Balanced number of both forms of breeding**
1984	1,329	49.1	1,377	50.9	2,706	
1985	1,454	50.8	1,408	49.2	2,862	
1986	1,441	49.1	1,495	50.9	2,936	
1987	1,377	44.7	1,705	55.3	**3,082**	**Number of bison exceeded 3,000 individuals for the first time**
1988	1,412	44.9	1,734	55.1	3,146	
1989	1,452	45.6	1,735	54.4	3,187	
1990	1,477	42.3	1,855	55.7	3,332	
1991	1,512	44.4	1,895	55.6	**3,407**	
1992	1,403	41.4	**1,990**	58.6	3,393	
1993	1,450	42.2	1,831	55.8	3,281	
1994	1,481	45.5	1,778	54.5	3,259	
1995	1,229	39.6	1,870	60.4	3,099	
1996	1,162	38.7	1,840	61.3	3,002	
1997	1,096	37.5	1,829	62.5	2,925	
1998	1,133	39.9	1,709	60.1	2,842	
1999	1,177	40.4	1,738	59.6	2,915	
2000	1,154	40.2	1,710	59.7	2,864	
2001	1,155	39.0	1,809	61.0	2,964	
2002	1,186	38.5	1,893	61.5	3,079	

(continued)

Table 23.1 (continued)

Years	Captive breeding		Free-ranging populations		Total 100 %	Remarks
	Number	%	Number	%		
2003	1,166	38.2	1,887	61.8	3,053	
2004	1,200	38.0	1,955	62.0	3,155	
2005	1,305	38.0	2,134	62.0	3,439	
2006	1,376	39.0	2,163	61.0	3,539	
2007	1,443	37.9	2,367	62.1	3,810	
2008	1,392	34.5	2,640	65.5	**4,032**	**Number of bison exceeded 4,000 individuals**
2009	1,444	34.0	2,787	66.0	4,231	
2010	1,475	33.0	**2,956**	67.0	**4,431**	**Maximum number of bison in free-ranging populations and in world population**

Data from the European Bison Pedigree Book. *Source* Krasińska and Krasiński (2007), updated

living free. In turn, the whole period 1965–2001 saw the percentage of all bison living in the wild increase from the aforementioned 28 % in 1965 to around 60 % in 2001. Between 1997–2003, there was a slight decline in the number of bison worldwide. Equally, the population living free looks set to increase further if the plan to create two large (c. 1,000-strong) populations in Russia comes to fruition. The first of these would be in the area around Vologda, and the second in the extensive forests between Bryansk, Orel and Smolensk (Belousova 1999). For good or ill, the remaining countries inhabited by bison (Poland, Belarus, Ukraine and Lithuania) do not at present seem likely to play host to any more major increases in numbers of free-ranging populations, and may even witness population declines.

The definition of free-ranging populations needs to be confined to those in which breeding bison of both sexes are present, while the area of occurrence is in no major way confined by artificial boundaries and barriers.

The European Bison Pedigree Book includes, alongside captive-breeding centres and free-ranging populations, those that may be considered semi-free. This category is taken to include centres in which bison are within a fenced-off or otherwise contained area, but in such circumstances as make it impossible to know the pedigree of individuals born. The concept first emerged in the EBPB in 1997, in connection with the CHERGA Centre in the North Altai of Russia. In 1998, this was joined by the ZALISSKA Centre (Ukraine), while in 2007 two Centres in Romania known as NEAGRA BUSCANI and VANATORI NEAMT were added to the listing under this category. The year 2008 brought a further increase in the number of semi-free populations as a Centre in the south of France (the RESERVE BIOLOGIQUE DES MONT D'AZUR) was added, while a further augmentation of 2010 involved the populations named MOGILIEV in Belarus and DÖBERIZER HEIDE in Germany. There were thus some 163 bison in semi-free populations as of the end of 2010.

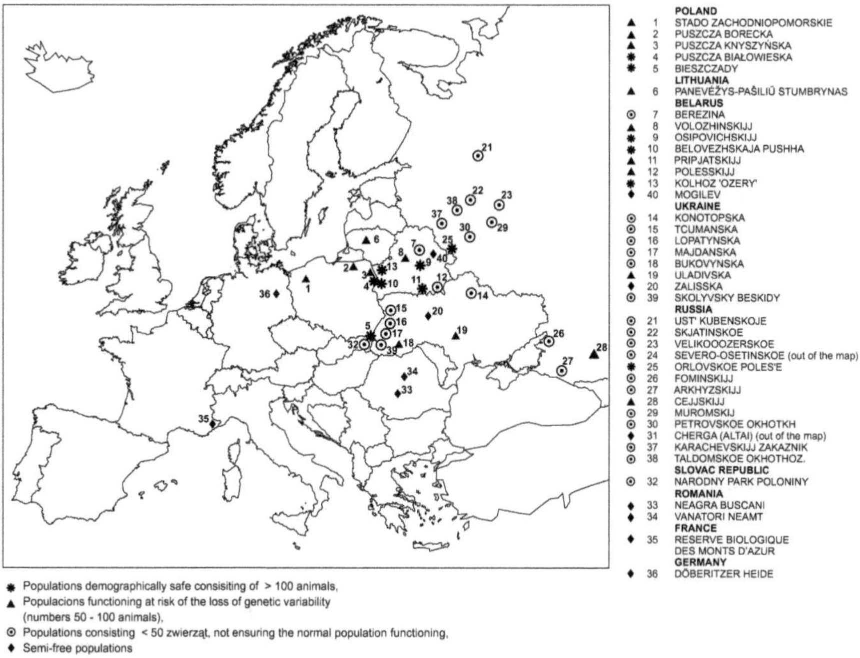

Fig. 23.1 The distribution of free-ranging populations of the European bison around the world as of 2010. *Source* Krasińska and Krasiński (2007), updated

As of 2004, there were 29 free-ranging populations of European bison in the world, plus two semi-free ones. By the end of 2010, the bison worldwide counted 4,431 individuals, including 2,956 (67 %) in 33 free-ranging and 7 semi-free populations and 1,475 (33 %) in captivity (Table 23.1, Raczyński 2010). These populations are located in the eastern part of the original range of the Lowland European bison: in Belarus, Lithuania, Poland, Russia, Slovakia and Ukraine (Fig. 23.1). It was as recently as in 2004 that the European Bison Pedigree Book registered the new free-ranging population in the Slovak Republic, while in 2009 two new Russian populations were listed, these having arisen thanks to migrations of individuals from existing free-ranging herds.

23.3 Strategies for the Management of Bison Populations

On the basis of the results of scientific work carried out in the Białowieża Forest, as well as the experience gained with other free-ranging populations, a macro-scale strategy for the management of the European bison population was drawn up during the Symposium of the International Union for the Conservation of Nature (IUCN) held in Międzyzdroje, Poland, in 1995 (Pucek et al. 1995). This was then extended and augmented by recommendations *vis-a-vis* the protection of *Bison bonasus* (Pucek et al. 2004).

More important prescriptions of this strategy are that:

- the Lowland and Lowland-Caucasian line bison should be managed as separate populations for as long as possible, or until such time as the two come into contact naturally;
- as populations of at least 100 individuals are deemed demographically secure, European bison should ideally be introduced into environments in which such a large number can ultimately be achieved (while it is true that even populations of 50+ individuals are not likely to be threatened with extinction from the demographic point of view, they are at increased risk of losing genetic variability);
- new populations should be formed from 15 to 20 individuals selected in relation to their origin, ages and sex;
- regulation of numbers based around knowledge of demographic structure is essential if balance between the animals and their environment is to be maintained (excessive exploitation and the consequent degradation of the environment have a negative impact on the state of the population);
- populations of 3,000 free-ranging individuals should be achieved for each of the breeding lines, giving 6,000 animals in all (this would denote a more than 3-fold increase in numbers as compared with the current situation and will only be possible to achieve if the size of the population in Russia is augmented very considerably);
- pathogens and diseases known and unknown pose a threat, and this is all the worse where the level of resistance to disease is low, and the vulnerability to different pathogens high;
- matings between the captive-bred and free-ranging populations should be permitted, especially where individuals representing precious genotypes are concerned;
- the process of reintroducing bison in forest and other ecosystems should be continued, including on extensive areas of abandoned farmland or on military training grounds (new places for the introduction of bison may be located within the historical range of the species, but also beyond it); a recent analysis by Kuemmerle et al. (2012) sees hope for the establishment of large new meta-populations in Eastern European areas like the Carpathians and Belarussian-Ukrainian borderland, as well as in several areas of European Russia;
- there should be facilitation of individual migrations between herds representing the existing genetic lines, albeit in respect of just a few locations like the Białowieża Forest (where such opportunities are lacking, males should be systematically transferred between herds with a view to genetic variability being enhanced, albeit with considerable care taken to avoid the transmission of disease);
- isolated populations should be linked into meta-populations through natural or assisted transfers of animals;

- American-European bison hybrids should be maintained, for research purposes for example, but these animals should be isolated from bison of the pure breeding lines;
- coordinating programmes must be drawn up, in order that genetic variability might be retained and enhanced, while inbreeding is confined, in the existing free-ranging populations (this objective may be achieved through exchanges of animals between herds, as well as the perfection of management methods);
- a gene bank should be established through the collection of semen, while conditions for the exchange of genes between herds should be safeguarded;
- the genes of all the founders represented in conditions of captive breeding will need to be introduced into the free-ranging populations;
- a start should be made to the founding of a more contiguous range for the species—in the case of bison of the Lowland-Caucasian line, opportunities for this have arisen in the Carpathians, in the borderland between Poland, Slovakia and Ukraine, as well as on the border between Ukraine and Romania (see also Perzanowski et al. 2006);
- the devising of models for the proper management of the existing free-ranging populations is indicated, with the creation of watering places, cultivated meadows or forest glades;
- there is an urgent need for the tasks fulfilled by the European Bison Pedigree Book to be expanded;
- there is a need to establish an International European Bison Breeding Centre that would have as its task the coordination of the restitution, reintroduction and monitoring process in respect of captive herds and free-ranging populations, as well as their management on the basis of genetic data. This recommendation has in fact been put into effect in recent years, with the 2007 founding of the European bison Conservation Center (EBCC) of the IUCN, whose main office—located at the Warsaw University of Life Sciences (SGGW)—is in contact with regional offices seated in various different countries.

In relation to their abundances, the free-ranging populations have been divided into three categories (Table 23.2). The most important of these from the point of view of the protection and restitution of *Bison bonasus* are the Category I populations numbering at least 100 individuals, since these are regarded as demographically stable (Pucek et al. 1995). There are only six of these, in fact, two being resident in the single natural complex of habitat formed by the Białowieża Forest in Poland and Belarus (Table 23.2). Populations of 50–100 individuals are capable of longer term functioning, although they are ultimately at risk of losing genetic heterogeneity if their further development is not assured. Of the nine populations assigned to this category, three are in Belarus and two in Poland, while there is one each in Lithuania, Russia and Ukraine (Table 23.2).

The category including the most populations is that with 50 individuals or fewer (Table 23.2). Where numbers are this low, isolation ensures the inviability

Table 23.2 Division of the free-ranging populations of European bison listed in the European Bison Pedigree Book of 2010

Name or number* of populations	Country	Number of animals
Populations demographically safe consisting >100 animals		
PUSZCZA BIAŁOWIESKA	Poland	473
BELOVEZHSKAJA PUSHHA	Belarus	415
	Total Białowieża forest	888
BIESZCZADY	Poland	304
ORLOVSKOE POLES'E NATIONAL'NIJJ PARK	Russia	217
OSIPOVICHSKIJJ LESKHOZ MOGILEVSKOJJ OBLASTI	Belarus	152
KOLKHOZ "OZERY" GRODNENSKOJJ OBLASTI	Belarus	139
Populations consisting <100 animals functioning at risk of the loss of genetic variability		
PUSZCZA KNYSZYŃSKA	Poland	94
ULADIVSKA	Ukraine	92
PRIPJATSKIJJ NATIONAL PARK	Belarus	87
PUSZCZA BORECKA	Poland	81
STADO ZACHODNIOPOMORSKIE	Poland	81
VOLOZHINSKIJJ LESKHOZ	Belarus	79
POLESKIJJ	Belarus	76
CEJJSKIJJ ZAKAZNIK (Kaukaz)	Russia	65
PANAVĖŽYS-PAŚILIŲ STUMBRYNAS	Lithuania	62
Populations consisting of < 50 animals not ensuring the normal population functioning during long period		
10 populations*	Russia	172
6 populations*	Ukraine	135
BEREZYNA	Belarus	36
NARODNY PARK POLONINY	Slovakia	9
Semi-free population		
CHERGA	Russia	42
NEAGRA BUSCANI	Romania	41
VANATORI NEAMT	Romania	22
RESERVE BIOLOGIQUE DES MONT' D'AZUR	France	32
MOGIIEV	Belarus	24
ZALISSKA	Ukraine	20
DÖBERITZER HEIDE	Germany	6
Together		*2,956*

Names of breeding centres after the European Bison Pedigree Book 2010. *Source* Krasińska and Krasiński (2007), updated
*Number of populations

of the population over the longer term. Unfortunately, only two such populations (in central Russia) have any chances of developing beyond this danger point. An obstacle to the development of three other populations of this size in northern Russia may be posed by the harsh climate there, most notably the severe frosts and thick snow cover of winter.

23.4 Free-Ranging Populations in Poland

There are five free-ranging populations in Poland, four in the east of the country and one in the north-west (see Fig. 22.1). The total number of bison present in the wild state in 2001 exceeded 600 individuals, while at the end of 2010 the figure was 1,033. There are no real prospects for increasing the numbers of bison beyond this, or for the establishment of any new populations in Poland. The largest population—that in the Białowieża Forest—numbered 473 individuals as of 31 December, 2010, leaving 560 animals spread around the remaining four populations. The rank and significance of these latter groups is thus put into perspective.

Poland's Bieszczady Mountains sustain a population of Białowieża-Caucasian line bison, while that in the Borecka Forest exemplifies a medium-sized (81-strong) population living in contiguous forest complexes of diverse environments. Two other populations—PUSZCZA KNYSZYŃSKA and STADO ZACHODNIOPOMORSKIE utilise a forest-field environment. In 2010, the numbers of bison in each of the latter exceeded 80 individuals (standing at 81 and 94 individuals, respectively).

Name and location: PUSZCZA BIAŁOWIESKA

In north-eastern Poland, on the border with Belarus and in the voivodship of Podlaskie and poviat of Hajnówka, this centre is some 80 km to the south of the city of Białystok (see Fig. 22.1). The free-ranging population of bison in the Białowieża Forest has been discussed exhaustively in Chap. 10.1.

Name and location: PUSZCZA BORECKA

Located in north-eastern Poland, within the Mazurian Lakeland and in the voivodship of Warmińsko-Mazurskie and the poviat of Giżycko (see Fig. 22.1).

Year of establishment: 1962.

Breeding line and founder group: Bison of the Lowland-Caucasian line were present here in the years 1962–1969, but these were recaptured and transferred to the Reserves in Niepołomice and Smardzewice, or else to zoos. Lowland (Białowieża) bison have in turn been present here since 1971, the founder group comprising 15 (7, 8) animals aged 1–4 were brought here in 1970—5 (2, 3) and 1971—10 (5, 5). These bison were derived from the captive breeding facilities at Białowieża—7 (3, 4), from the free-ranging PUSZCZA BIAŁOWIESKA population—7 (2, 5) and from the Reserve at Pszczyna—1 (1, 0).

Area of forest complex and natural conditions: The contiguous forest complex extends over some 180 km^2, the main strongholds for bison being in Borki Forest District, wherein some 77.3 % of the land is under forest. The dominant tree species here are Norway spruce (35.4 %), oaks (19.4 %), birches (16.3 %) and alders (11.9 %). The forests have a luxuriant understorey that mainly comprises hazel, raspberry and rowan, as well as wealth of herb-layer species typical for oak-lime-hornbeam forest. Water is accessible in the many forest streams. These forests are exploited for forestry, target areas being cut and then left for restocking without being fenced off. At most, trees are protected against debarking on a case-by-case basis.

Area of occurrence: the area intensively penetrated by bison covers around 107 km^2, or nearly 60 % of the forest complex as a whole (Fig. 23.2).

Population dynamics: as of 31 December, 2010, there were 81 bison here, and the numbers have fluctuated within the range 50–80. Inventorying is done by way of counts at winter refuges. In the years 1991–1993, the coefficient of births (ratio of calves born to overall population size) stood at 15.2 % on average. In the same period, the coefficient of mortality stood at 3.9 % on average, albeit with variations from year to year in the range 1.5–12.5 % (Krasiński and Krasińska 1994b). In contrast, in the case of the 1970–2006 period (of 36 years), the birth coefficient assumed a value of 16 % on average (range 5.5–27.9 %). In turn, the coefficients of fertility for the period 2005–2006 were of 60 and 55 %.

Status and level of human interference: the Borecka Forest is used for hunting of bison, carefully identified individuals being brought in specifically to be hunted. Winter is a time of supplementary feeding here—with hay, silage, and a mix of concentrates plus mineral supplements and tree bark.

Threats: bison from other Polish breeding centres are transferred into the Borecka Forest—in fact as many as 138 in the years 1974–1993 inclusive. With few exceptions, these bison did not participate in the further development of the population, being shot in the course of hunts. The core population did not change much, however. The transportation of bison from the different centres referred to above opens up possibilities for the transmission of pathogens and consequent spread of disease outbreaks.

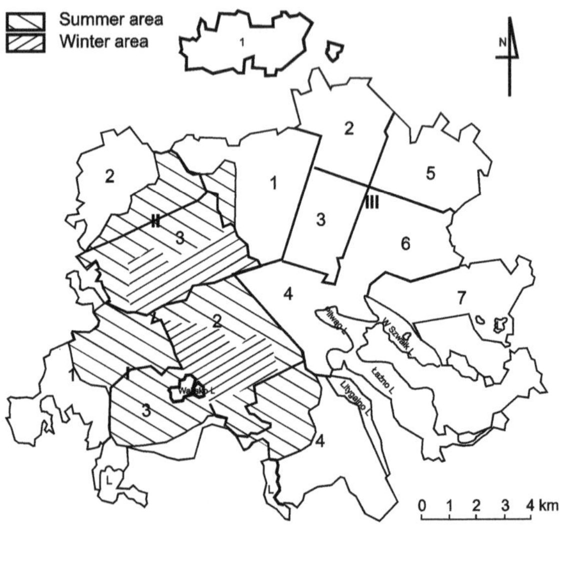

Fig. 23.2 The spread of the bison population in the Borecka Forest. I, II—area of Borki forest district, III—area of Czerwony Dwór FD, Arabic numbers denote subdistricts. *Source* Krasiński and Krasińska (1992)

There have been odd cases of individuals—mainly bulls—wandering off beyond the limits of the Borecka Forest, even covering distances of 30–200 km (see Chap. 16.4, Fig. 16.13). It is usually the fate of such bison to die in various kinds of accidents.

Prospects: a population size of around 50 is always associated with a risk of genetic diversity being lost. This risk is reduced by bringing in individuals from other centres in Poland, however. Furthermore, at 80, the size of the population in the Borecka Forest is at the optimal level for this forest complex.

Name and location: BIESZCZADY

Low mountains in south-eastern Poland on the border with Ukraine and Slovakia, and specifically in Poland's Podkarpackie voivodship (see Fig. 22.1).

Year of establishment: 1964. Bison were released from their acclimatisation enclosure on 4 May, 1964.

Breeding line and founder group: bison of the Lowland-Caucasian line. The first group of 7(5, 2) bison arrived in 1963 from Niepołomice and Pszczyna, being kept initially in an acclimatisation enclosure. Bull M 1128 PULON and young males M 1456 PURUS, M 1402 PUSZCZAR and M 1471 PUJAR derived from the line of bull M 879 STOR-PER, which was born in 1952 at Stockholm Zoo and was brought to Poland in 1956. Its ancestors in the third generation were M 163 BORUSSE and F 161 BISCAYA—along both the paternal and maternal lines. The founder group comprised a total of 19 (11, 8) bison, 11 (6, 5) of which came from Niepołomice and 8 (5, 3) from Pszczyna. In 1971, 8 males were brought in from Smardzewice, in 1975—12 (4, 8) individuals and in 1976—6 (1, 5) from Niepołomice. A total of 45 (20, 25) bison were thus introduced into the eastern part of the Bieszczady Mountains. Over the period 1976–1980, the acclimatisation enclosure within Komańcza Forest District in the western part of the Bieszczady Mountains held a group of five (1, 4) bison originating from the Niepołomice Reserve. A further 10 (5, 5) animals were brought from the Reserve in Pszczyna in 1980. All 15 bison were released into the wild in 1980. Two years later, two males were imported from Smardzewice.

The European Bison Pedigree Book treats the populations of the eastern and western Bieszczady Mountains together, and this is the only place in Poland where bison of the Białowieża-Caucasian line are present.

Area of forest complex and natural conditions: An area of no less than 1,446.79 km^2 includes the 292 km^2 protected within the Bieszczadzki National Park. The Bieszczady Mountains are in turn part of the Eastern Carpathian mountain massif—a chain of moderate altitude (400–1,300 m a.s.l.) overgrown with mixed forests in which beech, spruce and fir dominate the canopy. The understorey includes hazel, elder and red-berried elder, sallow, rowan and hornbeam, as well as raspberry and blackberry en masse. The many valleys combine with the open grasslands at higher altitude to offer a varied food base. The different ranges are linked by accessible passes, ensuring that there are no natural barriers to migratory movements here. Regular forestry management is pursued within the Forest Districts, but the forests within the National Park are, of course, protected.

Area of occurrence: there are two populations in the Bieszczady Mountains (Fig. 23.3):

1. **The eastern population** inhabits an area along the Polish-Ukrainian border. The eastern boundary of the area occupied is formed by the River San, while main strongholds of the bison are within Lutowiska and Stuposiany Forest Districts, as well as the Bieszczadzki National Park itself. This is the oldest part of the population. Having left the acclimatisation enclosure in May 1964, the bison here spent most of their time in the vicinity of the Widełki massif at altitudes of 759–900 m a.s.l. As the population grew, they then extended their area of occurrence steadily, although it long remained the case that the Lutowiska and Stuposiany Forest Districts provided the main refuge (for 97 % of bison in 1973 and for 59 % in 1999). In 1979, a small mixed group was noted in Brzegi Dolne FD, to the north of the main stronghold, for the first time. It was in this area that a separate group took shape, its size increasing to 18 animals (or 11 % of the whole Bieszczady population).
2. **The western population** was founded in 1980, through the release into the wild of bison held in an enclosure since 1976. The latter was located in the western part of the Bieszczady Mountains, close to the village of Wola Michowa and some 20 km from Komańcza, near the Slovakian border. Bison then colonised the Chryszczata and Krąglica mountain massifs to the north of the original enclosure. They spend the summer in the more elevated parts of the mountains, in fir-beech stands where meadows and clearings are also present nearby. The area occupied in summer extends to around 65 km^2. From December to March, the bison are present at lower levels, the winter stronghold of around 12 km^2 being located in alder-pine stands in which supplementary food is also on offer. The eastern limits of the range here are set by the Lesko–Cisna road. Animals present in the western part of the Bieszczady are not in contact with the eastern population.

Fig. 23.3 Areas of European bison population in the Bieszczady Mountains (*grey colour*). The eastern population is located along the Ukrainian border, while the stronghold for the western population lies within Baligród forest district. *Source* Krasińska and Krasiński (2007)

The introduction to the western part offers a good example of just how important the choice of individuals for release can be. In fact, the animals brought in were selected more or less randomly, and their physical condition was poor, ensuring that various health problems were encountered during the first phase of the introduction.

It is worth noting the record-breaking movement made by the young bull M 1287 PULPIT in 1966. In the course of 28 days, the animal covered 300 km, before being captured at Żabno near Tarnów (see Fig. 16.13).

Population dynamics: As of 31 December, 2010, there were 304 individuals. The population of bison grew in a natural way up to 1983, in which year it reached 184. The years 1984–1986 then brought some decline on account of a higher level of culling. By 1989, there were just 74 animals—a radical reduction also linked with migrations across the border into Ukraine and Slovakia. In 1997, the population once again exceeded 150. According to data for the years 1979–1987 (9 years), the coefficient of births was of 17.1 % on average (range 12–22 %), while the coefficient of natural mortality was of 2.9 % (range 0.5–6 %). In the same period, some 12.8 % of the population was culled on average, the figures for individual years ranging from 3.1 % in 1982 to 25.9 % in 1986.

Status and level of human interference: Selected individuals are culled by way of hunting. In the winter, bison receive supplementary hay, root crops and silage as food, though it is generally recognised that the animals in this population make only limited use of this available food source.

Threats: A serious threat to this population is posed by the tuberculosis, noted for the first time in 1996 among bison within Brzegi Dolne Forest District. A lack of appropriate legal regulations also ensures that the fates of any bison migrating into Ukraine or Slovakia remain unknown. Single cases of bison being attacked by wolves and bears are also now on record here.

Prospects: Cooperation and coordination regarding the breeding of bison should take in all administrative units active in the areas inhabited by the bison, i.e. the Lutowiska, Stuposiany and Brzegi Dolne Forest Districts and Bieszczadzki National Park. There are also plans for enhanced cross-border cooperation between Poland, Ukraine and Slovakia with a view to the area occupied by bison in the region being expanded. At the beginning of December 2002, three males from Sweden and one female from Denmark were brought into the Bieszczady population with the aim of enriching the gene pool. The bison in question were also of the Białowieża-Caucasian line.

Name and location: PUSZCZA KNYSZYŃSKA

The Knyszyn Forest is located in north-eastern Poland, in the Podlaskie voivodship, 30 km east of Białystok (see Fig. 22.1).

Year of establishment: 1973 (from 1969 on there had been a single male in the area, this having "emigrated" from the Białowieża Forest).

Breeding line and founder group: these are Lowland (Białowieża) bison. On 2 April 1973, two males from captivity at Białowieża were brought in, these being M 2247 PODBIAŁ (a 5-year-old) and M 2383 PODKOLAN (a 4-year-old). They were supplemented, on 18 April 1973, by a further three females aged 4–7 that had been captured from the free-ranging population in the Białowieża Forest.

Area of forest complex and natural conditions: the Knyszyn Forest is a 1,235 km^2 area located to the east and north of Białystok. It stands out in its relatively well-preserved state and environmental diversity. However, 80 % of the forest complex is of coniferous (pine-spruce) stands, while only 15 % is either deciduous or mixed forest. The most important forest-forming species are Scots pine (71.8 % by area) and Norway spruce (12.6 %), while birch occupies 5.1 % of the area, oak 4.8 % and alder 3.8 %. There is also a very limited admixture of hornbeam (1.1 %), aspen (0.7 %) and ash (0.1 %). A Landscape Park established in 1988 takes in the greater part of the Knyszyn Forest (745 km^2), its area being subject to certain restrictions on forest exploitation. Still stricter protection is also extended to the most valuable elements of nature within the Landscape Park area, thanks to the creation of as many as 18 Nature Reserves.

Area of occurrence: By the end of 1993, bison occupied an area of 63 km^2 located within Waliły and Krynki Forest Districts. In 2005–2007, the area increased to 250 km^2, of which 74 km^2 (i.e. 6 % of the Knyszyn Forest) constitute woodlands. In summer, bison utilize mainly forest area, while in autumn–winter period they head off to the north, to cultivated fields (Fig. 23.4). Food is then obtained on fields sown with winter cereals and rape, and bison tend to locate wherever these crops can be found (Fig. 23.5).

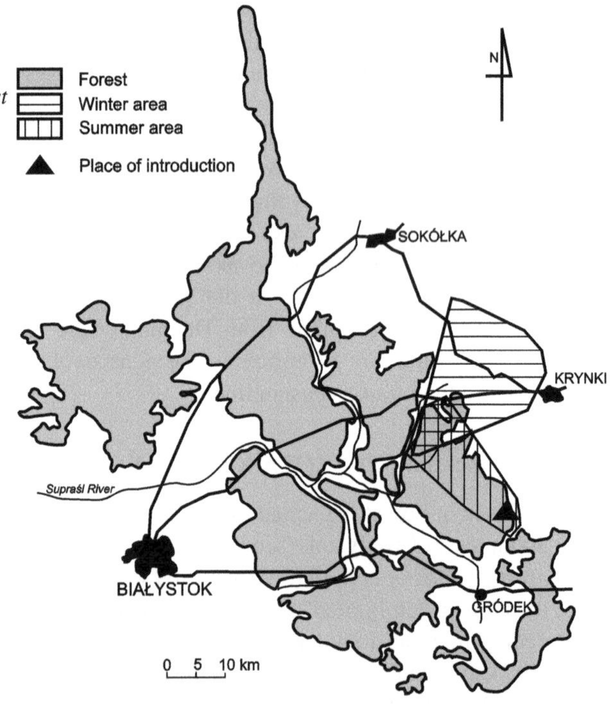

Fig. 23.4 The distribution of the European bison population within the Knyszyńska Forest: 1—*forest* boundary, 2—*Winter area* 3—*Summer area*, 4—places at which bison feed on cultivated fields, 5—introduction site, *L*—name of forestry sub-district, *Rz*—name of river. *Source* Krasiński et al. (1994b)

Fig. 23.5 Bison inhabiting highly fragmented or small forest complexes often leave to feed on farmers' fields. Those in the Knyszyńska Forest, the Polish Western Pomeranian herd and at Borosovskijj Leschoz in Belarus or Panavėžys-Pašilių Stumbrynas in Lithuania all live in a forest-field landscape. Bison in the Knyszyńska Forest, March 2011. *Photo* T. Kamiński

Population dynamics: As of 31 December, 2010, there were 94 bison here. However, in the 20-year period 1983–2003, there were never fewer than 11, and never more than 26. However, from 2004 on, there was a distinct and steady increase in the size of this population. Data on causes of death here are lacking, although several known cases resulted from collisions with vehicles.

Status and level of human interference: Neither culling nor hunting are practised. Care over the bison is extended by the State Forests' units administering the areas occupied. There is periodic (although irregular) supplementary feeding with hay, provided in winter.

Threats: numbers are so small that related individuals tend to mate and there is a high coefficient of inbreeding. The impact of poaching on mortality has not been determined. There are many times more red deer than bison here, although there is no tangible evidence that interspecific competition takes place.

Prospects: Due too its small size, this population was long in an unsafe state from the demographic point of view, thanks to the low rate of natural increase and unsuitable structure. However, the herd was reinforced with two females from the Białowieża Forest in 2006, and if its high rate of development is maintained for a while longer it will then be possible to assign this population to the group regarded as secure from the demographic point of view.

Name and location: STADO ZACHODNIOPOMORSKIE

Called the NADLEŚNICTWO WAŁCZ population to 2002.

Occurring in north-western Poland, in the voivodship of Western Pomerania (Zachodniopomorskie), between the towns of Wałcz and Mirosławiec (see Fig. 22.1).

Year of establishment: 1980. On 4 August, 1980, the bison left the acclimatisation enclosure built in Międzyrzecz Forest District, in which they had spent 5 months. The free-ranging population of NADLEŚNICTWO WAŁCZ was first noted in the Pedigree Book for 1983.

Breeding line and founder group: Lowland (Białowieża) bison, the founder group of which comprised 8 (4, 4) bison aged 2–5 that had been captured on 28 February, 1980 from the free-ranging herd in the Białowieża Forest.

Area of forest complex and natural conditions: The Forest Districts of Wałcz and Mirosławiec—where the bison occur most often—cover more than 300 km^2, and of this more than 90 % is forest. Coniferous forest habitats prevail (accounting for 89 % of the total). The most important forest-forming species is Scots pine (over 87 % of the area). Deciduous forests habitats—above all with birch, oak and beech—cover just 9 % of the area. Both Forest Districts are cut across by river valleys and ribbon lakes, and there are also meadows in forest clearings.

Area of occurrence: In the first years of reintroduction, the bison here occupied a 67 km^2 area in the vicinity of the acclimatisation enclosure. However, by 2000 they were present over about 55 km^2 of Wałcz Forest District and 50 km^2 of Mirosławiec FD, giving a total range of 105 km^2. The main stronghold was located in a forest-field area (Fig. 23.6). Today, this population comprises two

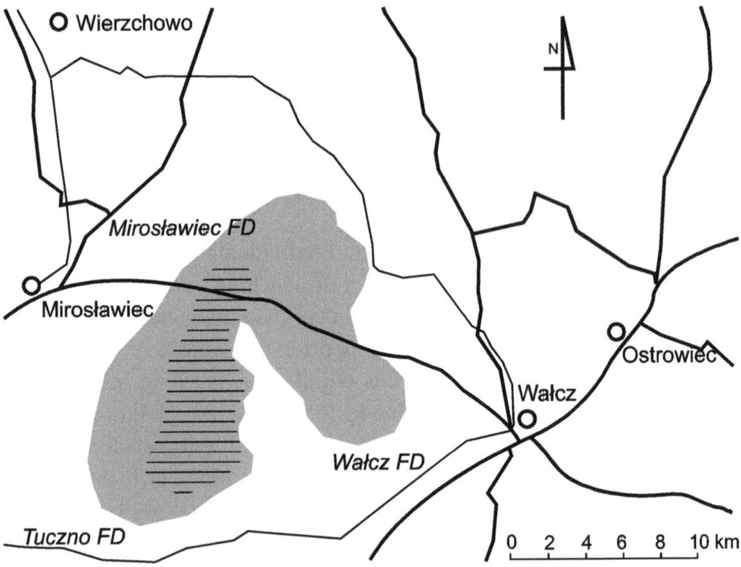

Fig. 23.6 The area inhabited by the Western Pomeranian herd (STADO ZACHODNIOPOMORSKIE). The hatched areas are those most frequented by the bison

distinct herds. One of these came into existence in 1980, its current 55 individuals inhabiting the Forest Districts of Mirosławiec, as well adjacent Tuczno, Człopa, Świerczyna and Wałcz, extending a total range of 150 km^2. The second herd occupies some 77 km^2 of the old Drawski military training ground, which thus lacks cultivated fields and has only scattered forest. The home range of the first herd is only 30 % forest, the remaining area being of cultivated fields. In turn, the second herd has within its range at least 20 % accounted for by open land.

Population dynamics: As of 31st December 2010, there were 81 animals. In 1981, a 5-year-old female brought in from Białowieża gave birth to a bull-calf and this was followed in 1982 by the birth of no fewer than 4 (1, 3) young ones. The year 1982 can thus be regarded as the start of the natural breeding. The population remained at the level of 11–13 individuals in the years 1987–1997, but rose into 21 in 1998. Where numbers are this small, even single deaths, cases of poaching or migrations out of the area are of considerable significance to the status of the population. However, since 2007, it has been possible to note a steady increase in numbers thanks to the reinforcement of the population with animals transported in, as well as the creation close to the main herd of satellite herds arising out of implementation of the programme for the protection of this population drawn up in 2005.

Status and level of human interference: In the first phase, this population was regarded as a scientific experiment. The aim was to determine whether it was possible for a small population to function largely within a single selected forest complex. The assumption was that numbers would be regulated by the culling of selected individuals (Graczyk 1987). When the experiment ended, the forest administration undertook to look after the bison that remained.

Threats: Inbreeding depression can only intensify in such a small population. Furthermore, bison living in a field-forest landscape often go out into fields of crops to feed. The owners of those fields often apply drastic methods to scare the bison off. A major threat to these bison is posed by the passing through the central part of their range of National Road No. 10—on which several are killed each passing year. Cases of poaching are also known. In 1983, a bull moved some 80 km out of the forest into Nieświady and Tychowo FDs. In turn, in 1985, 2 (1, 1) bison emigrated 40 km north-eastwards.

Prospects: A programme for the active protection of bison in the voivodship of Western Pomerania was drawn up in 2005. This *inter alia* envisaged the introduction of new individuals with a view to genetic diversity being increased and the overall size of the population likewise.

The above descriptions of the Polish populations draw on sources listed as follows: Deus (2000), Graczyk et al. (1978), Graczyk (1987), Janicka et al. (2008), Jaroński and Pepera (1969), Kosiba (1964), Krasińska and Krasiński (2000), Krasiński (1963, 1967, 1968a, 1978a, b, 1990, 1994a, b, 1997, 1992), Krasiński et al. (1994b), the European Bison Pedigree Book; Olech and Perzanowski (2002, 2007), Paszkiewicz 2004), Paszkiewicz and Ryba (2000), Perzanowski and Kozak (2000), Perzanowski and Paszkiewicz (2000), Polakowski (1961), Ścibor (1965), Tarcz and Tracz (2010), Żoch (2007), Żórawski and Lipiec (1997, 1998);

23.5 Free-Ranging Populations in Lithuania

Name and location: PANEVEŽYS-PAŚILIŲ STUMBRYNAS, called NAUJAMESTIS until 1993

Located in the central part of Lithuania, near the town of Panevėžys.

Year of establishment: 1973.

Breeding line and founder group: bison of the Lowland line. Over the period 1969–1972, 10 (2, 8) bison were transferred here from the Prioksko-Terrasnyjj Reserve (Russia) and placed in an enclosure. In 1973, 5 (2, 3) bison were released into the wild, giving a beginning to a free-ranging population. In subsequent years, further bison were released from their enclosures to join the herd living wild.

Area of forest complex and natural conditions: Bison inhabit a forest-field landscape. While they take refuge in mixed deciduous forest covering some 50 km^2, they mainly feed out on cultivated fields and meadows.

Area occupied: the population occupies some 100–120 km^2 in total, but the main stronghold covers 20 km^2.

Population dynamics: While there were between 5 and 21 individuals in the years 1973–1998, there were 28 in 1999 and 62 in 2010.

Status and degree of human interference: The European bison is in the Red Book of Protected Species in Lithuania. Changes in land ownership (privatisation) combine with a lack of legal regulations on compensation for damage done by bison to necessitate changes in the way the population is managed. Since 1980, numbers have been regulated by culling of the "excess" population and/or selected individuals. Licences to hunt bison are issued by the Ministry of the Environment, at the rate of about 4–5 (individuals) per year. Periodic but irregular supplementary feeding takes place in winter.

Threats: As migratory individuals are met with across Lithuania—even several hundred kilometres from the reintroduction centre—bison mostly die unnatural deaths due to poaching or different types of accident. A further threat is posed by the high level of inbreeding due to small population size. There was no success with an attempt, in 1974–1981, to introduce bison in the Širvintai area to the south and east of Panevežys. The principle reasons for this were due to anthropogenic pressure, via conflicts with local farmers in connection to damage done to crops. There was a lone male in this region in the years 1992 and 1993.

Prospects: It is proposed that a population of around 30 animals be maintained, with numbers being regulated along commercial lines (via hunting licences). There remains a need for compensation for the damage bison do to be brought under legal regulation.

Sources: Baltrunaite (1999), Balčiauskas (1999, 2000), European Bison Pedigree Book (1969–2010).

Table 23.3 Geographical areas in which free-living populations of bison are located

Explan. on Fig. 23.1	Region and name of population	Habitat, location	Number on 31 Dec. 2010 (**altogether**)	Remarks
	1. BIAŁOWIEŻA FOREST	**Lowland mixed and deciduous forests**	**(888)**	
4	PUSZCZA BIAŁOWIESKA Western part	Polish-Belarussian border Poland	473	
10	BELOVEZHSKAJA PUSHHA Eastern part	Polish–Belarussian border Belarus	415	
	2. CARPATIANS	**Hills**	**(355)**	
5	BIESZCZADY	Polish–Ukrainian border Poland	304	
18	BUKOVYNSKA	Ukrainian–Romanian border Ukraine	31	
17	MAJDANSKA	South-western Ukraine	11	
32	NÁRODNÝ PARK POLONINY	South-eastern Slovak Republic	9	From 2004
	3. CAUCASUS	**High mountains**	**(92)**	
28	CEJJSKIJJ	In the Northern Caucasus Russia	65	
27	ARKHYZSKIJJ	In the Northern Caucasus Russia	19	
24	SEVEROOSETINSKOE	In the Northern Osetia Russia	2	
39	SKOLIVSKY BESKIDY	South-western Ukraine	6	From 2010
	4. LOWLAND FOREST OF POLAND		**(256)**	Excluding the Białowieża Forest
2	PUSZCZA BORECKA	In the north-eastern Poland	81	
1	STADO ZACHODNIOPOMORSKIE	In the north-eastern Poland	81	
3	PUSZCZA KNYSZYŃSKA	In the north-eastern Poland	94	
	5. LOWLAND FOREST OF LITHUANIA		**(62)**	
6	PANEVĖŽYS-PAŠILIU STUMBRYNAS	Central part of Lithuania	62	
	6. LOWLAND FOREST OF BELARUS		**(569)**	
7	OKHOTKHOZJAJJSTVO “BEREZINA”	North-east of Minsk Belarus	36	To 2002 BORISOVSKIJJ
8	VOLOZHINSKIJJ	West of Minsk Belarus	79	
9	OSIPOVICHSKIJJ	South-east of Minsk Belarus	152	
11	PRIPJATSKIJJ	South of Belarus	87	
12	POLESSKIJJ	South-eastern Belarus	76	
13	KOLHOZ “OZERY”	South-western Belarus	139	

(continued)

Table 23.3 (continued)

Explan. on Fig. 23.1	Region and name of population	Habitat, location	Number on 31 Dec. 2010 (**altogether**)	Remarks
	7. LOWLAND FOREST OF UKRAINE		**(179)**	
14	KONOTOPSKA	Northern Ukraine	41	
15	TSUMANSKA	Western Ukraine	25	
16	LOPATYNSKA	Western Ukraine	21	
19	ULADIVSKA	Central Ukraine	92	
	8. LOWLAND FOREST OF RUSSIA		**(238)**	
25	ORLOVSKOE POLES'E	Central Russia	217	
37	KARACHEVSKIJ ZAKAZNIK	Central Russia	8	From 2009
30	PETROVSKOE OKHOTHOZ	Central Russia	13	From 2008
	9. BEYOND THE ORIGINAL RANGE OF THE EUROPEAN BISON		**(130)**	
21	UST' KUBENSKOJE	Northern Russia	39	
22	SKNJATINSKOE	Northern Russia	37	
23	VELIKOOZERSKOE	Northern Russia	20	
26	FOMINSKIJJ	Southern Russia	1	
29	MUROMSKIJ	Northern Russia	26	From 2002
38	TALDOMSKOE ROOR	Northern Russia	7	From 2009
	10. SEMI-FREE POPULATIONS		**(187)**	
31	CHERGA	Ałtai Russia	42	
20	ZALISSKA	Central Ukraine	20	
33	NEAGRA BUSCANI	Southern Romania	41	From 2007
34	VANATORI NEAMT	Northern Romania	22	From 2007
40	MOGILEV	Western Belarus	24	From 2010
36	DÖBEROTZER HEIDE Germany	Eastern Germany	6	From 2010
35	RÉSERVE BIOLOGIQUE DES MONTS D'AZUR France	Southern France	32	From 2008
	Altogether		**2,956**	

Situation as of 31st December 2010. *Sources of data* The European Bison Pedigree Book 2010. Explan: explanations. Dec.: December. *Source* Krasińska and Krasiński (2007), updated

23.6 Free-Ranging Populations in Belarus

As of 31 December, 2010, there were seven free-ranging populations and one semi-free population in Belarus, the total number of animals being 1,008 (Table 23.3, Fig. 23.1). This reflects a steady increase in numbers here. Among the three large populations with more than 100 individuals each, the largest is obviously that of the BELOVEZHSKAJA PUSHHA in the eastern part of the Białowieża Forest. Three others have more than 50 individuals, while one is small, with just 35 (Table 23.2). In line with the strategy drawn up for Belarus, all the populations are deemed to form a single metapopulation, exchanges of individuals occurring to limit inbreeding. The bison resident in the National Parks and Reserves of Belarus are designed members of the primary herd. The protection and utilisation of this herd's genetic resources will resemble those for species in the Red Book of Belarus. A special licence will thus be required to obtain any animals from this herd. In contrast, all other bison in the country form part of the reserve herd, in relation to which the level of protection is less rigorous, to the extent that hunting is permitted.

Name and location: BELOVEZHSKAJA PUSHHA

The eastern (Belarussian) part of the Białowieża Forest (Fig. 23.1). The free-ranging population here was first noted in the Pedigree Book for 1958. It has been discussed exhaustively in Chap. 10.2.

"BEREZINA" OKHOTKHOZJAJJSTVO BEREZINSKOGO BIOSFERNOGO ZAPOVEDNIKA from 2002

Located in the centre of the country, to the north-east of Minsk. The name applying in the years 1974–1986 was **BEREZINSKIJJ ZAPOVEDNIK**, and it was to this that the first bison were brought. The name after that was **BORISOVSKIJJ LESKHOZ MINSKOJJ OBLASTI**. In 1974, 5 (1, 4) bison were translocated to the Berezina Reserve—from the main breeding centre at the Prioksko-Terrasnyjj Reserve (Russia). In April 1974, these bison contrived to break down the fence and made it into the wild. Six more bison (4, 2) were translocated in 1974, although some of these died. Overall, then, the population has its origin in 6–7 individuals of the Lowland line. In autumn 1975, the bison moved out into the neighbouring Borisov area, coming under the care of the Borisov Forest District. The area occupied by today's population is of some 28 km^2 of forest and field landscape. The bison make frequent use of the latter habitat. As of the end of December 2010, there were 36 individuals here. The history and current situation of the animals here make it clear that a free-ranging population may not be established in areas unsuited to them—in this case marshland. The bisons' emigration moved them out of such habitat into what was a densely populated and urbanised area, and thus led to human–bison conflicts, especially in respect of animals' tendency to feed on crops and the related problem of poaching.

PRIPJATSKIJJ NATIONAL PARK

There are two populations in the same region of southern Belarus: the "PRIPJATSKIJJ" population along the southern side of the River Pripjat', and the LJASKOVICHI population on the north bank. The free-ranging "PRIPJATSKIJJ" population has been noted in the European Bison Pedigree Book since 1990. Its founder group was made up of eight Lowland line bison translocated between 1988 and 1989 from the Belarussian part of Białowieża Forest and the Prioksko-Terrasnyjj Reserve, and held initially in an 80 ha enclosure. The animals were freed in 1990. A National Park of 650 km^2 is located on the southern bank of the River Pripjat'; however, marshes account for more than 70 % of the area, and in fact bison have colonised a small (around 6 km^2) area of marshland there. In the years 2000–2004, all of the animals in this population left the National Park to occupy land on the "Polesskij Leskhoz" and "Agro-Pripjat" farms. The size of the population did not exceed 30 for many years. However, as at the end of December 2010, there were as many as 87 bison here. The bison receive supplementary food in winter.

The "**LJASKOVICHI**" population was founded in 2000 from 13 Lowland bison brought in from the Belarussian part of Białowieża Forest. The bison are present in the Najdo-Belevskij Forest covering c. 40 km^2. This area is dominated by mixed/deciduous stands with large numbers of oaks. Meadows present along river valleys combine with clearings within the forest to account for 11.4 % of the total area. At the end of 2001, there were 17 individuals here. Care over both populations is exercised by the PRIPJATSKIJJ NATIONAL PARK. Because of this, at that moment both populations are included in the European Bison Pedigree Book under common name PRIPJATSKIJJ NATIONAL PARK.

OSIPOVICHSKIJJ LESHOZ MOGILEVSKOJJ OBLASTI

This free-ranging population lives close to the confluence point of the River Svisloch with the Berezina, some 80 km south-east of Minsk. The population was founded in 1997, the founder group being of 15 Lowland European bison brought in from the Belarussian part of Białowieża Forest. Bison make occasional use of the supplementary feeding opportunities provided here in, but a large part of the range is accounted for by fields that may well have forest on three sides. There were 152 animals in this population as of the end of 2010.

VOLOZHINSKIJJ LESHOZ

The name derives from the town of Volozhyn. The bison live in the Naliboki Forest, a 240 km^2 area some 70 km west of Minsk. The stronghold is within Rumski Forest District, wherein some 36 % of the land is deciduous forests, while clearings account for 5.2 % of the total area. At present, the range occupied by the population in the winter period is of around 70 km^2, though this extends to 400 km^2 in the growing season, and thereby accounts for almost the entire area of the Naliboki Forest. Population size has increased slowly, such that there were 79 bison here by the end of 2010. In winter, the bison make sporadic use of supplementary feeding.

POLESKIJJ RADIACJONNO-EKOLOGICHESKI ZAPOVEDNIK

This population is located within the region known as Mozyrskoe Polesie, north-west from the Ukrainian town Chernobyl and the Belarussian-Ukrainian border, along the River Pripjat'. The left-bank areas along the river that are inhabited were very much contaminated by the Chernobyl nuclear disaster, so the population is viewed as an experimental one with no prospects for development. The founder group here was of 17 bison from the BELOVEZHSKAJA PUSHHA population. A further 18 individuals were brought in 1996. This population occupies a range of some 15 km^2, translating into a population density of 3.9 individuals per 1,000 ha. As of 31 December, 2010, there were 76 animals here.

KOLKHOZ "OZIERY"

This population resides within an area located some 25 km north-east of Hrodna. It was founded in 1998 from a founder group comprising 18 Lowland bison brought in from the Belarussian part of Białowieża Forest. The best conditions for bison within the above Forest Districts extend over some 350 km^2 in a continuous forest complex, with no villages or other human settlements in the interior. The dominant forest trees are Scots pine, alder, birch and spruce, while there are also some 500 ha of meadows here. Bison make sporadic use of available supplementary food in the winter period. In the spring-summer period, the population ranges over an area of some 12 km^2, as compared with 8–9 km^2 in autumn and 6–7 km^2 in winter. This population is growing gradually, reaching 139 at the end of 2010 this making it the third largest Belarussian population, and including it within the small group that comprises over 100 specimens.

In 2010, the semi-free **MOGILEV** population was registered for the first time in the European bison Pedigree Book.

The descriptions of the Belarussian populations presented above are derived from: Bunevich (1991), Bunevich and Kochko (1988), Korochkina (1969a, 1973, 1974, 1975), Korochkina and Kochko (1983), the European Bison Pedigree Book; Kozlo (1999, 2001), Kozlo and Bunevich 2009, Kozlo and Uglyanec (1999), Kozlo et al. (1997, 1999), Krasiński et al. (1994a, 1999), Kuchmiel et al. (1999), Lavov and Voronova (1982), Romanov and Kozlo (2001), and Uglyanec (1993, 1994, 1999).

23.7 Free-Ranging Populations in Ukraine

Ukraine supports seven free-ranging populations and one semi-wild one (Fig. 23.1, Table 23.3). According to the Pedigree Book, there were 247 European bison within the country as of the end of 2010. The free-ranging populations here are mainly present in game-managed areas, and are thus available for hunting. The dramatic reduction of the numbers of European bison, observed in Ukraine in recent years results from the cessation of its protection, increased poaching and commercial hunting conducted under the guise of selective shooting (Parnikoza et al. 2009).

The free-ranging herds derive from individuals of Belarussian origin (from the BELOVEZHSKAJA PUSHHA) and Russian origin (the Prioksko-Terrasnyjj and Okskijj Reserves). The two populations in the Ukrainian Carpathians are the BUKOVYNSKA and MAJDANSKA populations, while northern Ukraine supports the DANIVSKA and KONOTOPSKA centres, and the west has two more named TSUMANSKA and LOPATYNSKA. Finally, there are wild animals in the

ULADIVSKA population in central Ukraine. Only the LOPATYNSKA group has Lowland line bison originating from Lithuania, while the remaining populations comprise Lowland-Caucasian line bison. The NADVIRJANSKA population was closed down in 2002.

TSUMANSKA (until 2002 CUMANSKA). This oldest of free-ranging populations was established in 1965 in the Volhynia region of Ukraine, near Tsuman', between the cities of Lutsk and Rivne. The founder group comprised 15 (6, 9) Lowland-Caucasian line bison, deriving from the BELOVEZHSKAJA PUSHHA population. The population peaked at 203 in 1991, but had fallen to just 25 by the end of 2010. This reflects the fact that harvesting by hunting is carried out here.

BUKOVYNSKA is a population founded in 1970 in the Ukrainian Carpathians, close to the border with Romania, not far from the city of Chernivtsi. The Pedigree Book originally mentioned two populations at the BUKOVINSKOE and ZHUBROVICA Game Reserves. However, from 1994 only one population has been taken to exist—the BUKOVYNSKA population. The founder group comprised 19 (8, 11) bison of the Lowland-Caucasian line, originating in the BELOVEZHSKAJA PUSHHA in Belarus and the Russian Prioksko-Terrasnyjj and Okskijj Reserves. Numbers rose steadily, to peak at 225 in 1994, but there was then a decline to 31 by late 2010. Selective culling is carried out here. Interesting information from this centre includes the news that four bison were killed by bears here. Migrations by single individuals are also on record. The area occupied by bison takes in 590 km^2 (340 km^2 at Bukovynskoe and 290 km^2 at Zhubrovica). The size of the population is sufficient to safeguard genetic diversity, provided that the downward trend to numbers is arrested.

MAJDANSKA population, established in 1974, is in the Carpathians, not far from Majdan, within an area managed for hunting. The founder group comprised 10 (4, 6) bison from the Belarussian part of Białowieża Forest. As of the end of 2010, there were 11 bison here. The Majdan centre lies just 50–60 km east of the free-ranging population in the Polish Bieszczady Mountains, but there are so far no reports of any natural exchanges of individuals between them. Bison currently occupy an area of some 270 km^2.

ULADIVSKA this population dating back to 1979 is in the central Ukraine, to the north of Vinnytsia. Bison are present within two units of territorial administration, hence the earlier habit of the Pedigree Book's referring to the separate ULADIVSKA and PODILSKA populations. However, since 1994, these have been noted together as ULADIVSKA. Reference in the years 1979–1993 was also made to the VINNICKOE game holding. The founder group here was of 13, originating from the Prioksko-Terrasnyjj Reserve and the free-ranging TSUMANSKA population. Bison range over some 120 km^2 here, and the population at the end of 2010 was 92 individuals. Hunting culls are carried out.

Insert V. 1, 2 The Belarussian part of the Forest features many carvings making reference to local wildlife. *Photo* ZA Krasiński.

Insert V. 3, 4 A "Friend of the European Bison" medal of diameter 6.5 cm, as funded for the 75th anniversary of the commencement of attempts to restitution the bison in the Białowieża Forest. The medal went out to its first 6 recipients on February 20, 2004, these people being honoured for their contribution to rescuing the species from extinction. The medal's designer is Maciej Kiernicki. *Photo* ZA Krasiński A life-sized statue of a European bison stands in the square in front of the Poviat Office in Hajnówka. Unveiled on September 16, 2001, it bears the plinth inscription *żubr puszcz imperator* ("the European Bison—Forest Emperor"). This comment is inspired by the words of Adam Mickiewicz in Book IV of *Pan Tadeusz*. The idea was also used previously as the title of an information booklet authored by Zbigniew A. Krasiński and published by Białowieża National Park in 1999. The monument is the work of Jan Siuda. *Photo* ZA Krasiński

Insert V. 5. 6 Lithograph of a male and female bison by F. Kostecki. *Source* Jarocki (1930)

Insert V. 7, 8 A silver one-rouble coin issued by the Bank of Russia in 1997. A commemorative silver coin with a nominal value of 100 zł was previously issued by the Polish State Mint in 1977

DANIVSKA. A population in northern Ukraine founded in 1979. The centre is 50 km east of Chernigiv. The founder group comprised 13 (6, 7) Lowland-Caucasian bison from the Prioksko-Terrasnyjj Reserve. The area of occurrence is 312 km^2. As of 2001, there were 70 European bison here, compared with 22 at the end of 2004. This population was closed down in 2007.

LOPATYNSKA. This population was founded in 1980–1981 and is in western Ukraine, 60 km east of Lvov and south of Brody. The founder group of 12 (7, 5) bison came from Lithuania's Panevéžys-Paśilių Stumbrynas Breeding Centre. The area of occurrence is 160 km^2. At the end of 2010, there were 21 individuals here, while the desirable target figure is of 60.

KONOTOPSKA. Established in 1985, this centre is located to the north of Konotop. The founder group of 14 (3, 11) bison came from the semi-free ZALISSKA population. There were 41 Lowland-Caucasian line bison here in 2010. This is the only population in Ukraine that has not declined in recent years. Bison inhabit a forest complex of 50 km^2, surrounded by fields and meadows. Limited harvesting by hunting is carried out.

SKOLIVSKY BESKIDY—a new free-ranging population, created in 2010, into which part of the bison herd from the "Skolivsky Beskidy" enclosure centre were released.

ZALISSKA. This population enjoys "semi-free" status, on a Reserve founded in 1996 in an area of central Ukraine 40 km north of Kiev (Kyiv). The founder group of eight (4, 4) came from the Belarussian part of the Białowieża Forest and there were 20 individuals here at the end of 2010.

The descriptions of the Ukrainian populations offered above are based on the sources:

Bandarenko and Haiecki (1999), Perzanowski and Kozak (2000), Parnikoza et al. (2009), Pucek et al. (2004), the European Bison Pedigree Book.

23.8 Free-Ranging Populations in Russia

The old Soviet Union held the greater part of the world population of European bison, but its collapse in 1991 saw the Russian Federation left with free-ranging bison in the Caucasus Mountains only. An important role in the restitution of the European bison in Russia—but also in Belarus and Ukraine—was played by two closed breeding centres, i.e. the Prioksko-Terrasnyjj and Okskijj Reserves. Around 450 bison from these centres were used to supply other breeding centres and populations, bringing about their establishment and development.

The Prioksko-Terrasnyjj Reserve was founded in 1946, when bison were brought in from Poland. A total of 19 (10, 9) were sent in the years 1946–1948. The Reserve is about 100 km south of Moscow, close to the town of Serpuhov. The Bison Reserve is a separated-off part of the Biosphere Reserve located along terraces of the right bank of the River Oka. The presence of American bison being bred there represented a major threat to any maintenance of the purity of *Bison bonasus*. However, according to the statements made by the Reserve's long-time Director Mikhail Zablocky, the American bison were kept entirely separately, such that no crossing took place.

The Okskijj Reserve is also located along the River Oka, but further to the south-east. It was founded in December 1959, when the first Lowland-Caucasian line European bison were transferred here from the Prioksko-Terrasnyjj Reserve. In the years 1959–1962, a total of 11 (2, 9) bison came in.

Overall, Russia now has 12 free-ranging populations and one of semi-wild status, the number of bison present at the end of 2010 being 496 (Fig. 23.1, Table 23.3). A strategy for the conservation of the European bison in Russia, as accepted by the Ministry of the Environment, was published in 2002 (Flint et al. 2002).

The Caucasus

The mountain massif of the Northern Caucasus constitutes the natural area of occurrence of the Caucasian subspecies of the European bison *Bison* bonasus caucasicus, which became extinct in 1927. It was for this reason that the efforts to reintroduce the species began here (Table 23.3). Unfortunately, the first bison to come here had an admixture of American bison blood and were placed within the Caucasus Biosphere Reserve. A second group of such hybrids were released into the eastern part of the Caucasus range, gaining the name NALCHIK after a town not far away. A further threat to bison in this area was posed by the widespread gun ownership among local people. Periods of political instability were in turn associated with an increase of poaching. The effect of this was the total removal of the ASSINSKIJ Game Reserve population in Ingushetia, the SEVERO-OSETINSKOE population in North Ossetia and the SUNŽENSKOE population in Chechnya, as well as a collapse of numbers at both the CEJJSKIJJ and ARCHIZSKIJ centres.

CEJJSKIJJ ZAKAZNIK

This population requires fuller discussion, since it is the largest and oldest free-ranging herd in the Caucasus.

Name and location: the name comes from the Cejjskijj Game Reserve Holding and the region's highest peak—Mount Tsey at an impressive 4,646 m a.s.l. The Reserve is located in the Republic of North Ossetia, within the Ardon Valley.

Year of establishment: the first bison were translocated in December 1964, being placed in an acclimatisation enclosure from which they were released less than a year later (in 1965).

Breeding line and founder group: Lowland-Caucasian line bison, the first 14 being transferred here from the Prioksko-Terrasnyjj Reserve and the Belarussian part of Białowieża Forest. The founder group thus comprised some 46 animals.

Area of forest complex and natural conditions: The Reserve covers 730 km^2 and represents a very diverse area including everything from foothills forest through to ice-covered bare rock on mountain peaks. Forests account for 1/3 of the area and there is a large contiguous block under tree cover in the northern part of the Reserve. Around 200 km^2 is suitable for bison. Scrubby foothills, bordering with cultivated fields and meadows, are covered mainly with hazel. At higher altitudes, the deciduous forest has a prevalence of beech and hornbeam, but there are also mountain birches, maples, limes, willows, wild apples and pears. The southern slopes also support oak woodland.

Area of occurrence: c. 150 km^2

Population dynamics: Numbers rose steadily, exceeding 100 in 1965 and 200 in 1975. The peak population was the 270 noted in 1990 and 1992. However, by 1993, the numbers had declined drastically to 117, and this decline continued subsequently. By the end of 2010, just 65 bison were left.

Threats: If the downward trend observed for numbers is maintained, another free-ranging population in this region may become extinct.

ARKHYZSKIJ UCHASTOK TEBERDINSKOGO ZAPOVEDNIKA

Bison were brought here in the years 1968–1969. They came from the Prioksko-Terrasnyjj and Okskijj Reserves, from which 12 (4, 8) animals were transferred. As of 2010, the population numbered 19, with all animals being confined to a single valley. The desirable target size of the population has been set at 80. There is some hope for this, since this population is currently regarded as the safest anywhere in the Caucasus region.

"Mountain bison" in the Caucasus Biosphere Reserve

The central part of the main mountain massif of the Caucasus—between the Rivers Belaya and Laba—is the location for the Caucasus Biosphere Reserve with its so-called "mountain bison". Due to the inclusion of an admixture of American bison blood, these are not noted in the European Bison Pedigree Book. Rautian et al. (2000) proposed that the animals here be distinguished as a separate "subspecies" under the name *Bison bonasus montanus*. All these animals originate from five (1, 4) American–European bison hybrids brought here in 1940 from the Askania Nova steppe reserve (now southern Ukraine). The animals were kept in an enclosure, and, from 1949 on, the share of European bison blood in the population was increased by bringing in Lowland-Caucasian line males. Among these were as many as 15 bulls arriving in the Caucasus from Poland. The bison were released into the wild from 1954, and the population rose rapidly from then on—reaching an impressive 449 by 1965 (Kalugin 1968). By 1984, there were no fewer that 1,300 "mountain bison" (Yazan and Nemtsev 1985). Unfortunately, poaching set in and numbers declined thereafter, but there were still an estimated 550 in 1999 (Pucek et al. 2004).

The Caucasus Mountains sustain yet one further small population of American–European bison hybrids, in the area to the north of Nal'chik. Fortunately, the existence of natural barriers prevents the hybrids from mating with individuals from the pure-blood European bison populations present elsewhere in the Caucasus Mountains.

Lowland forests of Russia

Extensive forest areas between Bryansk, Kaluga, Orel and Smolensk in the western part of Russia were chosen for the founding of two free-ranging populations (Table 23.3). In the future, these might even constitute a metapopulation with a total of 800–1,000 individuals.

ORLOVSKOE POLES'E NACIONAL'NIJJ PARK

In 1996, eight (3, 5) bison were brought into the south-eastern part of the "Orlovskoe Poles'e" National Park from the Prioksko-Terrasnyjj Reserve, as well as four from the Okskijj Reserve. The potential area that could be inhabited by bison here is of 500 km^2, taking in the National Park "Orlovskoe Poles'e" and adjacent forests to the north in the Kaluga region. The forests across the area have a prevalence of Scots pine and spruce, with an admixture of birch, aspen, oak and black alder. The understorey has sallow, alder buckthorn, rowan, raspberry and hazel and the growing season lasts for about 180 days. Snow cover persists here from the first third of December through to the end of March, its greatest depth being 60–70 cm. Agriculture is extensive, and poaching may pose a threat. As of 2010, there were 217 bison here.

ZAPOVEDNIK BRJANSKIJJ LES

Three (1, 2) young bison were brought to this area located to the south of Bryansk in the Nerussia Valley in April 1997. As of the end of 2001, there were still only three animals present. An excessively small founder group continued to obstruct the development of this bison herd, which was abandoned in 2002, the three animals being transferred to the "Orlovskoe Polesie" National Park.

Populations in Russia beyond the original range of the species

Three free-ranging populations of an experimental nature have been established in northern Russia (Table 23.3). The authors of the strategy for the protection of bison in Russia (Flint et al. 2002) maintain that a c. 3,000 km^2 area in the north could theoretically support a population 500–1,000 strong. In fact, however, the poor development of numbers in practice—or even declines—may attest to the difficulties with European bison acclimatising there.

SKNJATYNSKOE OKHOTKHOZJAJJSTVO. Founded in 1986, this population is also known in Russia as Tverska—on account of its being close to the city of Tver. As of 2010, there were 37 individuals here.

VELIKOOZIERSKOE OKHOTKHOZJAJJSTVO. Established in 1989, and also referred to after the city of Vladimir. The population here declined from 15 in 2000. As of 2010, there were 20 individuals here.

UST'KUBENSKOE OKHOTKHOZJAJJSTVO. Established in 1992, and also referred to after the nearest town of Vologda. This is a small population, with just 39 bison present in 2010.

In the south of Russia, a new population was established close to Rostov on the Don in 1993. The centre is known as FOMINSKIJJ ZAKAZNIK (Table 23.3). However, there is lack of information on the situation this group of bison finds itself in. As of 2010, there were only one animals here.

In 2009, two new free-living populations were registered with the European bison Pedigree Book for the Russian Federation. Known as KARACHEVSKIJJ ZAKAZNIK and TALDOMSKOE ROOR, these populations have arisen through migration into an area by individuals originating with other wild populations.

Semi-free populations

The presence of a population of European bison in the Upper Altai has offered the chance of extending the area the species occupies far to the east. Its presence over there is thus a reflection of an experiment carried out by the Siberian Branch of the Russian Academy of Sciences at its Cherga SO RAN (CHERGA) Experimental Station. Population **CHERGA** is deemed "semi-free" on account of the fact that animals live in a fenced-off area, but pedigrees are not conducted here (Table 23.3). Animals were brought here from the Prioksko-Terrasnyjj Reserve in three stages: in December 1982—4 (1, 3), in June 1983—four (1, 3) and in April 1984—three (1, 2). One female died while being moved, so the founder group was made up of 11 (3, 8) Lowland line bison. These were placed in a 150 ha Reserve divided into two enclosures. The Reserve was enlarged to 400 ha in 1990. Both enclosures support large meadows overgrown with very luxuriant vegetation on fertile soil (see colour insert). Bison prove surprisingly efficient at crossing the hilly terrain with its thick woody vegetation. The forested part of the Reserve has birch, larch, aspen and Scots pine (see colour insert). The understorey supports alder buckthorn, pea tree and bird cherry. The climate is rather harsher than at Białowieża, with permanent snow cover for around 148 days, although the depth rarely exceeds 30 cm. The first calves in the Altai were born in 1983. Bison receive supplementary feeding of hay during the winter period, and numbers have remained at around 30 (42 at the end of 2010). The establishment of a free-ranging population of the species here seemed very desirable, since bison were found to have acclimatised very well. However, poaching may pose a serious threat to the persistence of this population. The exchange of Lowland bison between the centre in the Altai and others is hindered by its distance from European breeding centres.

In compiling data on the free-ranging populations of bison in Russia, use has been made of source materials contained in the studies by Flint et al. (2002), Gusarov (2011), Mizin (1999), Popov (1989), Pucek et al. (2004), Tyapougin and Gusarov (2004), Zablocky (1965), Sipko (unpublished), the European Bison Pedigree Book.

23.9 Free-Ranging Populations in Slovakia

Name and location: NÁRODNÝ PARK POLONINY

North-eastern Slovakia, close to the borders with both Poland and Ukraine.

Year of establishment: 2004.

Breeding line and founder group: Bison of the Białowieża-Caucasian line. The founder group comprised five (2, 3) individuals originating from the following captive-breeding centres:

1 (1, 0) Artis Zoo Amsterdam, The Netherlands
2 (0, 2) Parco Natura Viva, Bussolengo, Italy
2 (1, 1) Natur und Tierpark Goldau, Switzerland

These European bison spent 1.5 years in an acclimatisation enclosure, before being freed on 10 December, 2004. The founder group was augmented by a further 2 (1, 1) animals from Chomutow Zoo in the Czech Republic, these being released on 10 October, 2005. These bison are capable of being joined by bulls originating in the free-ranging population in Poland's Bieszczady Mountains. As of 2010, nine individuals were present here.

Area of forest complex and natural conditions: the overall area occupied by bison in Slovakia is around 100 km^2. The bison most often spend time in low mountains at altitudes of 500–800 m a.s.l. The dominant tree species here is beech (Fagus silvatica), while other broadleaved trees include sycamore (Acer pseudoplatanus), large-leaved lime (Tilia platyphyllos), ash (Fraxinus excelsior) and elms (Ulmus sp.). The conifers present in turn include fir (Picea abies), artificially planted spruce (Picea abies), Scots pine (Pinus sylvestris) and larch (Larix decidua). The non-forest vegetation in the region is diverse and mosaic-like, with a mix of fields and meadows crossed by streams.

Area occupied: Radio telemetric data point to an area of 78.24 km^2 being used by the bison.

Prospects: plans envisage a metapopulation arising within the Eastern Carpathians, in the area where Poland, Slovakia and Ukraine meet.

Source materials: Adamec et al. 2006, Pčola et al. 2006, Pčola and Gurecka 2008.

Chapter 24
Hybrids Between the European Bison and Domestic Cattle

The crossing of such species belonging to the same genus as the European and American bison poses no difficulty, and the existence of these hybrids in fact represented a major problem in the first years of restitution of the European bison, since the purity of *Bison bonasus* (L.) came under threat. In contrast, the obtainment of hybrids between genera, e.g. *Bison* and *Bos* (the genus including domestic cattle) is more problematical. It is worth recalling that domestic cattle were grazed in the Białowieża Forest for many years at the same time as bison were present there, yet no births of hybrids in natural conditions were noted.

However, man-made hybrids of American bison and cattle have long been known in North America, under the name "cattalo". Indeed, uncontrolled formation of hybrids may threaten the purity of the species: when investigated, the herd of "buffalo" in Custer State Park was found to contain sequences of mtDNA originating from cattle (Polziehn et al. 1995), a fact supporting the supposition that the founders of the population included back crosses indistinguishable from American bison in their outward appearance.

The first attempts at crossing European bison with cattle were pursued by Leopold Walicki at his estate near Grodno, in the years 1847–1859. Walicki's aim was to breed up strong working cattle (Karcov 1903). In turn, the experiments done in the years 1958–1976 at Białowieża, by the Mammal Research Institute PAS, had as their theoretical goal the tracing of the mechanisms of inheritance, the determination of hybrids' fertility and the practical utilisation of heterosis, i.e. hybrid vigour. First-generation (F_1) hybrids outclass their progenitors (bison and cattle) in both body mass and size. The largest F_1 males achieved body masses of 1,030 kg, cf. 880 kg for females (Krasińska 1988). A total of 71 hybrids were ultimately obtained at Białowieża, 11 of these being first-generation. The father of all the hybrids was the bull M 1077 POKORNY (Fig. 24.1), while female F 1073 PONĘTNA gave birth to the world's first hybrid, sired by a bull named Filon (Dehnel 1961) (Fig. 24.2). The bison–cattle hybrids had no special name, but after various proposals had been considered it was the Polish name "żubroń" that was decided upon. The raising of hybrids for commercial purposes in fact continued

M. Krasińska and Z. A. Krasiński, *European Bison*,
DOI: 10.1007/978-3-642-36555-3_24, © Springer-Verlag Berlin Heidelberg 2013

Fig. 24.1 The bull M 1077 POKORNY was the father of all the specimens of the hybrids (bison–cattle) obtained in the course of experimentation at the Mammal Research Institute of the Polish Academy of Sciences in Białowieża in the years 1958–1976. *Photo* ZA Krasiński

Fig. 24.2 Filon was the world's first hybrid born to a bison female at Białowieża in 1960. *Photo* Z. Pucek

Fig. 24.3 The first-generation hybrid "FAKIR" was obtained at Białowieża by crossing a domestic cow of the black-and-white lowland breed with a bison (specifically the 4-years-old POKORNY). *Photo* ZA Krasiński

on the Polish State Farms until 1984, artificial insemination of domestic cows with bison semen being applied to that end. The hybrids resulting from crosses between bison and different breeds of cattle were distinguishable on the basis of their very variable colouration (see colour insert). As first-generation (F_1) males (Fig. 24.3) are infertile, further crossing was facilitated by having hybrid females covered by either male bison or bulls (by way of back-crossing). The information on F_1 male fertility obtained by Walicki is not fully reliable. Contemporarily, a fertile male has only appeared where the cross was of 75 % bison blood. More information on the hybrids and the centres keeping them may be obtained from the monograph entitled Hybrids of European bison and domestic cattle (Krasińska 1988).

Chapter 25
The American Bison as a Close Relative of the European Bison

As the names suggest, American bison *Bison bison* (popularly "buffalo") and European bison *Bison bonasus* live on two separate continents. They nevertheless both arise from a common ancestor living in south-east Asia (where there are no bison at all at present). The ancestors of the American bison reached the Americas via a land bridge between the continents spanning what is today the Bering Sea. Final and full separation of the two species was assured when the ice sheets retreated once more some 10,000 years ago (Shapiro et al. 2004).

Bison populations eventually came to extend over huge areas of both North America and Eurasia. Many authors contend that European bison roamed as far as Europe's Atlantic coast, and their populations certainly extended as far as the Caucasus Mountains in the east (Heptner et al. 1966; Pucek 2004). American bison in turn ranged over a huge area from the Gulf of Mexico to Alaska.

Even allowing for the relatively short span of evolutionary time involved, these vast ranges with varied habitat were sufficient to encourage a diversification of populations. Two recognisable subspecies of European bison arose—i.e. the Lowland or Białowieża bison *B. bonasus bonasus*—as an inhabitant of the huge European forests, plus the Caucasian bison *B. bonasus caucasicus*, which was confined to the relatively limited areas of the Caucasus massif. Perhaps unsurprisingly, the latter subspecies of restricted range did not survive through to our times in pure form, since the last free-living specimens were killed in the Caucasus in 1927 (Bashkirov 1939; Nemtsev et al. 2003).

Nevertheless, the shrinking of contiguous areas of forest across Europe also resulted in ever-greater constriction of the Lowland bison's once very extensive range—to the point where just one wild population (in the Białowieża Primaeval Forest) remained in existence by the mid-19th century. And even there the last wild individuals succumbed at the beginning of the 20th century (Sztolcman 1926). Reinstatement of the species was thus only possible using animals that had survived in captivity.

The story of the American bison is rather different, since—as recently as in the 15th century—there were still simply huge herds of prairie bison (*Bison*

M. Krasińska and Z. A. Krasiński, *European Bison*,
DOI: 10.1007/978-3-642-36555-3_25, © Springer-Verlag Berlin Heidelberg 2013

bison bison) roaming across America's still-pristine grasslands, while the wilds of Canada continued to provide a habitat for the forest-dwelling "wood buffalo" *B. bison athabascae*.

However, notwithstanding initial populations running into the tens of millions, intensive hunting, disease and the development of cattle grazing were all able to first affect, and then drastically reduce, the numbers of American bison. By the end of the 19th century, there were just a couple of hundred of the great beasts left, this situation motivating conservation-minded individuals to capture the wild animals and bring them into breeding enclosures. That left just 23 bison in the wild in the USA—all in Yellowstone National Park on the borders of Wyoming, Montana and Idaho, as well as an estimated 300 wood buffalo in the vicinity of Canada's Great Slave Lake. Alas, even the latter population was not safe, since the area it inhabited was made subject to the importation of 6,673 prairie bison in the years 1925–1928, the result being extensive interbreeding of individuals from the two subspecies (Banfield and Novakowski 1960).

The abundance of European bison around the world is known with near-total precision, thanks to registration in the European Bison Pedigree Book. The grand total in fact exceeded 4,400 in 2010, with bison present in zoos, in reserves and in various other kinds of animal collection. Nevertheless, a majority (60 %) of animals are now back in the wild state, moving freely around larger forest complexes, and even areas of fields and forest, in Belarus, Lithuania, Poland, Russia, Slovakia and Ukraine.

The number of American bison in the USA and Canada now stands at an estimated 500,000. However, most bison are in private hands, living in enclosures and on farms, and thus being raised for commercial purposes. The two largest and best-known wild populations remain the US population at Yellowstone (some 4,000 strong) and the Canadian population in the huge Wood Buffalo National Park, which nevertheless supports "just" 2,500 of the animals.

As membership of the same genus *Bison* ought to suggest (though does not always do so), European and American bison are close relatives. Nevertheless, these are visibly distinct animals, as can be well seen if their outlines and key body features are compared (as in Figs. 25.1 and 25.2). The line followed by the back differs for a start. Beyond the head of a American bison, there is a clear change of angle that sees the silhouette slope steeply upwards. The line is in turn a much more gentle one in the case of European bison, and is (justifiably) considered "carp-like". Not unnaturally, it is in the large adult males of the two species that these differences are best seen (Reynolds et al. 1982; Krasiński 2010).

Hence the heads of both kinds of bison are low-set. However, while the horns of an American bison extend upwards and only slightly inwards, those of the European species are mostly characterised by greater curvature. The front parts of a "buffalo" have long, curly hair in a mass that comes to rather an abrupt end beyond the shoulders. Furthermore, the hair on the head and neck is of a dark chocolate hue, as compared with that along the back and sides which is paler brown. There is also longer dangling hair covering the forelegs of an American bison, to the point where an animal can seem to be "wearing puffed-out

Fig. 25.1 Comparison of silhouette shape of European bison and American bison. Figure by T. Samojlik. *Source* Krasiński (2010) modified

pantaloons". The hindquarters of an American bison have short hair of a coppery-brown colour. In turn, the hide of a European bison is more grey-brown in colour, while the beard borne by its American cousin is both longer and more marked (Reynolds et al. 1982; Krasińska 2010).

Both species are characterised by sexual dimorphism. The males (bulls) are larger and more massively built, with a more distinct "hump". The smaller females (cows) are more delicately built and with less of a hump. Potentially achievable sizes and body weights in the two species are similar, though ultimately the largest male European bison may be larger than American specimens (maximum 840 kg, as compared with 770), with longer and thicker horns also. Calves of both species are relatively small at birth, rarely exceeding 30 kg in weight (Halloran 1960; McHugh 1972; Krasińska and Krasiński 2002).

The two species interbreed readily and their offspring are fertile. Indeed, in the first years of work to reinstate the European bison, such hybridisation posed a serious threat to the purity of the species. Fortunately, work on individuals' pedigrees allowed for the elimination of hybrids from the work to recreate a *Bison bonasus* population. The early period of the reinstatement (1929–1930) for example saw three European-American bison hybrids brought into the breeding reserve at Białowieża, along with pure European bison. When the status of these animals was discovered, they were transported away again with all speed. In contrast, other breeders reacted less decisively, the most notorious example involving the herd at Woburn Abbey in the UK, whose hybrids obstructed the work to bring back the European bison.

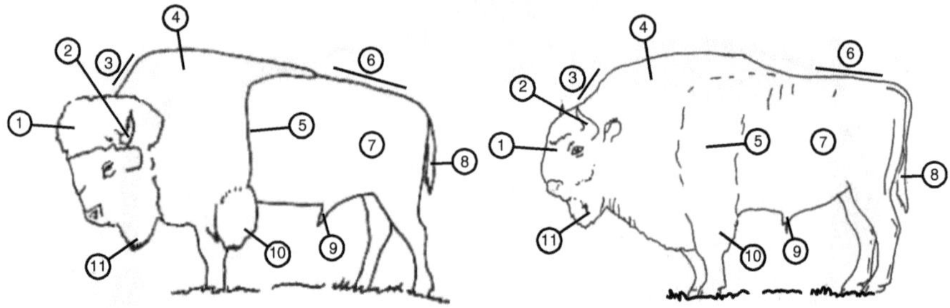

Fig. 25.2 Comparison of body composition of European bison and American bison. Silhouette of American bison after: Reynolds et al. (1982). Figure by T. Samojlik. *Source* Krasińksi (2010) modified

American bison

1. Hair on head longer than in European bison, dark brown, clear "fringe" on forehead.
2. Horns short, often concealed in hair, ends pointing upwards.
3. Back slopes sharply upwards behind the head.
4. Front part of the body covered with longer, curly hair, chocolate brown on the head and neck, pale brown on the back.
5. Distinct boundary between the longer "mantle" of hair around the foreparts and the shorter hair further back.
6. Gentle downward slope to the hind part of the back.
7. Hindquarters featuring a short, smooth pelt copper-brown in colour.
8. Tail short and thin, with smooth hair, ending in a brush of longer hairs.
9. In males, the tuft of hair surrounding the prepuce is very visible against the smooth underbelly.
10. Longer hair covering the upper part of the forelimbs ends near the carpus, giving the impression that a bison is dressed in "puffed-out pantaloons".
11. Beard distinct, triangular in shape and goatee-like.

European bison

1. Shapely head set higher than in American bison.
2. Horns longer and more massive, curving inwards, especially in females.
3. Back following a gentle concave curve behind the head, followed by a convexity, hence rather "carp-like"
4. Hair covering the anterior parts of the body is shorter than in American bison, grey-brown in colour, but with a hint of ruddy
5. Much less clear boundary between the hair covering the fore and hind parts of the body.
6. Rear part of the back level, not downward sloping.
7. Pelt of the hind parts of the body with much shorter hairs than on the fore parts.
8. Tail reaches to the tarsal joint, covered by longer hairs than in American bison.
9. In adult males, a distinctly (if slightly) protruding belly to the rear of the prepuce.
10. Upper part of the forelimbs covered with shorter hair than in American bison.
11. Beard shorter than in American bison, gently merging with the upper part of the chest.

Another mistake of this kind was made in the 1940s, when an admixture of American bison genes entered the population of European bison inhabiting the central part of the Caucasus Mountains, as later included in the Biosphere Reserve there. The consequences of this are severe, since they ensure that the best habitat for (what should be) European bison, is actually inhabited by hybrids that may not be registered in the Pedigree Book. This notwithstanding the intriguing fact that what now occurs there is a "mountain bison" that may represent a new breed or even a new "species", in the view of certain Russian researchers (Rautian et al. 2000).

A fashion developing in recent years has involved the on-farm breeding and raising of American bison in Europe. This has involved a large number of firms and private breeders in the importation to Europe (for example France, Belgium and Germany, and latterly also Russia and Poland) of "buffalo". And, as bad luck would have it, some of those farmed American bison live in uncomfortably close proximity to free-ranging populations of European bison, with potentially dangerous consequences for the purity of the latter species, should individuals for example escape from less-well-fenced farm enclosures.

References

Adamec M, Perdikova Z, Pčola Š et al (2006) Priestorova aktiva zubrov hrivnatých (*Bison bonaus*) v NP. Poloniny (Spatial activity of the European bison in Poloniny N.P.). In: Olech W (ed) Perspektywy rozwoju populacji żubrów. Wyd Artisco, Goczałkowice-Zdrój, pp 55–62 (In Slovak)

Aleksandrov VN, Golgovskaya KJ (1965) Kormovye ugodia zubrov Kavkazskogo Zapovednika (Food base of European bison in Caucasus Reserve).Tr Kavkaz Gosud Zapoved 8:129–154 (In Russian)

Aleksandrowicz NP (1999) Zubr (*Bison bonasus bonasus* Linnaeus 1758) v okhotnichei dobyche naseleniya sredniovekovych beloruskich gorodov (Bones remains of European bison in the hunt trophies of medieval Belarussian towns). In: Proceedings of conference: Nauchno-prakticheskaya konferentsiya posvyashchenna 60-letiyu so dnia obrazovanya Gosudarstvennogo zapovednika "Belovezhskaya Pushcha", Minsk, 22–24 Dec 1999 [In Russian]

Aleksandrowicz K, Perzanowski K, Cătănoiu S et al (2009) Browse and ground flora supply in selected tree stands of Vanatori Neamt Nature Park. Studies Res Vanatori Neamt Nature Park 2:40–46

Anusz Z, Walkowiak E, Krupa J et al (1986) Occurence of *Coxiella burneti* antibodies in bison (*Bison bonasus*) from Białowieża Primeval Forest and cattle and sheep from the villages around it. In: Abstracts of the annual meeting of the European association for animal protection, Hungary, 1–6 Sept 1986

Aulak W (1976) Development and production of undergrowth of *Tilio-Carpinetum* Tracz. 1962 as one of the basic elements of the trophy level in forest ecosystems. Zesz nauk SGGW-AR, Rozpr nauk 60: 1–151 (In Polish with English summary)

Babik W, Kawałko A, Wójcik JM et al (2012) Low major histocompatibility Complex Class I (MHC I) variation in the European Bison (*Bison bonasus*). J Hered 103(3):349–359. doi:10.1093/jhered/ess005

Balčiauskas L (1999) European bison (*Bison bonasus*) in Lithuania: status and possibilities of range extension. Acta zool Lith Biodivers 9:3–18

Balčiauskas L (2000) Restoration of European bison to Lithuania: achievements and problems. In: Balčiauskas L (ed) Proceedings of international symposium "European Bison—yesterday, today and tomorrow", 9–10 Dec 2000, Šiaulai University, Šiaulai

Baltrunaite L (1999) State and prospect of ungulate animals breeding in enclosure in Lithuania. Acta Zool Lith 9:55–60

Banfield AWF, Novakowski NS (1960) The survival of the wood bison (*Bison bison athabascae* Rhoads) in the Northwest Territories. Nat Hist Pap Natl Museum Canada 8:1–6

Bandarenko VD, Haiecki PB (1999) Reaklimatizatsiya zubra u zakhodnim regionie Ukrainy (Introduce of European bison in western Ukraine) In: Abstracts of conference:

DOI: 10.1007/978-3-642-36555-3,

Nauchno-prakticheskaya konferentsiya posvyashchenna 60-letiyu so dnya obrazovaniya Gosudarstvennogo zapovednika “Belovezhskaya Pushcha”, Kamenyuki, 22–24 Dec 1999 (In Russian)

Baranov AS, Pucek Z, Kiseleva EG et al (1997) Developmental stability of skull morphology in European bison, *Bison bonasus*. Acta Theriol 42(Suppl 4):79–85

Bashkirov VS (1939) Kavkazskii zubr. Monograficheskii ocherk (Caucasian European bison monographic review) In: Kulagin NK (ed) Kavkazskii zubr. Glac Uprav po zapoved, zoopark zoosad Moskva, pp 7–22 (In Russian)

Belousova IP (1986)Prioksko-terrasnyi biosfernyi zapovednik, pos. Danki (Prioksko-terrasnyi Biosphere Reserve) In: Proceedings of conference: Piervoe Vsesoyuznoe Soveshchanie po problemam zookul’tury. Tezisy dokladov chast’ 2. Zookul’tura Moskwa 1986 (In Russian)

Belousova IP (1993) Obosnovanie neobhodimosti razrabotki v programu sokhraneniya geneticheskogo raznoobraziya vida zubr (*Bison bonasus* L.) (The necessity to develop European bison conservation program). In: Proceeding of conference: Sovremennye problemy okhrany zubra, Minsk (In Russian)

Belousova IP (1999) Znachenie i ocenka pokazatelei geneticheskogo raznoobrazya dla resheniya problemy sokhraneneniya evropeiskogo zubra (The importance and assessment of genetic diversity indicators for the conservation of European bison). Prioksko-terrasnyi gosud biosf zapov, Pushchino (In Russian)

Benecke N (2005) The Holocene distribution of European bison—the archeozoological record. Munibe (Antropoliogie-Arkeologie) 57:421–428

Bojanus LH (1827) De Uro nostrate eiusque sceleto commentatio. Nova Acta Physico-Medica. Academiae Cesareae Leopoldino-Caroline Naturae Curiosum 32:1–184

Van den Brink WJ (1980) The behaviour of wisent and bison in larger enclosures. Acta Theriol 25:115–130

Brincken J (1826) Memoire Descriptif sur la Foret Imperiale de Białowieża, en Lithuanie N. Glücksberg, Varsovie

Borchers K, Brackmann J, Wolf O et al (2002) Virologic investigations of free-living European bison (*Bison bonasus*) from the Białowieża Primeval Forest. Poland J Wildlife Dis 38:533–538

Borowski S, Kossak S (1972) The natural food preferences of the European bison in seasons free of snow cover. Acta Theriol 17:151–169

Bouzat JL, Giovambattista G, Golijow GD, Lojo MM, Dulout FN (1998) Genética de la Conservación de Razas Autóctonas: El ganado Criollo Argentino. Interciencia 23:151–157

Brewczyński P, Welz M (2011) Threat of tuberculosis of European bison in the Bieszczady. Euro Bison Conserv Newslett 4:63–70

Burzyńska B, Olech W, Topczewski J (1999a) Phylogeny and genetic variation of the European bison *Bison bonasus* based on mitochondrial DNA D-loop sequences. Acta Theriol 44:253–262

Burzyńska B, Topczewski J (1995a) Genotyping of *Bison bonasus* β-casein gene following DNA sequence amplification. Anim Genet 26:335–336

Büchner E (1900) Powolne wymieranie żubra w Puszczy Białowieskiej (The gradual extinction of European bison in the Białowieża Forest Łowiec Polski 7 (19):2–4 ((20):2–3, (21):1–2, (22):2–3, (23):3–4 [In Polish])

Bunevich AN (1989a) Polovovozrastnoi sostav selektsionnogo-vybrakovannych zubrov Belovezhskoi Pushchi (Age and sex composition of culled European bison in Białowieża Forest). In: Proceedings of conference: Nauchno-prakticheskaya konferentsiya, posvyashchenna 50-letiyu regulyarnych issledovanii v Belovezhskoi Pushche, 19–21 Dec 1989 (In Russian)

Bunevich AN (1989b) Teritoryal’noe razmeshchenie zubrov v Belovezhskoi Pushchi (Spatial distribution of European bison in Białowieża Forest) In: Proceedings of conference: Nauchno-prakticheskaya konferentsiya, posvyashchenna 50-letiyu regulyarnych issledovanii v Belovezhskoi Pushche, 19–21 Dec 1989 (In Russian)

Bunevich AN (1989c) Kharakteristika mesta obitaniya razlichnykh grup zubrov Belovezhskoi Pushchi v vegetatsyonnyi sezon (Description of habitats used by European bison in growing season in Belovezhskaya Pushcha). In Abstracts of conference: Vsesoiuznoe soveshchanie po probleme kadastra i ukhota zhivotnogo mira, Ufa (In Russian)

Bunevich AN (1991) Itogi rasseleniya zubrov po territorii Belovezhskoi Pushchi (Distibution of European bison on area of in Białowieża Forest). Zapov Bel Issled 15:98–109 (In Russian)

Bunevich AN (1994) Sostoyanie populatsii zubrov v vostochnoi chasti Belovezhskoi Pushchi (Status of European bison population in eastern part of Białowieża Forest). In: Proceedings of conference: Mezhdunarodnyi simpozium po zubru "Sovremennye problemy okhrany zubra", 26–27 Marta 1992, Kamieniuki, Minsk 1994 (In Russian)

Bunevich AN (1999a) Povrezhdaemost' zubrami drevostoev v mestach zimnei kontsentratsii (Damage of trees by European bison in area of winter concentrations). In Abstracts of conference: "Problemy izucheniya sokhraneniya i isopol'zovaniya biologicheskogo raznobraziya zhivotnogo mira" Minsk 1999 (In Russian)

Bunevich AN (1999b) Mortality cases of European bison in Belarussian part of the Białowieża Forest. In: Abstracts of international scientific conference on health protection of free-ranging *Bison bonasus* in Białowieża Forest, Warszawa, 26–27, Nov 1999

Bunevich AN (1999c) Bioritmy reproduktsii zubra (*Bison bonasus bonasus*) (Rhythm of reproduction of European bison in eastern part of Białowieża Forest). In Abstract of conference: Mezhdunarodnoaya nauchno-prakticheskaya konferentsiya "Biologicheskie Ritmy". Brest 1999 (In Russian)

Bunevich AN (1999d) Prichiny smertnosti zubrov v Belovezhskoi Pushchi i ikh dinamika (Dynamic of mortality reasons of European bison in eastern part of Białowieża Forest). In: Abstracts of conference: Nauchno-prakticheskaya konferentsiya "Belovezhskaya Pushcha na rubiezhe tret'ego tysyachletiya" posvyashchennoi 60-letiyu so dnia obrazovaniya Gosudarstvennogo zapovednika "Belovezhskaya Pushcha", Kamenyuki, 22–24 Dec 1999 (In Russian)

Bunevich AN (2003a) Dinamika areala populatsii zubra w vostochnoi chasti Belovezhskoi Pushchi (Dynamic of area of European bison population in eastern part of Białowieża Forest. In: Abstracts of conference "Znaczenie badań naukowych dla ochrony żubrów w Polsce", Katedra Genetyki i Ogólnej Hodowli Zwierząt Wydział Nauk o Zwierzętach, SGGW Warszawa, 13 June 2003 (In Russian)

Bunevich AN (2003b) Analiz formirowania populatsii zubra v beloruskoi chasti Belovezhskoi Pushchi (Analysis of development of population of European bison in Belarussian part of Białowieża Forest). Belov Pushcha Issled 11:178–204 (In Russian with English summary)

Bunevich AN (2003c) Dynamika i struktura areala populatsii zubra v Beloviezhskoi Pushche (Dynamic and range of structure of European bison population in Białowieża Forest). Belov Pushcha Issled 11:160–177 (In Russian with English summary)

Bunevich AN (2006) Massa tela i ekster'nye pokazateli zubrov Belovezhskoi Pushchi (Body mass and measurements of the European bison in the Belovezhskaya Pushcha). Belov Pushcha Issed 12:121–141 (In Russian with English summary)

Bunevich AN (2010) Occurrence of uro-genital disease (balanoposthitis) in males of the European bison in the Belorussian part of the Białowieża Forest. Park nar rez Przyr 29(3):77–87 (In Polish with English summary)

Bunevich AN, Kochko FP (1988) Dinamika chislennosti i struktura populatsii zubrov Belovezhskoi Pushchi (Dynamic of numbers and structure of population of Euroopean bison in eastern part of Białowieża Forest). In: Populatsyonnye issledovanniya zhivotnykh v zapovednikakh. Sbornik nauchnykh trudov. Nauka, pp 96–114 (In Russian)

Bunevich AN, Krasińska M, Daleszczyk K (2006) Creation and growth of the free-ranging population of lowland European bison, *Bison bonasus bonasus* (L.) in the Belarussian part of Białowieża Forest). Parki nar Rez Przyr 25(4):101–118 (In Polish with English summary)

Burzyńska B, Topczewski J (1995b) Genotyping of *Bison bonasus* β-casein gene following DNA sequence amplification. Animal Genet 26:335–336

Burzyńska B, Olech W, Topczewski J (1999b) Phylogeny and genetic variation of the European bison *Bison bonasus* based on mitochondrial DNA D-lop sequences. Acta Theriol 44:253–262

Cabaj W, Moskwa B, Pastusiak K et al (2005) Antibodies to *Neospora caninum* in the blood of European bison (*Bison bonasus bonasus* L.) living in Poland. Vet Parasitol 128:163–168

Cabaj W, Moskwa B, Pastusiak K et al (2006) *Neospora caninum* in European bison in Poland. In: Olech W (ed) Perspektywy rozwoju populacji żubrów, Wyd Artisco, Goczałkowice Zdrój, pp 89–108 (In Polish with English summary)

Cabaj W, Bień J, Goździk K et al (2009) (*Neospora caninum* in European bison in Poland—the current research problem). Eur Bison Conserv Newslett 2:102–111 (In Polish with English summary)

Cabaj W, Bień J, Goździk K et al (2010) *Neospora caninum* in European bison living in Białowieżą. Poland. Eur Bison Conserv Newslett 3:63–68 (In Polish with English summary)

Cabaj W, Goździk K, Bień J et al (2012) Neospora caninum in European bison eliminated at Białowieża in 2010–2012. Eur Bison Conserv Newslett 5:26–32

Caboń-Raczyńska K, Krasińska M, Krasiński ZA (1983) Behaviour and daily activity rhythm of European bison in winter. Acta Theriol 28:273–299

Caboń-Raczyńska K, Krasińska M, Krasiński ZA et al (1987) Rhythm of daily activity and behaviour of European bison in Białowieża Forest in the period without snow cover. Acta Theriol 32:335–372

Celtis K (1502) Quattour libri amorum (Translation from Latin into Polish), see: Ziembicki W 1934)

Czudek A (1930) Rozwój przyszłości żubrów śląskich (Future of European bison on Silesia). Sylwan 48:1–15 (In Polish)

Czykier E, Krasińska M (2004) Early spermatogenesis in European bison. Acta Theriol 49:543–554

Czykier E. and Krasińska M (2006) The concentration of free testosterone in blond serum of European bison males depending on age, body mass and the weight of testes – preliminary study. Parki nar Rez Przyr 25 (4):119–131 (In Polish with English summary)

Czykier E, Sawicki B, Krasińska M (1999) Postnatal development of the European bison spermatogenesis. Acta Theriol 44:77–90

Dackiewicz J (1999) Oddziaływanie żubrów na las w miejscach ich zimowej koncentracji w Puszczy Białowieskiej (Influence of European bison upon the forest at winter feeding sites of Białowieża Forest). M.Sc.Thesis, Faculty of Forestry, Warsaw University of Life Sciences-SGGW (In Polish)

Daleszczyk K (2002) Behawior rozrodczy *Bison bonaus* (L.) (Behaviour of breeding of European bison (L.)). PhD Thesis, Mammal Research Institute PAS, Białowieża (In Polish)

Daleszczyk K (2004) Mother-calf relationships and maternal investment in European bison *Bison bonasus*. Acta Theriol 49:555–566

Daleszczyk K (2005) Activity pattern of calf and its integration into herd in European bison, *Bison bonasus*. Fol zool 54:359–363

Daleszczyk K, Bunevich AN (2009) Population viability analysis of European bison populations in Polish and Belarussian parts of Białowieża Forest with and without gene exchange. Biol Conserv 142:3068–3075. doi:10.1016/j.biocon.2009.08.006

Daleszczyk K, Czykier E (2010) Do European bison bulls in Białowieża Forest differ in their rutting behaviour depending on age? Eur bison Conserv Newlet 3:5–16

Daleszczyk K, Krasińska M, Krasiński ZA, Bunevich AN (2007) Habitat structure, climatic factors and habitat use by European bison (*Bison bonasus*) in Polish and Belarussian parts of Białowieża Forest, Poland. Can J Zool 85:261–272. doi:10.1139/Z06-209

Daleszczyk K, Krasiński ZA (2001) Parturition behaviour of European bison living in reserves. Folia Zool 50:75–78

Dehnel A (1961) Die ersten hybriden zwischen *Bos taurus domesticus* und *Bison bonasus* (L.). Acta Theriol 5:45–50

Demiaszkiewicz AW (1988) (Onchocercosis of European bison and cattle in Białowieża Forest). Med Weter 44:343–345 (In Polish with English summary)

Demiaszkiewicz AW (1991) (A case of chorioptosis in the bison (*Bison bonasus*) in the Białowieża Forest). Med Weter 64:547–548 (In Polish with English summary)

Demiaszkiewicz AW (1995) Studies on tissue nematodes from subfamily *Onchocercinae* Leiper, 1911 in free living and domestic ruminants in Poland). Rozp nauk Mon, Wyd SGGW, Warszawa (In Polish with English summary)

Demiaszkiewicz AW, Dróżdż J, Lachowicz J (1999a) Occurence of lung nematodes in red deer in Białowieża Forest. Med Weter 55:519-520 (In Polish with English summary)

Demiaszkiewicz AW, Lachowicz J (1991) *Trypanosoma theileri* Laveran 1902—a parasite of cattle in Poland. Med Weter 47:112–113 (In Polish with English summary)

Demiaszkiewicz A, Osińska B, Bielecki W (1999b) A case of onchocercosis in European bison with disease of genital organs. Med Weter 55:321–322 (In Polish with English summary)

Demiaszkiwicz AW, Pyziel AM (2010) Forming of European bison helminthfauna in Białowieża Forest. In: Kowalczyk R, Ławreszuk D, Wójcik JM (eds) European bison conservation in the Białowieża Forest. Threats and prospects of the population development. Mammal Research Institute PAS, Białowieża (In Polish with English summary. http://krainazubra.pl/files/ochrona_zubra.pdf)

Demiaszkiewicz AW, Pyziel AM, Lachowicz J (2008) Helminthological status of European bison in Białowieża Forest in the winter 2007/2008. Eur Bison Conserv Newslett 1:42–53 (In Polish with English summary)

Demiaszkiewicz AW, Pyziel AM, Lachowicz J et al (2009) Dictyocaulosis of European bison in Białowieża Forest. Eur Bison Conserv Newslett 2:112–118 (In Polish with English summary)

Demiaszkiewicz AW, Pyziel AM, Kuligowska I et al (2010) Parasitoses of the European biaon caused by nematodes located in the klarge intestine. Eur Bison Conserv Newslett 3:69–74 (In Polish with English summary)

Dmowski K (1998) Lead, cadmium and zinc in tissues of European bison (*Bison bonasus* L.) from Białowieża National Park. In: Proceedings of 2nd international conference on trace element-effect on organisms and environment, Cieszyn, Poland, 23–26 June 1998

Drożdż A, Weiner J, Gębczyńska Z et al (1975) Some bioenergetics parameters of wild ruminants. Pol Ecol Stud 1:85–101

Dróżdż J (1961) A study on helminthes and helminthiases in bison, *Bison bonasus* (l.) in Poland. Acta Parasit Pol 9:55–95

Dróżdż J (1967) The state of research on the helminthofauna of European bison. Acta Theriol 12:377–384

Dróżdż J (2000) Influence of the migration of animals on formatting of their parsitofauna. Med Weter 56:154–157 (In Polish with English summary)

Dróżdż J, Demiaszkiewicz AW, Lachowicz J (1989) The helminth fauna of free ranging European bison, *Bison bonasus* (L.). Acta Parasit Pol 34:117–124 (In Polish with English summary)

Dróżdż J, Demiaszkiewicz AW, Lachowicz J (1990) Nematodes in large intestine of European bison. Wiad Parazyt 36:35–38 (In Polish with English summary)

Dróżdż J, Demiaszkiewicz AW, Lachowicz J (1994) The helminth fauna of free-ranging European bison, *Bison bonasus* (L.) studied again 8 years after reduction of bison in the Białowieża Forest. Acta Parasit 39:88–91

Dróżdż J, Demiaszkiewicz AW, Lachowicz J (1998a) *Ashworthius sidemi* (Nematoda, Trichostrongylidae) a new parasite of the European bison *Bison bonasus* (L.) and the question of independence of *A. gagarini*. Acta Parasit Pol 43:75–80

Dróżdż J, Demiaszkiewicz AW, Lachowicz J (1998b) The helminth fauna of free-ranging European bison, *Bison bonasus* (L.). Acta Parasit Pol 34:117–124

Dróżdż J, Demiaszkiewicz AW, Lachowicz J (1999) Formating of helminthofauna of free-living European bison in three different forest biotopes (Białowieża Forest, Borecka Forest and Bieszczady Mountains). In: Abstracts of international scientific conference on health protection of free-ranging *Bison bonasus* in Białowieża Forest, Warszawa, 26–27 Nov 1999

Dróżdż J, Demiaszkiewicz AW, Lachowicz J (2000) Ashwortiosis a new parasitic disease of wild ruminants. Med Weter 56:32–35 (In Polish with English summary)

Dróżdż J, Demiaszkiewicz AW, Lachowicz J (2002a) Forming of gastro-intestinal nematodes fauna of free ranging European bison in Białowieża Primeval Forest during last 17 years (1984–2001). Wiad Parazytol 48:375–381 (In Polish with English summary)

Dróżdż J, Demiaszkiewicz AW, Lachowicz J (2002b) Exchange of gastrointestinal nematodes between roe and red deer (Cervidae) and European bison (Bovidae) in the Bieszczady Mountains (Carpathians, Poland). Acta Parasitol 47:314–317

Dróżdż J, Demiaszkiewicz AW, Lachowicz J et al (2003) Rola jeleniowatych w rozprzestrzenieniu aswortiozy wśród wolno żyjących żubrów w Polsce (The role of red deer in spreading of ashwortiosis among free-living European bison in Poland.In: Abstracts of conference "Znaczenie badań naukowych dla ochrony żubrów w Polsce", Katedra Genetyki i Ogólnej Hodowli Zwierząt, Wydział Nauk o Zwierzętach SGGW Warszawa, 13 June 2003

Dvorak LE, Tolkach VN (1989) Zapas kormovych rastenii napochvennogo pokrova v osnovnych tipach lesa Belovezhskoi Pushchi (Supply of food plants in forest floor in the main types of forest in Belovezhskaya Pushcha). Zapov Bel 13:104–112 (In Russian)

Dzięciołowski RM (1991) Ecological niches of five big ungulates in forest tracts. Folia Pol Ser A Forestry 33:56–70

Dzięciołowski R, Kossak S, Borowski S, Morow K (1975) Diets of herbivorous mammals. Polish Ecol Stud 1:33–50

Eichwald EK (1830) Naturhistorische Skizze von Lithuanien, Volhynien und Podolien. J. Zawadzki, Wilno

Empel W (1962) Morphologie des Schädels von *Bison bonasus* (Linnaeus 1758). Acta Theriol 6:53–111

Empel W, Roskosz T (1963) Das Skelett der Gliedmassen des Wisents *Bison bonasus* (Linnaeus, 1958). Acta Theriol 7:259–300

European Bison Pedigree Book see Giergiel J, Groeben G, Krysiak K, Mohr E, Pilarski W, Raczyński J, Żabiński J

Falińska K (1973) Seasonal dynamics of herb layer in forest communities of Białowieża National Park. Phytocoenosis 2:1–120

Faliński JB (1968) Położenie, granice i nazewnictwo fizjograficzne Puszczy (Location, boundaries and physiographic nomenclature of Białowieża Forest). In: Faliński JB (ed) Park Narodowy w Puszczy Białowieskiej. Państ Wyd Roln Leśne, Warszawa, pp 13–24

Faliński JB (1973) Herb layer filling by plant cormus in the *Querco-Carpinetum* community in the Białowieża National Park. Phytocoenosis 2:123–142

Faliński JB (1986) Vegetation dynamics in temperate lowland primeval forest. Ecological studies in Białowieża Forest. Geobotany (8), Dr W. Junk Publishers, Dordrecht

Faliński JB (2001) Phytophenological atlas of the forest communities and species of Białowieża National Park. Phytocoenosis 13:1–176

Fedyk S, Sysa P (1971) Chromosomes of European bison, domestic cattle and their hybrids. Acta Theriol 16:465–470

Flerov KK (1932) Obzor diagnostichestich priznakov belovezhskogo i kavkazskogo zubrov (Diagnostic features of Białowieża bison and Caucasian bison). Izvestiya Akad Nauk SSSR. Otd matemat i estestv nauk 10:1579–1589

Flerov KK (1979) Sistematika i evolutsiya (Sistematics and evolution of European bison). In: Sokolov VE (ed) Zubr. Morfologiya, sitematika, evolutsiya, ekologia, Izdatel'stvo Nauka, Moskva, pp 1–112 (In Russian)

Flint VE, Belousova IP, Pererva VN et al (2002) Strategiya sokhraneniya zubra v Rosii (Strategy od conservation of European bison in Russia). Rosiiskaya Akademiya Nauk, Moskva

Flisiak R, Załężny W, Prokopowicz D (2000) Evaluation of the relationship between humoral immunological response against *Borrelia burgdorferi* and exposure to ticks. Med Weter 56:579–581 (In Polish with English summary)

Floeriche K (1930) Wisent und Elch zwei urige Rechen. Franckh'sche Verlagshandlung Stuttgart

Frankham R, Gilligan DM, Morris DR, Briscoe DA (2001) Inbreeding and extinction: effects of purging. Conserv Genet 2:279–284

Fuller WA (1959) The horns and teeth as indicator of age in bison. J Wildlife Manage 23:342–344

Gambaryan PP, Sukhanov VB (1979) K biomekhanike dvigatel'nogo apparata zubra (Biomechanics of motoric system in European bison). In: Sokolov VE (ed) Zubr. Morfologiya, sitematika, evolutsiya, ekologia, Izdatel'stvo Nauka. Moskva, pp 128–141

Gątkiewicz T, Zbrożek P (eds) (2003) Leśny Kompleks Promocyjny Puszcza Białowieska (Białowieża Forest). Media Art, Anna Włodarczyk, Białystok, Bydgoszcz

Geysztor J (1938) Żubr padł (Death of European bison). Łow pol 5:86–87

Gębczyńska Z, Gębczyński M, Martynowicz E (1991) Food eaten by free-living European bison. Acta Theriol 36:307–313

Gębczyńska Z, Kowalczyk J, Krasińska M et al (1974) A comparison of the digestibility of nutrients by European bison and cattle. Acta Theriol 19:283–289

Gębczyńska Z, Krasińska M (1972) Food preferences and requirements of the European bison. Acta Theriol 17:105–117

Gębczyński M, Tomaszewska-Guszkiewicz K (1987) Genetic variability in the European bison. Biochem Syst Ecol 15:285–288

Gill J (1968) The capacity and weight of the walls and digesta of the alimentary tract in the European bison. Acta Theriol 31:499–509

Giergiel J (ed) (1985-1986) European Bison Pedigree Book 1985–1986. PWN, Warszawa

Gill J (1989) Seasonal changes in the red blood cell system in the European bison (*Bison bonasus*). Comp Biochem Physiol 92A:291–298

Gill J (1990) Seasonal cyclicity in carbohydrate metabolism parameters in the European bison, *Bison bonasus* L. Comp Biochem Physiol 96A:435–439

Gill J (1992a) Seasonal changes in the white blood cell count and blood cell sedimentation rate in the European bison *Bison bonasus*. Acta Theriol 37:279–290

Gill J (1992b) Seasonal changes in activity of some enzymes in the European bison *Bison bonasus*. Acta Theriol 37:291–300

Gill J (1995) Serum lysozyme in the European bison, *Bison bonasus* (L.). Comp Biochem Physiol 110B:235–240

Gill J (1999) Zarys fizjologii żubra (Overwiev of European bison physiology). Wyd Severus, Warszawa

Gill J (2002) Skutki restytucji żubra, po 70 latach hodowli, na tle innych gatunków zagrożonych homozygotycznością (Effects of European bison restitution after 70 years of breeding, as compared to other species endangered with homozygosity). Kosmos 51:483–489 (In Polish)

Gill J (2006) Physiological results of the restitution of the European bison after 70 years of breeding at the present danger from the homozygosity. In: Kita J, Anusz K (eds) Health threats for the European bison particularly in free-roaming populations in Poland. The SGGW Publishers, Warszawa, pp 259–266

Glunz R (2008) History, present situation and future prospects of the European bison breeding centre in Hardehausen. In: Abstracts 1st international european bison congress, Hardehausen, 9–12 Sept 2008

Głowaciński Z (ed) (1992) Polska Czerwona Księga Zwierząt (Polish Red Book). PWRiL, Warszawa (In Polish)

Głowaciński Z (ed (2002) Czerwona Lista Zwierząt Ginących i Zagrożonych w Polsce (Red List of Endangered and threatened Animals in Poland). Oficyna Wyd Text, Kraków (In Polish)

Graczyk R (1987) European bison, *Bison bonasus bonasus* (Linnaeus, 1758) in Wielkopolska. Wyd A R w Poznaniu 173

Graczyk R, Sikora S, Wygralak W (1978) Żubr *Bison bonasus bonasus* (Linnaeus 1758) w lasach Nadleśnictwa Waliły (European bison Bison bonasus bonasus (L.) in Foret Dirtrict Waliły). Rocz A R w Poznaniu 24:67–71

Gralak B, Krasińska M, Niemczewski C et al (2004) Polymorphism of bovine microsatellite DNA sequences in the lowland European bison. Acta Theriol 49:449–456

Grochmalnicki J (1933) Results of ten years' bison-breeding in Poland. Polski Oddział MTOZ, Poznań (In Polish with English summary)

von der Groeben G (1932) Das Zuchtbuch. Berichte d Internat Ges z Erhaltung d Wisent 5(1):5–50

Grzegrzółka B, Olech W, Krasiński ZA (2004) The genetic structure of Lowland European bison free ranging herds in Poland. Parki nar Rez Przyr 23(4):665–677 (In Polish with English summary)

Gstalter A, Lazier P (1996) Le Bison d' Europe. Mythe et renaissance d'une espčce sauvage. Traces, E. C. Editions, Le Vigan

Guillaume P (2003) O wybawieniu od śmierci cześnika księcia polskiego (About the rescue of the death of Polish Prince). In: Gall Anonim Kronika polska (trans: Grodecki R). Introduction and working out M. Plezia. Ossolineum, DeAgostini, Wrocław, pp 167–169

Halloran AF (1960) American bison weights and measurements from the Wichita Mountains Wildlife Refuge. In: Proceedings of Oklahoma academy sciences, vol 41. pp 212–218

Hartl GB, Pucek Z (1994) Genetic depletion in the European bison (*Bison bonasus*) and the significance of electrophoretic heterozygosity for conservation. Conserv Biol 8:67–174

Hayward AW, Kowalczyk R, Krasiński ZA et al (2011) Restoration and intensive management have no effect on evolutionary strategies. Endangered Species Res 15:53–61. doi:doi:10.3354/esr00371

Gusarov I (2011) Condition of population of European bison in northern region of Russia. Eur Bison Conserv Newslett 4:117–120

Hedemann O (1939) Dzieje Puszczy Białowieskiej w Polsce przedrozbiorowej (w okresie do 1798) (History of Białowieża Forest in Poland before the partition (until 1798). IBL Rozpr sprawozd Series A 41, Warszawa

Hedrick PW (1994) Evolutionary genetics of the major histocompatibility complex. Am Nat 143:945–964

Heptner VG, Nasimovich AA, Bannikov AG (1966) Die Sägetiere der Sowietunion. I Paarhufer und Unpaarhufer. G. Fischer Verlag, Germany

Hofmann RR (1978) Die Stellung europeischän Wildwiederkäuer im System des Assungstypen. In: Hofmann RR (ed) Wildbiologische information für den Jäger, Jagd + Hege Verlag St.Gallen

Hofman E (2008) Wykorzystanie przez żubry (*Bison bonasus*) terenów otwartych przyległych do Puszczy Białowieskiej i Puszczy Knyszyńskiej w latach 2000-2007 (Utilisation of open areas adjoining the Białowieża Forest and Knyszyńska Forest in 2000–2007). M.Sc. Thesis. Lublin, Catholic University of Lublin (In Polish)

Hofman-Kamińska E, Kowalczyk R (2010) What are bison searching for in agricultural fields? In: Kowalczyk R, Ławreszuk D, Wójcik JM (eds) European bison conservation in the Białowieża Forest. Threats and prospects of the population development. MRI PAS Białowieża, pp 135-146 (In Polish with English summary. http://krainazubra.pl/files/ochrona_zubra.pdf)

Hofman-Kamińska E, Kowalczyk R (2012) Farm crops depredation by European bison (Bison bonasus) in the vicinity of forest habitats in northeastern Poland. Environ Manage 50:530–541. doi:10.1007/s00267-012-9913-7

Holodova MV, Belousova IP (1989) Potreblenie i usvoenie pitatel'nykh veshchestv i energii zubrami (*Bison bonasus*) (Digestibility of nutrients by European bison (*Bison bonasus*). Zool Zhurnal 68:107–117 (In Russian)

Hoppe S (1981) Słownik języka łowieckiego (Hunting dictionary). PWN, Warszawa [In Polish]

Hussowski M (1523) Pieśń o Zubrze jego postaci, dzikości i o polowaniu na niego w przekładzie Jana Kasprowicza (Song of European bison). Białowieski Park Narodowy (1994), Białowieża (In Polish)

Izdebska JN (1999) Pasożyty zewnętrzne żubrów w stadzie wolnym w Puszczy Białowieskiej (External parasites of European bison in free-ranging population in Białowieża Forest). In: Abstracts of conferenceon Międzynarodowa konferencja naukowa Ochrona zdrowia żubra w stanie wolnym w Puszczy Białowieskiej, Warszawa, 26–27 Nov 1999

Izdebska JN (2000) Stawonogi pasożytnicze żubra jako potencjalny wektor patogenów (European bison arthropods as a potential vector of pathogens). In: Buczek A, Błaszak C (eds) Stawonogi pasożytnicze i alergogenne. Wyd KGM, Lublin, pp 57–64

Izdebska JN (2001a) European bison arthropods from closed polish breeding facilities. Acta Parasit 46:135–137

Izdebska JN (2001b) The occurence of parasitic arthropods in two groups of European bison in the Białowieża Primeval Forest. Wiad Parazyt 47:801–804

Izdebska JN (2001c) (Parasitic arthropods in bison from Bieszczady Moutains). Scientific Messenger of Lviv State Academy of Veterinary Medicine named after S.Z. Gzhytskyj 3(2):208–211

Izdebska JN (2003) *Bisonicola sedecimdecembrii (Mallophaga, Trichodectidae)* wszoł, który przetrwał? (Biting louse, specific parasite of European bison *Bisonicola sedecimdecembrii (Mallophaga, Trichodectidae)*. In: Buczek A, Błaszak C (eds) Stawonogi i żywiciele]. Liber, Lublin, pp 105–115 (In Polish)

Izdebska JN (2005) Porównanie infestacji stawonogami pasożytniczymi żubrów ze stad wolnych i hodowli w Polsce (Comparison of inferstation by parasitic arthropods of European bison from free-living populations and captivity in Poland). In Olech W (ed) Ochrona żubrów zachodniopomorskich,Wydawca ZODR, Barzkowice, Ińsko, pp 38–42 (In Polish)

Izdebska JN (2006a) Studies on parasitic arthropods of European bison (*Bison bonasus*) in Poland. In: Kita J, Anusz K, (eds) Health threats for the European bison particularly in free-roaming populations in Poland, The SGGW Publishers, Warszawa, pp 139–140

Izdebska JN (2006b) Skin mites (Acari: Demodecidae, Psoroptidae, Sarcoptidae) in the European bison, *Bison bonasus*. Biol Lett 43:169–174

Izdebska JN (2011) *Bisonicola sedecimdecembrii* (Phthiraptera: trochodectodae] from European bison – redescription of adults and description of juvenile stages. Entomologica Fennica 22:69–77

Izdebska JN, Cydzik K (2010) Analysis of the reasons for differences in topical specificity among various species of tick (Acari, Ixodidae) infesting European bison. Eur Bison Conserv Newslett 3:75–84

Jakob W, Schröder HD, Rudolph M et al (2000) Necrobacillosis in free-living males European bison in Poland. J Wildlife Dis 36:248–256

Jaczewski Z (1958) Reproduction of the European bison, *Bison bonasus* (L.) in reserves. Acta Theriol 1:333–376

Jaczewski Z (1959) Spostrzeżenia z zakresu opieki wetererynaryjnej w rezerwatach żubrów w latach 1952–1954 (Veterinary care of the European bison in reserves in 1952–1954). Roczn Nauk roln. Series E 69:297–318 (In Polish)

Jaczewski Z, Korona Z (1994) Bibliography of the genus Bison (H.Smith, 1827) (until 1950). Polska Akademia Nauk, Instytut Genetyki i Hodowli Zwierząt, Warszawa

Janicka D, Hołubczat E, Karnecki R et al (2008) Restitution of the European bison *Bison bonasus* (L.1758) as a component of active conservation within Natura 2000 sites of Western Pomerania. Eur Bison Conserv Newslett 1:132–139 (In Polish with English summary)

Jarocki FP (1830) O Puszczy Białowieskiej i o celniejszych w niej zwierzętach (About Białowieża Forest and its more important animals). Komisja Rządowa Wyznań Religijnych i Oświecenia Publicznego, Warszawa (In Polish)

Jarońcki A, Pepera W (1969) Utworzenie ośrodka hodowli żubrów w Bieszczadach (Creation of European biosn breeding center in Bieszczady). Post restyt żubra. Państ Wyd Nauk, Warszawa, pp 193–204 (In Polish)

Jaroszewicz B, Pirożnikow E (2008) Diversity of plant species eaten and dispersed by the European bison *Bison bonasus* in Białowieża forest. Eur Bison Conserv Newslett 1:14–29

Jędrzejewska B, Jędrzejewski W (1998) Predation in vertebrate communities. The Białowieża Primeval Forest as a case study. Ecological Studies 135. Springer, Berlin

Jędrzejewska B. Okarma H, Jędrzejewski W et al (1994) Effect of exploitation and protection on forest structure, ungulate density and wolf predation in Białowieża Primeval Forest. J Appl Ecol 31:664–675

Jickeli K (1927) Die prahistorischen und historischen Wisentvorkommen in Siebenbürgen. Berichte d Internat Ges z Erhaltung d Wisent 2:163–170

Joly DO, Leighton FA, Messier F (1997) Tuberculosis and brucellosis infection of bison in Wood Buffalo National Park, Canada: Preliminary results. In: Proceedings of international symposium on bison ecology and management in North America, Montana, 4-7June, 1997

Jurkevich ID, Geltman WS (1965) Geografiya, typologiya i ratsionizatsiya lesnoi rastitelnosti Belorusi (Geography and typology of foret vegetation of Belarus). Nauka Tekhnika, Minsk (In Russian)

Kadulski S, Izdebska J (1996) *Demodex bisonianus* sp. nov. (*Acari Demodicidae*) a new parasite of the bison (*Bison bonasus L.*). Wiad parazyt 42:103–110

Kalugin CG (1968) Vostanovleniye zubra na severo-zapadnom Kavkaze (Restoration of European bison in north-western Caucasus). Trudy Kavkazskogo Zapovednika 10:3–94

Kamiński T, Hofman-Kamińska E, Kowalczyk R (2010) European bison diet in Białowieża Forest. In: Abstracts of conference VIII Miedzynarodowa Konferencja 'Zubr w Puszczy Królewskiej', Stowarzyszenie Miłośników Żubrów i Nadleśnictwo Niepołomice. Niepołomice 9–10 września 2010 (In Polish with English summary)

Kamiński S, Zabolewicz T (1997) DDE I restriction fragment length polymorphism in the *Bison bonasus* untranslated region of kappa-casein gene. Act Acad Agricult AC Technicae Olstenensis, Zoot 47:147–153

Kamiński S, Olech W, Oleński K et al (2012) Single nucleotide polymorphisms between two lines of European bison (*Bison bonasus*) detected by the use of Illumina Bovine 50 K BeadChip. Conserv Genet Resour 4(2):311–314. doi:10.1007/s12686-011-9535-z

Kania BF, Teuchmann JK, Piwowarczyk S et al (1973) Badania nad immoblilizacyjnym działaniem chlorowodoru etorfiny (M 99) na *Bison bonasus* oraz antagonizowanie tego działania chlorowodorkiem cyprenorfiny (M 285) (Sudies on an immobilise effect of etorphine hydrochloride (M99) and antidote cyprenorphine (M 285). Prz Zool 17:242–247 (In Polish)

Karcov G (1903) Belovezhskaya Pushcha (Białowieża Forest). AF Marks, St. Petersburg (In Russian)

Kaplan LO, Krumhardt KM, Zimmermann N (2009) The prehistoric and preindustrian deforestration of Europe. Quaternary Sci Rev 28:3016–3034

Kashtalian A (2009) Formation of Borisov–Berezinsky European bison population and its state. Eur Bison Conserv Newslett 2:17–29

Kaz'min VD, Smirnov KA (1992) Zimnee pitanie, kormovye resursy i troficheskoe vosdeistvie zubra na lesnye fitocenozy central'nogo Kavkaza (Winter feeding, food abundance and European bison influence upon the forest habitats of central Caucasus). Biuletin Moskovskogo obshchestva ispitatelei prirody. Odl Biol 97 (2):26–35 (In Russian)

Kelterborn T, Zenter F, Zacharias SK (2009) 52 years of European bison breeding on the Wisent-Island in the hart of Meklemburg-Vorpommern. Eur Bison Conserv Newslett 2:172–181

Kerley GIH, Kowalczyk R, Cromsight PGM (2012) Conservation implication of the refugee species concept and the European bison: king of the forest or refugee in a marginal habit? Ecography 35:519–529. doi:10.1111./j.1600-0587-2011.07146.x

Kingston N, Dróżdż J, Rutkowska M (1987) Occurrence of *Trypanosoma* in free living ruminants in Poland. Wiad parazyt 33:219–220 (In Polish with English summary)

Kingston N, Dróżdż J, Rutkowska M et al (1992) Redescription of *Trypanosoma (Megatrypanum) wrublewski* Wladimiroff et Yakimoff 1909 from the European bison, *Bison bonasus* L., from Puszcza Białowieska (Poland). Acta Parasit 37:163–168

Kiseleva EG (1969) Breeding of the Caucasus-Białowieża European bison in the Okski Reservation. Postępy Restytucji Żubra In: Proceedings of conference on Konferencja Polsko-Radziecka. Białowieża-Kamieniuki, 18–21 April 1967 (In Polish and Russian with English summary)

Kiseleva EG (1974) Razvedenie zubrov v pitomnike Okskogo Zapovednika (Breeding of European bison in Oksi Reserve). Berezin Zapoved Issled 3:103–138 (In Russian)

Kiseleva EG (1986) Zubrovoi pitomnik v Okskom Zapovednike: Sovremennoe sostoyanie i problemy. In Abstracts of conference on Pervoe vsesoiuznoe soveshchane po problemam zookul'tury. Tezisy dokladov, chast' II. (In Russian)

Kita J, Anusz K (eds) (2006a) Health threats for the European bison particularly in free-roaming populations in Poland. The SGGW Publishers, Warszawa

Kita J, Anusz K (2006b) Infectious diseases in *Bison bonasus* in the years 1910–1988. In: Kita J, Anusz A (eds) Health threats for the European bison particularly in free-roaming populations in Poland. The SGGW Publishers, Warszawa, pp 17–26

Kita J, Anusz K, Zaleska MM et al (2003) Realtionship among ecology, demography amd diseases of European bison (*Bison bonasus*). J Vet Sci 6(4):261–206
Kita J, Dziąba K, Piusiński W et al (1990) (A disease of genital organ of free-roaming male of the European bison in the Białowieża Primeval Forest, Polad). Med Weter 44:474–476 (In Polish with English summary)
Klevezal G, Pucek Z (1987) Growth layers in tooth cement and dentine of European bison and its hybrids with domestic cattle. Acta Theriol 32:115–128
Kobryń H (2006) Anatomy of European bison. In: Kita J, Anusz K (eds) Health threats for the European bison particularly in free-roaming populations in Poland. The SGGW Publishers, Warszawa, pp 293–299
Kobryńczuk F (1976) Joins and ligaments of hind-limbs of the European bison in its postnatal development. Acta Theriol 21:37–100
Kobryńczuk F (1985) The influence of inbreeding on the shape and size of the skeleton of the European bison. Acta Theriol 30:379–422
Kobryńczuk F (1999) Białowieski Skrzat Królewski (The Białowieża royal elf). Ortdruk, Białystok (In Polish)
Kobryńczuk F (2000) Żubr w "Pieśni" Mikołaja Hussowskiego i w polskich opracowaniach morfologicznych (cz. II) (European bison in song of Mikolai Hussowski and in morphological studies). Agricola, Pismo SGGW 47:16–19 (In Polish)
Kobryńczuk F, Krasińska M (1987) Taxonomic studies on skulls of European bison and domestic cattle. Acta Theriol 32:203–218
Kobryńczuk F, Krasińska M (1991) Shape of the mandible in European bison and domestic cattle hybrids. Acta Theriol 36:171–177
Kobryńczuk F, Krasińska M, Szara T (2008a) Polarization of skull shapes in adult Lowland European bison, *Bison bonasus bonasus*. Ann Zool Fenn 45:341–346
Kobryńczuk F, Krasińska M, Szara T (2008b) Sexual dimorphism in skulls of the lowland European bison, *Bion bonasus bonasus*. Ann Zool Fenn 45:335–340
Kobryńczuk F, Roskosz T (1980) Correlation of skull dimensions in the European bison. Acta Theriol 25:349–363
Koch W (1956) Jahreszeitlische Schwankungen der Fruchtbarkeit bei Wildrindern. Fortpflanzung, Zuchthygiene und Haustierbesamung 6:85–87
Kochko Ju (2003) The results of the study on the fauna of helminths of Belovezhskaya Pushcha in 20th century. Belov Pushcha Issled 11:205–223 (In Russian with English summary)
Korochkina LN (1969a) O utworzeniu wolnego stada żubrów białowieskich na terenie Puszczy Białowieskiej (Restoration of free-ranging populationof European bison in estern part of Białowieża Forest). Postępy Restytucji Zubra. In Proceedings of conference: III Konferencja Polsko-Radziecka, Białowieża-Kamieniuki, 18–21 April 1967 (In Polish)
Korochkina LN (1969b) O sozdaii volnogo stada zubrov v Belovezhskoi Pushchi (Restoration of free-ranging population of European bison in estern part of Białowieża Forest). Belov Pushcha Issled 5:28–40 (In Russian)
Korochkina LN (1969c) Drevesnaya rastitel'nost' v pitanii zubrov Belovezhskoi Pushchi (Woody plants in feeding of the European bison in Białowieża Forest). Belov Pushcha Issled 2:120–126 (In Russian)
Korochkina LN (1969d) Vidovoi sostav lesnoi travyanistoi rastitel'nosti v pitanii zubrov Belovezhskoi Pushchi. Belov Pushcha Issled 2:204–223 (In Russian)
Korochkina LN (1971a) Kormovaya baza i nekotorye itogi reaklimatyzatsii zubrov v Belovezhskoi Pushchi (Food resources and adaptation of European bison to the conditions of eastern part Białowieża Forest). Belov Pushcha Issled 5:164–176 (In Russian)
Korochkina LN (1971b) Vzaimootnosheniya zubrov s drugimi kopytnymi (Relationships between European bison and other ungulates). Belov Pushcha Issled 5:176–184 (In Russian)
Korochkina LN (1972a) Sutochnyi ritm aktivnosti zubrov Belovezhskoi Pushchi (Daily activity rhythm of European bison in Białowieża Forest). Belov Pushcha Issled 6:125–131 (In Russian)
Korochkina LN (1972b) Travjanistaya rastitel'nost' v pitanii zubrov Belovezhskoi Pushchi (Monocotyledonous plants in feeding of European bison in Białowieża Forest). Belov Pushcha Issled 6:110–124 (In Russian)

Korochkina LN (1973) Rayon obitaniya i statsional'noe razmieshchenie zubrov v Belovezhskoi Pushchi (Population home range and spatial distribution of European bison in Białowieża Forest) Belov Pushcha Issled 7:148–165 (In Russian)

Korochkina LN (1974) Voprosy reaklimatyzatsii zubrov i emkost' kormovych ugodii Belovezhskoi Pushchi (Adaptation of European bison and food abundance in Białowieża Forest). Belov Pushcha Issled 8:118–133 (In Russian)

Korochkina LN (1975) Metodika ucheta i dinamika chislennosti vol'nozhivushchikh zubrov Belovezhskoi Pushchi (Methods of counting and dynamic of numbers of European bison in Białowieża Forest). Belov Pushcha Issled 9:91–120 (In Russian)

Korochkina LN (1978) Vliyanie snezhnogo pokrova na osobennosti zimnego pitaniya olen'ikh (Influence of snow cover on the specificity of winter foraging of deer). Zapoved Bel Issled 2:100–110 (In Russian)

Korochkina LN, Bunevich AN (1980) Znachenie kormovych polei v pitanii kopytnykh Belovezhskoi Pushchi (Importance of open areas for the deer feeding). Zapoved Bel Issled 4:109–116 (In Russian)

Korochkina LN, Kochko FP (1982) K voprosu o smertnosti zubrov v estestvennych usloviyakh Belovezhskoi Pushchi (Mortality o f European bison in Białowieża Forest). Zapoved Bel Issled 6:96–103 (In Russian)

Korochkina LN Kochko FP (1983) Dinamika chislennosti vol'nozhivushchich zubrov Belovezhskoi Pushchi (Dynamic of number free-ranging population of European bison in Białowieża Forest). Zapoved Bel Issled 7:52–59 (In Russian)

Kosiba W (1964) Żubry w Bieszczadach (European bison in Bieszczady Mountains). Łow pol 20 (1239):2–3 (In Polish)

Kossak S (2001) The Białowieża Forest saga. Muza SA, Warszawa

Kośla T (1993) The contents of macroand microelements in fodder, blood serum and hair of European bison. Ann War Agric Univ. SGGW-AR Veterinary Medicine 17:79–85

Kośla T, Skibniewski M, Skibniewska E et al (2004) The zinc status in free living Euroepan bisons. Acta Allimentaria 33:269–273

Koulischer L, Tijskens T, Mortelmans J (1967) Mammalian cytogenetics. I. The chromosomes of three species of Bovidae: *Bos taurus, Bison bonasus, Cephalophus grimmi.* Acta Zoologica Pathologica Antverpiensia 43:135–141

Kowalczyk J, Gębczyńska Z, Krasińska M (1976) The digestibility of nutrients of natural diet by European bison in different seasons. Acta Theriol 21:141–146

Kowalczyk R (2010) European bison – king of the forest or meadows and river valleys? In: Kowalczyk R, Ławreszuk D, Wójcik JM (eds) European bison conservation in the Białowieża Forest. Threats and prospects of the population development. MRI PAS, Białowieża, pp 123–134(In Polish with English summary. http://krainazubra.pl/files/ochrona_zubra.pdf)

Kowalczyk R, Krasińska M,Kamiński T et al (2013) Movements of European bison *(Bison bonasus)* beyond the Białowieża Forest (NE Poland): range expansion or partial migrations? Acta Theriol doi:10.1007/s13364-013-0136-y

Kowalczyk R, Ławreszuk D, Niedziałkowski K et al (2010) Guidelines for conservation and management of European bison in Białowieża Primeval Forest. In: Kowalczyk R, Ławreszuk D, Wójcik JM (eds) European bison conservation in the Białowieża Forest. Threats and prospects of the population development. MRI PAS, Białowieża, pp 211–215 (In Polish with English summary)

Kowalczyk R., Schneider TC, Krasiński ZA et al (2008) Influence of winter supplementary feeding on space use by European bison in Białowieża Primeval Forest (Poland). In: Abstracts of 4th international conference on European bison in natura 2000 network. European bison Friends Society, Cisna, 15–16 Sept 2008

Kowalczyk R, Taberlet P, Coissac E et al (2011) Influence of management practices on large herbivore diet-case of European bison in Białowieza Primeval Forest (Poland). Forest Ecol Manag 261:821–828. doi:10.1016/j.foreco.2010.11.026

Kowalski K (1967) The evolution and fossil remains of the European bison. Acta Theriol 12:325–338

Kozlo PG (1999) Reintroduktsya belovezhskogo zubra (*Bison bonasus* L.) v Belarusi (Reintroduction of European bison in Belarus). Belovezhskaya Pushcha na rubiezhe tret'ego tysyachletiya. In: Abstracts of conference: Nauchno prakticheskaya konferentsiya posvyashchenna 60-letiyu so dnia obrazovaniya Gosudarstvennogo zapovednika "Belovezhskaya Pushcha", 22–24 Dec 1999 (In Russian)

Kozlo PG (2001) Zubr Belarusi. Natsyonal'naya programma sokhraneniya i ispol'zovaniya (European bison of Belarus. State program of conservation and management). In Proceedings of conference: Mezhdunarodnaya nauchnaya konferentsiya "Raznoobrazie zhiviotnogo mira Belarusi itogi izucheniya i perspektywy sokhraneniya", Minsk, 29–30 Nov 2001 (In Russian)

Kozlo PG, Bunevich AN (2009) Zubr v Belarusi (The European bison in Belarus). Belaruskaya Navuka, Minsk (Belarus) (In Russian)

Kozlo PG, Bunevich AN, Stavrovskii DD et al (1996) *Bison bonaus* v Belarusi: analiz sostoyaniya populatsii i strategiya okhrany vida (*Bison bonasus* in Belarus, status of populations and strategy of species conservation). In: Proceeding of conference: "Sokhranenie biologicheskogo raznoobraziya lesov Belovezhskoi Pushchi", Kamenyuki 1996 (In Russian)

Kozlo PG, Bunevich AN, Stavrovsky DD et al (1997) The European bison (*Bison bonasus bonasus*) in Belarus. Analysis of its population status and conservation strategy of species. In: Luchkov A. Tolkach V, Berwick S, Brylski Ph (eds) Belovezhskaya Pushcha Forest Biodiversity Conservation. Beloruskii Dom Pechati, Minsk, pp 196–208

Kozlo PG, Kuchmel SV, Emel'yanova LG et al (1999) Otsenka kormovykh uslovii Khajdo-Belevskogo uchastka ELOKH "Lyaskovichi" dla reintroduktsii zubrov (Estimation of food resources of Khajdo-Belevskogo uchastka ELOKH Lyaskovichi for the reintroduvtion of European bison). In: Proceedings of conference: VIII Zoologicheskaya nauchnoaya konferentsiya "Strukturno-funktsyonal'noe sostoyanie biologicheskogo raznoobraziya zhivotnogo mira Belarusi" 1999 (In Russian)

Kozlo PG, Uglyanec AV (1999) Otsenka sostoyaniya pripyatskoi mikropopulatsii zubrov, uslovii obitaniya i puti ich usklucheniya (Status of Pripyat micropopulation of European bison and food resources). Biologicheskoe raznoobrazie Natsyonal'nogo parka "Pripyatskii" i drugikh osobo okhranyaemykh prirodnykh territorii. In: Proceedings of conference: 30-letie Pripyatskogo gosudarstvennogo lanshaftno-gidrologicheskogo zapovednika (1969-1999) Nacjonal'nyi park "Pripyatskii" (1996–1999) (In Russian)

Kozul'ko GA, Zhukov VP (1999) Gosudarstvennyi natsional'nyi park "Belovezhskaya Pushcha"—stareishii zapovednik v Europie. (State National Park Białowieża Forest, the oldest reserve in Europe). Belovezhskaya Pushcha na rubiezhe tret'ego tysyacholetiya]. In: Proceedings of conference: Nauchno-prakticheskaya konferentsiya, posvashchonna 60-letiyu so dnya obrazovaniya zapovednika "Belovezhskaya Pushcha", Minsk, 22–24 Dec 1999 (In Russian)

Krasińska M (1988) Hybrids of European bison and domestic cattle. Prace Hab PAN Zakład Badania Ssaków PAB, Ossolineum (In Polish with English summary)

Krasińska M, Caboń-Raczyńska K, Krasiński ZA (1987) Strategy of habitat utilization by European bison in the Białowieża Forest. Acta Theriol 32:147–202

Krasińska M, Giżejewski Z, Czykier E, Krasiński ZA, Matuszewska M (2009) Changes of weight and size of European bison testes during postnatal development. Acta Theriol 54(2):111–112

Krasińska M, Krasiński ZA (1994) Spatial structure of the European bison population in the Polish part of the Białowieza Forest 1976–1993. Parki nar i Rez Przyr 13(4):69–87 (In Polish with English summary)

Krasińska M, Krasiński ZA (1995) Composition, group size, and spatial distribution of European bison bulls in Białowieża Forest. Acta Theriol 40:1–21

Krasińska M, Krasiński ZA (1997) Utilization of the Białowieża National Park by the European bison. Parki nar i Rez Przyr 16(2):3–25 (In Polish with English summary)

Krasińska M, Krasiński ZA (2000) Functioning of the European bison population in the Białowieża Primeval Forest. In: Proceedings of international symposium on European bison—yesterday, today and tomorrow, Šiauliai, Lithuania, 2–10 Dec 2000

Krasińska M, Krasiński ZA (2002) Body mass and measurements of the European bison during postnatal development. Acta Theriol 47:85–106

Krasińska M, Krasiński ZA (2002a) Plan ochrony fauny Białowieskiego Parku Narodowego. Żubr (Conservation action plan of fauna of Białowieża National Park. European bison). Białowieża National Park, Manuscript (In Polish)
Krasińska M, Krasiński ZA (2007) European bison: the nature monograph. Mammal Research Institute, Białowieża
Krasińska M, Krasiński ZA (2010) Course and dispersion of the posthitis/balanoposthits in males of the European bison in the Polish part of Białowieża Forest. Parki nar Rez Przyr 29(2):107–128 (In Polish with English summary)
Krasińska M, Krasiński ZA, Bunevich AN (1997a) Differentiation of the size of mixed groups of European bison populations depending on the habitats utilized in Białowieża Forest. Parki nar Rez Przyr 16(1):55–56 (In Polish with English summary)
Krasińska M, Krasiński ZA, Bunevich AN (1997b) Functioning of the European bison, *Bison bonasus bonasus* population in Polish and Belarussian parts of the Białowieża Forest in 1976–1995. Chrońmy Przyr 53(3):28–40 (In Polish with English summary)
Krasińska M, Krasiński ZA, Bunevich AN (2000) Factors affecting the variability in home range size and distribution in the Polish and Belarussian part of the Białowieża Forest. Acta Theriol 45:321–334
Krasińska M, Krasiński ZA, Ławreszuk D (2010) Dispersion ot the population and use of open areas by European bison In Białowieża Forest. In: Kowalczyk R, Ławreszuk D, Wójcik JM (eds) Ochrona żubra w Puszczy Białowieskiej. Zagrożenia i perspektywy rozwoju, Zakład Badania Ssaków Polskiej Akademii Nauk, Białowieża, pp 111–122 (In Polish with English summary)
Krasińska M, Szuma E, Kobryńczuk F et al (2008) Morphometric variation of the skull during postnatal development in the Lowland European bison (*Bison bonasus* bonasus). Acta Theriol 53:193–216
Krasiński Z (1963) O żubrach na wolności w Puszczy Białowieskiej (European bison in Białowieża Forest). Chrońmy Przyr 19(1):3–7 (In Polish)
Krasiński Z (1967) Free living European bison. Acta Theriol 12:391–405
Krasiński Z (1968a) Żubr w Puszczy (European bison in Białowieża Forest). In: Falinski JB (ed) Park Narodowy w Puszczy Białowieskiej. Państ Wyd Roln i Leśn, Warszawa, pp 162–174
Krasiński Z (1968b) Rezerwaty zwierzęce (Animals reserves in Białoweiża Forest). In: Faliński JB (ed) Park Narodowy w Puszczy Białowieskiej. Państ Wyd Roln i Leśn, Warszawa, pp 267–272
Krasiński Z (1971) Farmakologiczne poskramianie dzikich kopytnych (Immobilisation of wild ungulates). Prz zool 15:379–387
Krasiński ZA (1976) Niektóre aspekty ekologii żubrów w Puszczy Białowieskiej (Some aspects of European bison ecology in Białowieża Forest). PhD Thesis, Mammal Research Institute PAS, Białowieża (In Polish)
Krasiński ZA (1978a) Dynamics and structure of the European bison population in the Białowieża Primeval Forest. Acta Theriol 23:3–48
Krasiński Z (1978b) Restytucja i ewolucja populacji żubrów w Polsce (Restitution and evolution of Europeasn bison in Poland). Człowiek i Nauka, Wiedza Powszechna, Warszawa, pp 207–229 (In Polish)
Krasiński ZA (1983) The first 10 years of free breeding of bison in Białowieża Foret, 1952–1961. Parki nar i Rez Przyr 4(2):38–50 (In Polish with English summary)
Krasiński ZA (1986) European bison breeding center in Borecka Forest. Parki nar i Rez Przyr 7(3):73–79 (In Polish with English summary)
Krasiński ZA (1990) Stan hodowli żubra w Polsce (Situation of European bison in Poland). In: Pucek Z (ed) Ekspertyza o stanie, perspektywach i potrzebach hodowli i restytucji żubra w Polsce. Manuscript (In Polish)
Krasiński ZA (1993) Bison a relict of ancient times. Wyd BPN, Białowieża
Krasiński ZA (1994a) Restitution of the European bison in the Białowieża reserve in the years 1929–1952. Parki nar i Rez Przyr 13(4):3–23 (In Polish with English summary)
Krasiński ZA (1994b) Zagrożenia populacji żubrów w Puszczy Białowieskiej (Threats to European bison population in Białowieża Forest). Łowiec Polski 8:6–7 (In Polish)

Krasiński ZA (1994c) Walki żubrów (Fights of European bison males). Łow Pol 9:22–23 (In Polish)

Krasiński ZA (1996a) European bison breeding center in Pszczyna. Parki nar i Rez Przyr 15(3):25–42 (In Polish with English summary)

Krasiński ZA (1996b) Polany karmowe w Puszczy Białowieskiej (Clearings and meadows of Białowieża Forest). Parki nar 4:20–21 (In Polish)

Krasiński ZA (1997) Perspektywy wolnej hodowli żubrów w Puszczy Knyszyńskiej (Perspectives of growth of European bison breeding in Knyszyn Forest). Biul inf Wydz Ochron Środ Urzędu Wojew Białymstok 1(17):4–7 (In Polish)

Krasiński ZA (1999a) The European bison forest emperor. Białowieski Park Narodowy, Białowieża

Krasiński ZA (1999b) The European bison show reserve. Białowieski Park Narodowy, Białowieża

Krasiński ZA (2000) The role of large herbivores in shaping the nature reserves in the Netherlands. Parki nar i Rez Przyr 19:107–110 (In Polish with English summary)

Krasiński ZA (2010) (European bison and his close relative from North America). In: Kowalczyk R, Ławreszuk D, Wójcik JM (eds) European bison conservation in the Białowieża Forest. Threats and prospects of the population development. MRI PAS, Białowieża, pp 85–92 (In Polish with English summary. htp://krainazubra.pl/files/ochrona_zubra.pdf)

Krasiński ZA, Bunevich AN, Krasińska M (1994a) Characteristics of the European bison populations in the Polish and Belarussian part of the Białowieża Forest. Parki nar i Rez Przyr 13(4):25–67 (In Polish with English summary)

Krasiński ZA, Caboń-Raczyńska K, Krasińska M (1982) Immobilizing and marking of the European bison. Acta Theriol 27:181–190

Krasiński ZA, Krasińska M (1992) Free ranging European bison in Borecka Forest. Acta Theriol 37:301–317

Krasiński ZA, Krasińska M (1993) Stosowanie telemetrii w badaniach nad żubrami w Puszczy Białowieskiej (The use of telemetry in studies of European bison). Parki nar 4:12 (In Polish)

Krasiński ZA, Krasińska M (1994a) Ekologia populacji zubrov v zapadnoi chasti Belovezhskoi Pushchi (Ecology of European bison on western part of Białowieża Forest). In: Proceedings of conference: "Sovremennye problemy okhrony zubra", Minsk, 26–27 Mart 1992 (In Russian)

Krasiński ZA, Krasińska M (1994b) Performance of European bison population in the Borecka Forest. Park nar i Rez Przyr 13(4):89–106 (In Polish with English summary)

Krasiński ZA, Krasińska M (1997) The European bison breeding centre in Smardzewice. Parki nar i Rez Przyr 16(1):47–53 (In Polish with English summary)

Krasiński ZA, Krasińska M (2006) Threats to free-ranging populations of the European bison in the Białowieża Primeval Forest, Poland on the background of current state of species. In: Kita J, Anusz K (eds) Health threats for the European bison particularly in free-roaming populations in Poland. The SGGW Publishers, Warszawa, pp 183–191 (In Polish with english summary)

Krasiński ZA, Krasińska M, Bunevich AN (1999) Free ranging population of lowland European bison in the Białowieża Forest. Parki nar i Rez Przyr 18(4):23–75

Krasiński ZA, Krasińska M, Leniec H (1994b) The European bison in the Knyszyn Forest. Parki nar i Rez Przyr 13(4):107–114 (In Polish with english summary)

Krasiński ZA, Olech W, Pezanowski K, Bielecki W, Bereszyńaski A (2010) Plan ochrony żubra (*Bison bonasus*) w Białowieskim Parku Narodowych (European bison conservation action palan in Białowieża National Park). Białowieża National Park, Białowieża, Manuscript (In Polish)

Krasiński Z, Raczyński J (1967) The reproduction biology of European bison living in reserves and freedom. Acta Theriol 12:407–444

Krasiński Z, Raczyński J (1969) Z badań nad rozmnażaniem się żubrów w polskich rezerwatach hodowlanych i na wolności w Puszczy Białowieskiej (The breeding of European bison in captivity and freedeom). Post Rest Żubra 1:203–231 (In Polish)

Krasochko PA, Krasochko IA, Shashenko AS, Kochko YuP, Gaevisky VI, Bunevich AN (1997) Propagation of infectious diseases in European bison in Belovezhskaya Pushcha. In: Luchkov A,

Tolkach V, Berwick S, Brylski P (eds) Belovezhskaya Pushcha Forest biodiversity conservation. Beloruskii Dom Pechati, Minsk, pp 209–214

Krasochko PA, Kozlo PG, Krasochko IA, Yakubovskij MV, Bunevich AN, Kochko YP (2004) Ekologicheskie i veterinarnye aspekty zubrov v Belarusi (The ecological and veterinary aspects of European bison in Belarus). Biznesofset, Minsk (In Russian)

Krawczyński W (1947) Łowiectwo. Podręcznik dla leśników i myśliwych (Haunting textbook for foresters and hunters). Wyd Spółdzielni "Las", Warszawa (In Polish)

Krischke N (1984) Beiträge zum Soicialverhaltung des Wisents (*Bison bonasus bonasus* L.). Zeitschrift für Sägetierkunde 49:43–54

Krupka J (ed) (1990) Łowiectwo (Hunting). Państ Wyd Roln i Leśn, Warszawa (In Polish)

Krysiak K (1951a) Więzadło karkowe, *ligamentum nuchae* żubra (*Bison bonasus*) (*Ligamentum nuchae* of European bison). Fol morph 2:271–283 (In Polish)

Krysiak K (1951b) Znalezisko dwóch czaszek prażubra—*Bison priscus* Bojan,. z ziem polskich (The skull remains of *Bison priscus* Bojan. from Poland). Wiad archeol 18:185–199 (In Polish)

Krysiak K (ed) (1974) European bison pedigree book. PWN, Warszawa

Krysiak K, Świeżyński K (1967) The present state of research on the morphology of the European bison. Acta Theriol 12:323–331

Kuchmel SV, Emel'yanova LG, Deryabina TG (1999) Sravnitel'nyi analiz sostava kormov zubra v natsional'nych parkakh: pripyatskii (Comparative analysis of the composition of the diet of European bison in the national parks: pripyatskii) Belovezhskaya Pushcha na rubiezhe tret'ego tysyachletiya. In: Abstracts of conference: Nauchno-prakticheskaya konferentsiya posvyashchenna 60-letiyu so dnia obrazovaniya Gosudarstvennogo zapovednika "Belovezhskaya Pushcha", 22–24 Dec 1999 (In Russian)

Kuemmerle T, Radeloff VC, Perzanowski K et al (2012) Predicting potential European bison habitat across its former range. Ecol Appl 21:830–843. doi:10.1890/10-0073.1

Kulagin NM (1919) Zubr Belovezhskoi Pushchi (European bison of the Białowieża Forest). Izdatelstvo Moskovskogo Nauchnogo Instituta, Moskva (In Russian)

Lavov MA, Voronova TN (1982) Zubry v Berezinskom Zapovednike i prilegayushchikh teritoriyach (European bison in Bernskii Reserve and surroundings). Zapoved Bel 6:109–112 (In Russian)

Lehnen A, Busse HJ, Frölich K et al (2006) *Arcanobacterium bialowiezense* sp. Nov. and *Arcanobacterium bonasi* sp. nov. isolated from the prepuce of European bison (*Bison bonasus*) suffering from balanoposthitis and emended description of the genus *Arcanobacterium*, Collins et al 1983. Int J Syst Evol Micr 56:861–866. Doi.10, 1099/ijs.063923-0

Lisle GW, Mackintosh CG, Bengis RG (2001) *Mycobacterium bovis* in free-living and captive wildlife, including farmed deer. Revue Scientifique et Technique de l'Office International des Epizooties 20:86–111

Luenser K, Fickel J, Lehnen A et al (2005) Low level of genetic variability in European bison *Bison bonasus* from the Białowieża National Park in Poland. Eur J Wildlife Res 51:84–87. doi:10.1007/s10344-005-0081-4

Lynch M, Walsh B (1998) Genetics and analysis of quantitative traits. Sinauer, Sunderland

Ławreszuk D, Kowalczyk R, Wójcik JM (2010) Project LIFE "European bison conservation in the Białowieża Forest—Bison Land"—activities, Project results and ecological effects. In: Kowalczyk R, Ławreszuk D, Wójcik JM (eds) European bison conservation in the Białowieża Forest. Threats and prospects of the population development. MRI PAS, Białowieża, pp 5–22 (In Polish with English summary. http://krainazubra.pl/files/ochrona_zubra.pdf)

Łukaszewicz K (1952) Tur (The aurochs). Ochrona Przyr, Rocz Komit Ochr Przyr Pol Akademii Umiejętności 20: 1-32. [In Polish]

Mackintosh C, Haigh JC, Griffin F (2002) Bacterial diseases of farmed deer and bison. Revue Scientifique et Technique de l'Office International des Epizooties 21:249–263

Matuszewska M, Olech W, Bielecki W et al (2004) The influence of inbreeding on the pathological changes occurence in European bison male reproduction tract. Parki nar i Rez Przyr 23(4):685–679 (In Polish with English summary)

Matuszewska M, Sysa P (2001) Epididymal defects in European bison. Folia Morph 60:145
Matuszewska M, Sysa P (2002) Epididymal cysts in European bison. J Wildl Dis 36:637–640
McDonald JN (1981) North American bison. Their classivication and evolution. Univeristy of California Press, London
McHugh T (1972) The time of the buffalo. Alfred A. Knopff, New York
Melander Y (1959) The mitotic chromosomes of some cavicorn mammals (*Bos taurus* L., *Bison bonasus* L. and *Ovis aries* L.). Hereditas 45:649–664
Mizin IA (1999) Landshaftnye predpochteniya v vybore mestoobitanii evropeiskogo zubra (*Bison bonasus* L.) pri reklimatyzatsii (na primiere zapovednika "Briyanskii les") (The choice of place of reintroduction of Europeasn bison on example of Briyanskii Forest). Belovezhskaya Pushcha na rubiezhe tret'ego tysyachletiya. In: Abstracts of conference: Nauchno-prakticheskaya konferentsiya posvyashchenna 60-letiyu so dnia obrazovaniya Gosudarstvennogo zapovednika "Belovezhskaya Pushcha", 22–24 Dec 1999 (In Russian)
Mohr E (ed) (1933) Das Zuchtbuch. Berichte d Internat Ges z Erhaltung d Wisent 5(2):53–60
Mohr E (ed) (1937) Das Zuchtbuch. Berichte d Internat Ges z Erhaltung d Wisent 5(3):61–79
Mohr E (1951) L'elevage du bison d'Europe au cours des derniers annes son etat actuel. Mammalia 15:138–142
Mohr E (1952) Der Wisent. Die Neue Brehm-Bücherei, Leiptzig, p 74
Mysterud A, Bartoń KA, Jędrzejewska B et al (2007) Population ecology and conservation of endangered megafauna: the case of European bison in Białowieża Primeval Forest, Poland. Animal Conserv 10:77–87. doi:10.1111/j.1469-1795.2006.00075.x
Nahlik A (1991) W sprawie transylwańskiego żubra (Transylvanian European bison). Sylwan 85:63–68 (In Polish)
Nei M, Maruyama T, Chakraborty R (1975) The bottleneck effect and genetic variability in populations. Evolution 29:1–10
Nemtsev AS, Rautian GS, Puzarenko A, Yu et al (2003) Zubr na Kavkaze (The European bison in Caucasus). In: Rautian GS (ed) Kachestvo, Moskva-Maikop (In Russian)
Niezabitowski-Lubicz E (1937) Szczątki kopalne żubra (*Bison bonasus* L na ziemiach polskich.) (The bones remains of European bison in Poland). Rocz Nauk Roln Leśn 36:1–44 (In Polish]
Okarma H (2000) Ryś (Lynx). Oficyna Edytorska Wyd "Swiat", Warszawa [In Polish)
Okołów C (1966) Kiedy padł ostatni żubr w Puszczy Białowieskiej? (When the last European bison was killed in Białowieża Forest?) Chrońmy Przyr 22:28–30 (In Polish)
Olech W (1987) Analysis of inbreeding in European bison. Acta Theriol 32:373–387
Olech W (1989) The participation of ancestral genes in the existing population in European bison. Acta Theriol 34:397–407
Olech W (1998) The inbreeding of European bison (*Bison bonasus* L.) population and its influence on viability. In: Abstracts of 49th EAAP meeting, Warszawa
Olech W (2003) (Influence of individual inbred and mother's inbred on calves survival in European bison *(Bison bonasus)*. Rozpr Nauk Mon, Wyd SGGW. Warszawa (In Polish and English summary)
Olech W (2006) The analysis of European bison genetic diversity using the pedigree data. In: Kita J, Anusz K (eds) Health threats for the European bison particularly in free-roaming populations in Poland. The SGGW Publishers, Warszawa, pp 205–210
Olech W (2009) The changes of founders number and their contribution to the European bison population during 80 years of species restitution. Eur Bison Conserv Newslett 2:54–60
Olech W, Perzanowski K (2002) A genetic background for reintroduction program of of the European bison *(Bison bonasus)* in the Carpathians. Biol Conserv 108:221–228
Olech W, Perzanowski K (2007) Strategia ochrony żubrów (*Bison bonasus)* w Polsce (Strategy of protection of European bison (*Bison bonasus*) in Poland). Ministerstwo Środowiska. Warszawa, Manuscript (In Polish)
Olszewski JL (1986) Rola ekosystemów leśnych w modyfikacji klimatu lokalnego Puszczy Białowieskiej. Prace Habilitacyjne, Polska Akademia Nauk Zakład Badania Ssaków, Ossolineum, Wrocław:1–222 (In Polish)
Olszewski JL, Żarnowiecki G (1992) Snow cover in Białowieża. Parki nar i Rez Przyr 11(2):131–141 (In Polish with english summary)

Osińska B, Piusiński W (1997) Sarcocistosis of cardiac muscle in bison (*Bison bonasus*) from Białowieża Forest. Wiad parazyt 43:393–398 (In Polish with English summary)

Osińska B, Demiaszkiewicz AW, Lachowicz J (2010) Pathological lesions in European bison (*Bison bonasus*) with infestation by *Ashworthius sidemi* (Nematoda, Trichostrongylidae). Pol J Vet Sci 13:63–67

Parnikoza I, Borecko V, Sesin V et al (2009) History, current state and perspectives of conservation of European bison in Ukraine. Eur Bison Conserv Newslett 2:5–16

Parusel JB (1996) Pszczyńskie Żubry (European bison of Pszczyna). Centrum Dziedzictwa Przyrodniczego Górnego Śląska, Katowice (In Polish)

Paszkiewicz (2004) The use of tree stands by the population of the European bison of lowland-caucasian line Western Bieszczady). Parki nar i Rez Przyr 23(4):647–656 (In Polish with English summary)

Paszkiewicz R, Ryba MS (2000) Był, jest...czy będzie? Żubr w masywie Chryszczatej i Krąglicy (The history and future of European bison in Bieszczady). Fundacja Cum Tacent Clamant, Warszawa (In Polish)

Pčola Š, Adamec M, Pčola Š, Jun (2006) Reintrodukcja żubra *Bison bonasus* w Parku Narodowym Połoniny (Reintroduction of the European bison, *Bison bonasus* to Poloniny National Park) In: Olech W (ed) Perspektywy rozwoju populacji żubrów Wyd Artisco. Goczałkowice Zdrój, pp 43–54 (In Polish)

Pčola Š, Gurecka J (2008) Najnovšie poznatky z reštitúcie zubra hrivnatého (*Bison bonasus*) v Národnom Parku Poloniny (The most recent information about the restitution of European bison (*Bison bonasus*) to Poloniny National Park). Eur Bison Conserv Newslett 1:146–152 (In Slovak)

Peinado VI, Celdrán JF, Palomeque J (1999) Basic hematological values in some wild ruminants in captivity. Comp Biochem Physiol A 124:199–203

Pertoldi C, Bijlsma R, Loeschcke V (2007) Conservation genetics in a globally changing environment: present problems paradoxes and future challenges. Biodivers Conserv 16:4147–4163. doi:10.1007/s10531-007-9212-4

Pertoldi C, Tokarska M, Wojcik JM et al (2009a) Depauperate genetic variability detected in the American and European bison using genomic techniques. Biol Direct 4:48. doi:10.1186/1745-6150-4-48

Pertoldi C, Wójcik JM, Tokarska M et al (2009b) Genome variability in European and American bison detected using the BovineSNP50 BeadChip. Conserv Genet 11:627–634. doi:10.1007/s10592-009-9977-y

Pertoldi C, Tokarska M, Wójcik JM et al (2010) Phylogenetic relationships among the European and American bisons and seven cattle breeds as reconstructed using the bovineSNP50 Illumina genotyping BeadChip. Acta Theriol 55:97–108. doi:10.4098/j.at.0001-7051.002.2010

Perzanowski K, Adamec M, Pčola Š (2006) Status and restitution of a transboundary Polish-Slovak population of the European bison. In: Adamec M, Urban P (eds) Proceedings of conference 'Vyskum a ochrana cicavcov na Slovensku' VII, Zvolen 2005, Štátna ochrana prírody SR, Banská Bystrica

Perzanowski K, Januszczak M (2004) Preliminary assessment of the home range dynamics of the European bison *Bison bonasus* in Bieszczady Mountains. Parki nar i Rez Przyr 23(4):639–646 (In Polish with English summary)

Perzanowski K, Januszczak M, Wołoszyn-Gałęza A (2011b) Użytkowanie terenów otwartych i zalesionych przez żubry należące do zachodniej subpopulacji w Bieszczadach (Utilisation of open and forested areas by European bison belonging to the western subpopulation of Bieszczady). Roczniki Bieszczadzkie 19:191–206 (In Polish)

Perzanowski K, Kozak I (2000) Żubr w Karpatach w przyszłości i ocena dalszych perspektyw gatunku (*Bison* in the Carpathians in the future and assess the future prospects of the species). Dział nauk PAN 9:66–68 (In Polish)

Perzanowski K, Paszkiewicz R (2000) Restytucja i współczesny stan populacji żubrów w Bieszczadach (Restitution and contemporary status of the bison population in the Bieszczady Mountains). Monografie Bieszczadzkie, Ustrzyki Dolne 9:119–229 (In Polish)

Perzanowski K, Wołoszyn A, Januszczak M (2005) Wielkość ugrupowań w nizinnych i górskich populacjach żubrów (Size of groups in lowland and mountain populations of European bison). In: Hędrzak WM (ed) Zmiany w populacjach ssaków jako pochodna dynamiki zmian środowiska, Akademia Rolnicza w Krakowie, pp 3–10 (In Polish)

Perzanowski K, Olech W (2004) Recommendations for the reintroduction program of the European bison population in Bieszczady Mountains Poland. Biosphere Conserv 6(1):19–23

Pilarski W (1956) Deformacje kostne obserwowane na 6 okazach żubra (*Bison bonasus* L.) (Skeletal deformations observed in six specimens of Europan bison). Fol Morph 7:301–396 (In Polish)

Pilarski W (1962) Skóra i pokrywa włosowa żubra—*Bison bonasus* Linnaeus 1758. Część I i II (The skin and coat of European bison, Part I and II). PhD Thesis, Wydział Wetererynaryjny SGGW, Warszawa (In Polish)

Pilarski W (ed) (1975–1984) European Bison Pedigree Book 1975–1984. Państ Wyd Nauk, Warszawa

Pilarski W, Roskosz T (1957) Zjawisko ukrzyżowienia (*sacralisatio*) ostatniego kręgu lędźwiowego u samic żubra (*Bison bonasus* L.) (*Sacralisatio* the last lumbar vertebra in female European bison). Fol morph 8:109–110 (In Polish)

Piusiński W, Bielecki W, Małecka E et al (1997) Pathomorphology and pathogenesis of diseased genital organs (prepuce and penis) of bisons in the Białowieża Forest. Med Weter 53:596–600 (In Polish with English summary)

Podgurniak Z (1967) Pathological lesions in the European bison caused by foot-and mouth disease in Polish reservation. Acta Theriol 12:323–331

Podkówka W, Olszewski T, Kalisiewicz A (1984) Technologia produkcji siana (Hay production technology). PWR iL, Warszawa

Polakowski B (1961) Zespoły leśne Puszczy Boreckiej ze szczególnym uwzględnieniem leśnictwa Walisko i Lipowo (The forest types in Boreka Forest especielly in forest district Walisko and Lipowo). Stud Soc. scient Toruniensis 5:1–146 (In Polish)

Polziehn RO, Strobeck C, Sheraton J et al (1995) Bovine mtDNA discovered in north American bison populations. Conserv Biol 9:1638–1643

Popov VA (1989) Bison in Siberia (Cherga, Gornyj Altai) (European bison in Sibiria, Alatai, Russia). Manuscript (In Russian)

Pucek Z (1967) The most important problems for futher studies on European bison. Acta Theriol 12:495–501

Pucek Z (1984) Księga Rodowodowa Żubrów (European bison Pedigree Book). Prz zool 28:426–435 (In Polish)

Pucek Z (1986) Bison bonasus (Linnaeus 1758)—Wisent. In: Niethammer J, Krapp F (eds) Handbuch der Säugetiere Europas. Band 2/II (Paarhufer), AULA-Verlag GmbH, Wiesbaden, pp 278–315

Pucek Z (1991) History of the European bison and problems of its protection and management. In: Bobek B, Perzanowski K, Regelin W (eds) Global trends in wildlife management. Trans.18th IUGB congress, Kraków 1987. Świat Press, Kraków-Warszawa, pp 19–39

Pucek Z (1994) Postępy i zagrożenia restytucji żubra (Progress and threats to European bison restitution). Kosmos 43:147–169 (In Polish)

Pucek Z (2004) European bison – history of a flagship species. In: Jędrzejewska B, Wójck JM (eds) Essays on mammals of Białowieża Forest, Mammal Research Institute PAS, Białowieża, pp 25–34

Pucek Z (ed); Pucek Z, Belousova IP, Krasińska M, Krasiński ZA, Olech W (comps.) (2004) European bison. Status Survey and Conservation Action Plan. IUCN/SSB Bison Specialist Group IUCN, Gland, Cambridge

Pucek Z, Seal, US, Miller PS (eds) 1995 Population and habitat viability assessment for the European bison (*Bison bonasus*). Draft report, IUCN/SSC Conservation Breeding Specialist Group, Apple Valley, Minnesota, USA

Pytel SM (1969) Morphology of digestive tract of the European bison. Acta Theriol 14:349–402

Pytel S, Węgrzyn M (1976) Morphology of the liver of European bison. Acta teriol 21:19–30

Pytel S, Wojnowska M (1980) Morphology of the pancreas in the European bison. Acta Theriol 25:141–159

Pytel S, Węgrzyn M, Kobryńczuk F (1988) Abnormal development of the caudal *vena cava* in European bison. Acta Theriol 33:361–365

Pyziel AM, Kowalczyk R, Demiaszkiewicz AW (2011) The Annual cycle of shedding *Eimeria* oocysts by European Bison (*Bison bonasus*) in the Bialowieza Primeval Forest, Poland. J Parasitol 97(4):737–739. doi: 10.1645/GE-2567.1

Raczyński J (ed) (1987–2010) European bison pedigree book 1987–2010. Białowieski Park Narodowy, Białowieża

Radwan J, Kawałko A, Wójcik JM et al (2007) MHC-DRB3 variation in a free-living population of European bison (*Bison bonasus*). Mol Ecol 16:531–540. doi: 10.1111/j.1365-294X.2006.03179.x

Radwan J, Demiaszkiewicz AW, Kowalczyk R et al W (2010) An evaluation of two potential risk factors, MHC diversity and host density, for infection by an invasive nematode Ashworthius sidemi in endangered European bison (*Bison bonasus*). Biol Conserv 143:2049–2053. doi:10.1016/j.biocon.2010.05.012

Rautian GS, Kalabushkin BA, Nemtsev AS (2000) A new subspecies of the European bison, *Bison bonasus montanus* ssp. nov. (Bovidae, Artiodactyla). Doklady Biol Sci 375:636–640

Reed DH, Lowe EH, Briscoe DA et al (2003) Inbreeding and extinction: effects of rate of inbreeding. Conserv Genet 4:405–410

Reynolds HW, Glaholt RD, Hawley AWL (1982) Bison *Bison bison*. In: Chapman IA, Feldhamer GA (eds) Wild mammals in North America. John Hopkins University Press, Baltimore, pp 972–1007

Rhayan JC, Gidlewski T, Roffe TJ et al (2001) Pathogeny of brucellosis in bison from Yellowstone National Park. J Wildlife Dis 37:101–109

Romanov VS, Kozlo PG (2001) Sostoyanie i programa sokhraneniya i ispol'zovaniya belovezhskogo zubra (*Bison bonasus*) v Belarusi. T Bel gosud tekhnol univ, Seria I, 9:3–11 (In Russian)

Roskosz T (1962) Morphoplogie der Wirbelsäule des Wisents, *Bison bonasus* (Linnaeus 1758). Acta Theriol 6:113–164

Roskosz T, Empel W (1961) The size of the head and the height of spinous processes in the region of withers of the European bison. Acta Theriol 5:63–71

Roskosz T, Empel W (1963) Obserwacje L. H. Bojanusa nad czaszką i kręgosłupem żubra, *Bison bonasus* (L.) w świetle badań późniejszych (The research of LH Bojanus on skull and skeleton of European bison in comparison to modern anatimical studies). Prz zool 7:18–21 (In Polish)

Roskosz T, Empel W (1965) Komentarz naukowy i analiza dzieła L.H. Bojanusa *De* uro nostrate eiusque sceleto commentatio (Scientific comment and analysis of the work L.H. Bojanusa *De uro nostrate eiusque sceleto commentatio*). Mem Zoologi 14(1):47–185 (In Polish)

Rouys S (2000) Activity and movements of European bison in the Białowieża Forest, Poland. MSc Thesis, University of Greenwich

Rouys S, Krasińska M (2001) Influence of supplementary winter feeding on the daily travel distance of European bison in the Białowieża Forest, Poland. In: Field R, Warren RJ, Okarma H, Sievert PR (eds) Wildlife, land and people: priorities for the 21st century. The Wildlife Society, USA, pp 115–116

Rutberg AT (1983) Factors influencing dominance status in American bison cows (*Bison bison*). Zeitschrift für Tierpsychologie 56:206–212

Sawla A, Anusz K, Welz M et al (2011) Analysis of epizootiological situation in livestock and free-living wildlife on the Bieszczady Mountains in relation to occurrence of tuberculosis in European bison (*Bison bonasus*). Eur Bison Conserv Newslett 4:71–80

Samojlik T (2006) Hunts and other stays of kings of Poland and grand dukes of Lithuania in the Białowieża Primeval Forest in the 15–16th century. Kwart Hist Kult 54 (3-4):293–305 (In Polish with English summary)

Samojlik T, Jędrzejewska B (2003) Altana do polowań króla Stanisława Augusta Poniatowskiego w zagrodzie Teremiska—przyczynek do dziejów Puszczy Białowieskiej (Hunting house of king Stanislaw August Poniatowski in enclosure Teremiska as contribution to the history of the Bialowieza Forest). Sylwan 166:105–109 (In Polish)

Schnabel RD, Ward TJ, Derr JN (2000) Validation of 15 microsatellites for parentage testing in North American bison (*Bison bison*) and domestic cattle. Anim Genet 31:360–366

Selva N, Jędrzejewska B, Jędrzejewski W (2003) Scavenging on European bison carcasses in Białowieża Primeval Forest (Eastern Poland). Ecoscience 10:303–311

Sewalem A, Kistemaker GJ, Miglior F et al (2006) Analysis of inbreeding and its relationship with functional longevity in Canadian dairy cattle. J Dairy Sci 89:2210–2216. doi: 10.3168/jds.S0022-0302(06)72291-3

Shabailo VE, Pererva VI (1989) Issledovanie zabolevanii samcov zubrov Belovezhskoi Pushchchi i Nadvirnyanskogo lespromkhoza (Disease of European bison males.in Białowieża Forest and Nadvirnyainskij Forest District). In: Abstracts of conference: "50-let issledovanii v Belovezhskoi Pushche", Kamenyuki 19–21 Dec 1989 (In Russian)

Shackleton DM (1968) Comparative aspects of social organization of American bison. The University of Western Ontario, London

Shakun V (2011) Formation of the free-ranging population of European bison *(Bison bonasus)* "Volozhinskaya. Eur Bison Conserv Newslett 4:31–36

Shapiro B, Drummond AJ, Rambaut A et al (2004) Rise and fall of the Beringian Steppe Bison. Science 306:1561–1565

Shashenko AS, Krasoshko PA, Krasochko IA et al (1994a) Patologoanatomicheskie izmeneniya u belovezhskich zubrov pri nekroticheskom balanopostite (Pathomorphological signs observed at balanoposthitis in European bison males). In: Abstracts of conference: Mezhdunarodnoe simpozium po zubru"Sovremennye problemy okhrany zubra", Minsk (In Russian)

Shashenko AS, Krasochko PA, Krasochko IA et al (1994b) Vliyanie antitel k vozbuditelyam virusno-bakterial'nykh infektsii u belovezhskikh zubrov (Antibodies induced by viral-bacterial diseases of European bison). In: Abstracts of conference: Mezhdunarodnyi simpozium po zubru "Sovremennye problemy okhrany zubra", Minsk 1994 (In Russian)

Siński E, Gill J, Rijpkema S (1996) Prevalence of antibodies to *Borrelia burgdorferi* in European bison (*Bison bonasus*) from Białowieża Primeval Forest. Rocz AM Białyst 41:111–116

Sipko TP, Udina IG, Badagueva YN, Suklimova GE (1994) Polymorphism of k- casein gene in the family *Bovidae*: comparative analysis. Russ J Genet 30:203–207

Sipko TP, Rautian GS, Udina IG et al (1995) Investigation of blood group polymorphism in European bison (*Bison bonasus*). Russ J Genet 31:93–100

Sipko TP, Rautian GS, Udina IG et al (1997) Conservation of genetic material from endangered and economically important species in the estalishement of cryobanks. In: Tupaev TM, Kolcov NK (eds) Physiology and general biology reviews, Vol 13(3). pp 35–98

Sipko TP, Puzachenko AJu, Bunevich AN et al (2002) Imenchivost' ekster'ernykh priznakov zubrov (Variability of body measurements of European bison). In Proceedings of conference: Mezhdunarodnaya nauchno-prakticheskaya konferenciya 5–6 dekabrya 2002 "Voprosy sovremennogo okhotovedeniya", pp 267–285 (In Russian)

Slatis HM (1960) An analysis of inbreeding in the European bison. Genetics 45:275–287

Sokolov VE (1979a) Kozhnyi pokrov zubra (Skin of the European bison). In: Sokolov VE (ed) Zubr. Morfologiya, sitematyka, evolutsiya, ekologia, Izdatel'stvo Nauka, Moskva, pp 128–141 (In Russian)

Sokolov VE (ed) (1979b) Zubr. Morfologiya, sitematika, evolutsiya, ekologia (European bison. Morphology, sistematics, evolution, ecology). Izdatel'stvo Nauka, Moskva (In Russian)

Sokołowski AW (1998) Scientific value of Białowieża Forest. Parki nar i Rez Przyr 17(3) Supl:5–15 (In Polish with English summary)

Sokołowski AW (2004) Lasy Puszczy Białowieskiej (Frest types of Białowieża Forest). Centrum Informacyjne Lasów Państwowych, Warszawa (In Polish)

Speck S, Krasińska M, Lehnen A (2006) Normal vaginal mikroflora of free-ranging Euroepan bison (*Bison bonasus*). Vet Rec 159:600–601

Stachowiak I (1979) Jak pomierzyć i wycenić poroże żubra (Hunting appraisal and measurements of European bison horns). Łow pol 2:3–4 (In Polish)

Stefański W (1963) Parazytologia Wetererynaryjna (Veterinarian parasitology). PWRiL,Warszawa (In Polish)

Szafer W (1919) Z Puszczy Białowieskiej Wrażenia z wycieczki odbytej w kwietniu br. (Impressions from a trip to Białowieża Forest in April 1919). Sylwan 37:125–139 (In Polish)

Szafer W (1920) Z Puszczy Białowieskiej. Wrażenia z wycieczki odbytej w 1919 i 1920 r. (Impressions from a trip to Białowieża Forest in 1919 and 1920). Orli Lot, Mies krajozn dla młodz 8:90–105 (In Polish)

Szalay B (1913) Bôlčnyadatok kritikája. Vadászlap 14:260

Sztolcman J (1924) Materiaux pour l'histoire naturelle et pour l'historique du Bison d'Europe (*Bison bonasus* Linn.). Ann zool Mus Polon Hist Nat 2:49–136

Sztolcman J (1926) Żubr, jego historia, obyczaje i przyszłość (European bison its history, future and behaviour). Nakładem Centralnego Związku Polskiego Stowarzyszeń Łowieckich, Warszawa (In Polish)

Szwejkowski H (1954) *Sarcocystosis* w mięśniu sercowym żubra (*Bison bonasus* Boj.) w Polsce (*Sarcocystosis* in heart' muscle of European bison). In Abstract of conference: IV Zjazd Polskiego Towarzystwa Parazytologicznego w Gdańsku 30 Oct to 1 Nov 1954

Ścibor J (1965) Hodowla żubra w Polsce. Hodowla Żubrów w Puszczy Białowieskiej (Breeding of the European bison in Poland and in Białowieża Forest). In: Proceedings of conference: I i II Konferencja Polsko-Radziecka, Warszawa (In Polish)

Świeżyński K (1962) The skeletal musculatural system of the European bison, *Bison bonasus* (Linnaeus 1758). Acta Theriol 6:165–218

Świeżyński K (1968) The male reproductive organs of the European bison. Acta Theriol 13:511–551

Świeżyński K, Pilarski W (1956) Umięśnienie skórne żubra (Skin musculature of European bison). Fol morph 7:133–138 (In Polish)

Sysa P, Matuszewska M (2006) Cytogenetic ultrastructure of seminiferous tubules and epididymal cysts in European bison *Bison bonasus* (L). In: KitaJ, Anusz K (eds) Health threats for the European bison particularly in free-roaming populations in Poland. The SGGW Publishers, Warszawa, pp 244–249

Thiede S, Spergser J, Rosengarten R et al (2002) Antobodies against *mycoplasma bovigenitalium* in free-living European bison (*Bison bonasus*) with balanoposthitis. J Wildl Dis 38:760–763

Tiedemann R, Nadlinger K, Pucek Z (1998) Mitochondrial DNA-RFLP analysis reveals low levels of genetic variation in European bison *Bison bonasus*. Acta Theriol 5:83–87

Tokarska M (2010) Genetic variability of the extant Lowland bison (Bison bonasus bonasus) in Bialowieza Forest. Suggestions for protection of its genetic potential. Mammal Research Institute PAS in Białowieża, Białowieża (In Polish with English summary)

Tokarska M, Kawalko A, Wojcik JM et al (2009a) Genetic variability in the European bison (*Bison bonasus*) population from Białowieza forest over fifty years. Biol J Linn Soc 97:801–809. doi: 10.1111/j.1095-8312.2009.01203.x

Tokarska M, Marshall T, Kowalczyk R et al (2009b) Effectiveness of microsatellite and SNP loci (BovineSNP50 BeadChip Illumina®) in parentage assignment and identity analysis in species with low genetic diversity: the case of European bison. Heredity 103:326–332. doi:10.1038/hdy.2009.73

Tokarska M, Pertoldi C, Kowalczyk R et al (2011) Genetic status of the European bison *Bison bonasus* after extinction in the wild and subsequent recovery. Invited review. Mammal Rev 41(2):151–162. doi: 10.1111/j.1365-2907.2010.00178.x

Tracz M, Tracz M (2010) Conservation of European bison population in West-Pomeranian voivodship. Eur Bison Conserv Newslett 3:119–123 (In Polish with english summary)

Traul DL, Bhushan B, Eldridge JA et al (2005) Characterization of *Bison bison* Major Histocompatibility Complex Class IIa haplotypes. Immunogenetics 57:845–854

Tyapougin YA, Gusarov IV (2004) Creating of population of European bison in condition of northern region of Russia. In: Krasińska M, Daleszczyk K (eds) Proceedings of the conference on European bison conservation Białowieża, Poland, Mammals Research Institute PAS, Białowieża 30 Sept to 2 Oct 2004

Udina IG, Shakhaev GO (1998) Restriction fragment length polymorphism (RFLP) of exon 2 of the MHCBibo-DRB3 gene in European bison *Bison bonasus*. Acta Theriol 43 (Suppl. 5):75–82

Udina IG, Sokolova SS, Sipko TP et al (1994) Comparative analysis of polymorphism for DQB and DRB DNA *loci* of the major histocompatibility complex in the representatives of the family *Bovidae*. Russ J Genet 30:313–317

Uglyanec AV (1993) Predvaritel'noe rezultaty reintroduktsii zubra v Pripyatsakom zapovednikie (Preliminary results of European bison reintroduction in Pripyat' National Park). Zapoved Bel Issled 16:73–179 (In Russian)

Uglyaniec AV (1994) K voprosu sokhraneniya i ispol'zovaniya zubrov Belarusi (Concervation and management of European bison in Belarus). In: Abstracts of conference "Problemy izucheniya, sokhraneniya i izpol'zovaniya biologicheskogo raznobrazya zhivotnogo mira" Minsk (In Russian)

Uglyanec AV (1999) Osobennosti godovogo zhiznenogo cikla pripjatskoi mikropopulacji (Charcteristic of annual life cycle of European bison inPripyat' National Park). In Abstract of conference "Biologicheskie ritmy", Brest (In Russian)

Vajner L (1980) Etologická studie stáda zubru evropských, *Bison bonasus* (Linnaeus 1758) v pražke ZOO (Ecological investigation of European bison herd in Prague ZOO). Gazella 2:59–69 (In Czech)

Vuure van C (2005) Retracing the aurochs. History, morphology and ecology of an extinct wild ox. Pensoft Publishers, Sofia-Moscow

Wachnik Z (1983) Zarys chorób zakaźnych zwierząt (Infectious diseases of animals). PWN, Warszawa (In Polish)

Welz K, Anusz K, Salwa A et al (2005) Bovinae tuberculosis in European bison in Bieszczady region. Med Weter 61:441–444 (In Polish with English summary)

Westemeier RL, Brawn JD, Simpson SA et al (1998) Tracking the long-term decline and recovery of an isolated population. Science 282:1695–1698

Węgrzyn M (1968) Blood-vascular system of the European bison. Morphology of the heart. Acta Theriol 12:177–218

Węgrzyn M, Kupczyńska M (1986) Shape, size and weight of the heart in European bison. Acta Theriol 31:327–342

Węgrzyn M, Serwatka S (1961) Ligamentum sacrotuberale latum bei *Bison bonasus* (Linnaeus 1758) und *Bos taurus* (Linnaeus 1758). Acta Theriol 5:73–97

Węgrzyn M, Serwatka S (1984) Teeth eruption in the European bison. Acta Theriol 29:11–121

Williams ES, Thorne ET, Andersen SLH, Erriges JD (1993) Brucellosis in free-ranging bison (*Bison bison*) from Teton County, Wyoming. J Wildl Dis 29:118–122

Wolf O, Jakob W, Lange A et al (2000) Balanoposthitis of the European bison (*Bison bonasus*). In: Proceedings of international symposium on European bison—yesterday, today and tomorrow, Šiauliai Lithuania, 9–10 Dec 2000

Wołk E (1983) The hematology of the free-ranging European bison. Acta Theriol 28:73–82

Wołk E (1990) Fizjologiczne i ekologiczne aspekty hematologii ssaków w warunkach naturalnych (Physiological and ecological aspects of hematology of wild animals). Centum Doskonalenia Nauczycieli im. Wł. Spasowskiego, Oddział w Toruniu, Toruń (In Polish)

Wołk E, Józefczak E (1984) Serum proteins of the free-ranging European bison. Comp Biochem Physiol 79A:597–600

Wołk E, Józefczak E (1988) Serum biochemistry of free-ranging European bison. Acta Theriol 33:47–56

Wołk E, Krasińska M (2004) Has the condition of European bison deteriorated over last twenty years? Acta Theriol 49:405–418

Wójcik JM, Kawalko A, Tokarska M, Jaarola M, Vallenback P, Pertoldi C (2009) Post-bottleneck mtDNA diversity in a free-living population of European bison *Bisonbonasus*. Implications for conservation. J Zool 277:81–87. doi: 10.1111/j.1469-7998.2008.00515.x

Wróblewski K (1927) Żubr Puszczy Białowieskiej (European bison of Białowieża Forest). Wyd Polskie, Poznań (In Polish)

Wróblewski K (1932) Czy możliwe jest obecnie i jaka drogą można odrodzić wymierającego żubra? (If is possible the recover of endangered species Europeasn bison). Rozpr biol Med Wet, Roln, Hod 10:54–117 (In Polish)

Yazan Ju, Nemcev A (1985) Kavkazkie chistoporodnye gornye zubry (Mountain European bison in Caucasus). Okhota Okhot Khoz 1:16–17 (In Russian)

Zablocky MA (1949) Tridcat' piat' let razvedeniya zubrov i ikh gibridov v Zooparke Askania Nova (1902–1937) (Thirty five years of breeding of European bison and their hybrids in Askania Nova reserve (1902–1937). In: Kulagin NM (ed) Kavkazskii zubr. Glav. Upravlen. po Zoopark. Zoosadam, Moskva, pp 73–124 (In Russian)

Zablocky MA (1956) Gosudarstvennaya plemennaya kniga zubrov i bizonov (European bison and American bison State Pedigree Book). Ministerstva Sel'skogo-Khozyaistva SSSR. Moskva (In Russian)

Zablocky MA (1957) Nekotorye biologicheskie osobennosti zubra i ikh izmenenie v usloviyakh zagonnogo soderzhaniya (Characteristic of biology of European bison and the change of its behaviour in captivity). T Priok-Terras Zapoved 1:5–65 (In Russian)

Zablocky MA (1965) Razvedenie chistokrovnykh zubrov v SSSR (Breeding of European bison in USSR). In: Proceeding of conference: Hodowla Zubrów w Puszczy Białowieskiej I and II Konferencja Polsko-Radzieckiecka, Warszawa (In Polish and Russian)

Zaręba R (1962) "Ogrody do polowań" w Puszczy Białowieskiej w czasach Stanisława Augusta Poniatowskiego (Hunting gardens in Białowieża Forest during times of king Stanisław August Poniatowski). Sylwan 106:81–83 (In Polish)

Ziembicki W (1934) Polskie żubry i tury w poemacie łacińskim z 1502 roku (European bison and aurochs in Poland - in the Latin poem from 1502). Łowiec Polski 35:695–696 (In Polish)

Żabinski J (ed) (1947–1973) European bison pedigree book. PWN, Warszawa

Żabiński J (1960) The European bison (*Bison bonasus*). State Council for Conservation of Nature Poland, Warsaw No 9, Kraków

Żoch K (2007) Wpływ huraganu na letni i zimowy areał występowania populacji żubra Bison bonasus (Linnaeus 1758) w Puszczy Boreckie (An influence of windstorm on summer and winter ranges of European bison in Borecka Forest). M.Sc. Thesis, Faculty of Forestry, Warsaw University of Life Sciences-SGGW (In Polish)

Żórawski C, Lipiec M (1997) [eneralized tuberculosis in a European bison. Med Weter 53:90–92 (In Polish with English summary)

Żórawski C, Lipiec M (1998) Dzikie zwierzęta jako rezerwuar *Mycobacterium bovis* w przyrodzie (Wild animal as reservoir of *Mycobacterium bovis*). Nowa Wetererynaria 3:21–25 (In Polish with English summary)

Timeline

M. Zichy (1861)

M. Krasińska and Z. A. Krasiński, *European Bison*,
DOI: 10.1007/978-3-642-36555-3, © Springer-Verlag Berlin Heidelberg 2013

1409 – A great hunt through Poland's Białowieża Forest (*Puszcza Białowieska*) is led by King Władysław Jagiełło during their preparations for what would become known as the Battle of Grunwald

1627 – The last aurochs dies in the Jaktorowska Forest, not far from Warsaw. These animals have often been confused with the European bison or wisent

1752 – King August III's hunting session in the Białowieża Forest is commemorated by an obelisk that still stands on the embankment between the ponds at the Białowieża Palace Park—42 European bison were killed

1860 – Another 28 European bison succumb in the course of a great hunting session led by Tsar Alexander I. The cast-iron statue funded to commemorate that occasion stands today in Spała

1865 – Four European bison from the Białowieża Forest are sent to Pszczyna (Pless), these giving start to the so-called "Pszczyna (Pless) Line". It was their descendents that would play such a major role in the reinstatement of the species

1894 – A hunt by Tsar Alexander III costing the lives of 7 European bison coincides with a visit to the newly-built palace at Białowieża

1897 – 37 bison are killed over 10 days when Tsar Nicholas II goes off hunting. The session coincides with the completion of the Hajnówka-Białowieża railway

1900 – Along with his high-profile guests, Nicholas II enjoys the greatest hunt of his life, at Białowieża between 18 and 29 August. The 690 game animals taken at that time include no fewer than 45 bison

1903 – Publication of the Russian-language monograph entitled "Belovezhskaya Pushcha" (Białowieża Forest) by A.N. Marks, St. Petersburg, under the authorship of Georgij Karcov. The work devoted to the Białowieża Forest of course includes a long chapter on the last natural population of European bison. It also boasts a wealth of fine illustrations. Original copies are to be seen at the libraries of the Białowieża National Park (Poland) and the Belovezhskaya Pushcha State National Park (Belarus). A facsimile of the monograph came out in 2008 in Minsk (Belarus), the publishing house being Masterskaya Literatura

1912 – Tsar Nicholas II hunts for the last time at Białowieża

1913 – The bison of the Białowieża Forest are hunted for the last time, *inter alia* by invited guest the Prince of Monaco

1914 – Just prior to the outbreak of the World War I, there are 727 European bison in the Białowieża Forest

1919 – The last European bison dies in the Białowieża Forest

1923 – The International Society for the Protection of the European Bison is established

1924 The different captive breeding centres around the world can together muster just 54 European bison

1927 – The last Caucasian bison (*Bison bonasus caucasicus*) dies in the Caucasus Mountains

1927 – Appearance of a Polish-language monograph entitled "Żubr Puszczy Białowieskiej" (European bison of the Białowieża Forest), as published by Wydawnictwo Polskie in Poznań under the authorship of Konrad Wróblewski. The book represents a unique source of information on the biology and ecology of the animals living within the last natural population of European bison at the beginning of the twentieth century, as based on the author's own observations

1929 – Work on the reintroduction of the European Bison to Białowieża commences, as two individuals are brought into the newly-constructed reserve

1932 – The first European Bison Pedigree Book is published

1939 – The reserve in the Białowieża Forest now has 16 European bison, of which 7 are of the Białowieża line and 9 of the Białowieża-Caucasian line

1945 – Changes in borders leave a western part of the Białowieża Forest in Poland, while the eastern part comes within the USSR. The reserves in the western part have 17 bison, including 4 of the Białowieża (Lowland) line

1946 – The first 5 European bison from the reserve in Białowieża are sent to the eastern (Soviet) part of the Białowieża Forest

1950 – The last bison of the Lowland-Caucasian line from the reserve in Białowieża are transferred out of the Białowieża reserves

1952 – The first two Lowland line European bison are released into the wild in the Polish part of the Białowieża Forest

1953 – The first efforts to reinstate a wild population of European bison in the Belarussian part of the Białowieża Forest are made

1957 – The free-living population of bison in the Polish part of Białowieża Forest is augmented by its first calf

1961 – The first Polish-Soviet conference on a joint bison reintroduction programme for the Białowieża Forest is convened

1962 – The first European bison go free in Poland's Borecka Forest

1964 – Following a period in an acclimatisation enclosure, bison brought in from Niepołomice and Pszczyna are released into the wild in Poland's Bieszczady Mountains (south-east Poland)

1965 – Establishment of the first free-ranging population of CEJSKIJJ in the Caucasus Mountains of Russia (previously the Soviet Union), as well as TSUMANSKA in Ukraine

1966 – The last European bison of the Lowland-Caucasian line are removed from the Belarussian part of the Białowieża Forest

1971 – The wild-living population in the Polish part of the Forest rises back above 200

1973 Five Białowieża Forest bison are transferred to Poland's Knyszyn Forest, where they are released into the wild, thus beginning a free-ranging population in the Knyszyn Forest

1973 – Founding of the first free-ranging population in Lithuania

1978 – The world population of European bison goes back over 2,000

1979 – Appearance of the Russian-language monograph entitled "Zubr. Morfologia, sistematika, evolutsya, ekologia" from Nauka Publishers, Moscow, and under the editorship of V. E. Sokolov. The content is as revealed in the subtitle, dealing with morphology, systematics, evolution and ecology of the species. A noteworthy part of the book is that concerned with evolution within the genus *Bison*, as written by C.C. Flerov. In turn, the wild populations were only just being established at the time the book came out, ensuring that little information on ecology could by definition be available

1980 – European bison brought in from Białowieża are released into the forests of Wałcz Forest District (north-east Poland). From 2004 on, this population has gone by the name of the STADO ZACHODNIOPOMORSKIE

2003 – Publication of the Russian-language monograph on "Zubr na Kaukazie" (The European bison in the Caucasus), from Maikop, Moscow, and under the authorship of Aleksandr S. Niemtsev and co-authors. This presents the history of the European bison in the Caucasus Mountains, in relation to both its extinction and reinstatement, and to the genetics and ecology of the wild populations living in the high mountain massifs

2004 – Publication in English of a strategy for the protection of the species under the title "European bison. Status Survey and Conservation Action Plan", from the IUCN/SSC Bison Specialist Group, IUCN, Gland, Switzerland and Cambridge, UK, and under the authorship of Z. Pucek (ed), Z Pucek, I.P. Belousova, M. Krasińska, ZA. Krasiński and W. Olech (comps.). It contains a plan for the protection of the species on the global scale, taking in the captive breeding populations, as well as the free-ranging and semi-free ones. The primary preservation-related tasks are considered to entail the further spread of the species and the protection of its gene resources

2004 – Publication of the contemporary Polish-language monograph entitled "Żubr. Monografia przyrodnicza", from Wyd. Hajstra, Warsaw-Białowieża, and under the authorship of Małgorzata Krasińska and Zbigniew A. Krasiński. The bison of Białowieża occupy a significant part of the book, which discusses systematics, contemporary history, the last wild population living in the Forest in the late nineteenth and early twentieth centuries, the extinction of the species in the wild in the early twentieth century and the reinstatement of the species, first by captive breeding and then in the wild. The build of the bison and its habits are also described, as are its food and reproduction, and the social and spatial structures applying to the population. The final part of the book is devoted to the restoring of wild populations in different areas

- English-language translation (2007) entitled "European bison. The Nature Monograph" from the Mammal Research Institute, Białowieża
- German-language translation (2008) entitled *Der Wisent*, from Die Neue Brem-Bücherei Bd. 74.Westarp Wissenschaften Hohenwarsleben

2004 – The first free-ranging population of European bison in Slovakia comes into existence

2008 – The global population of European bison exceeds 4,000

2009 – Appearance of a Russian-language monograph entitled "Zubr w Belarusi"(The European bison in Belarus), from Belaruskaya navuka in Minsk, and authors Peter G. Kozlo and Aleksei N. Bunevich. The book discusses the biology and ecology of the bison in Belarus's free-ranging populations, with particular attention paid to the animals from the Belovezhskaya Pushcha

- A second augmented and corrected version of this book comes out in 2011

2010 – The world population of European bison stands at 4431 individuals, of which 2956 (67 %) are in the 33 free-ranging and the 7 semi-free populations, while 1475 (or 33 % of the total) are in captivity

2011 – The Polish and Belarussian parts of the Białowieża Forest are home to 909 European bison, and hence to more animals than at the beginning of the twentieth century

Appendices

M. Krasińska and Z. A. Krasiński, *European Bison*,
DOI: 10.1007/978-3-642-36555-3, © Springer-Verlag Berlin Heidelberg 2013

Appendix 1 The history of the restitution of the European bison in enclosed breeding conditions within the Białowieża Forest, 1929–1952

Years	Birth			Bison transported to the breeding centre					Bison transferred outside the breeding centre					Mortality				Number on 31 Dec.
	Sex	Ped No.	Name	Sex	Ped. No.	Name	Date	From	Sex	Ped. No.	Name	Date	Where	Sex	Ped. No.	Name	Date	
1	2	3	4	5	6	7	8	9	10	11	12	13	14	15	16	17	18	19
1929				M	163	BORUSSE	19.09	ZOO Warsaw										2 (1. 1) [2]
				M		Kobold	19.09	ZOO Warsaw										
				F	93	**BISERTA**	10.10	ZOO Warsaw										
				F		Faworyta	10.10	ZOO Warsaw										
1930	M	189	PUHACZ	F	161	**BISCAYA**	29.08	ZOO Warsaw										4(2, 2) [2]
				F		Stolce	29.08		M		Kobold		ZOO Warsaw					
1931	F	211	PUSTÓŁKA	M	101	HAGEN	27.06	ZOO Poznań										7 (3, 4) [4]
	M		Huron	F	96	**Gatczyna**	27.06											
	F		Popala															
1932	F	226	PUSTELNICA						M	189	PUHACZ	05	ZOO Warsaw	F	96	**Gatczyna**	8.05	6 (2, 4) [5]
	F													F	227	PULPA	20.05	
	F	227	PULPA															
	–		Puszcza															
1933	M	228	PUK											M	101	HAGEN	13.09	6 (2, 4) [5]
1934	F	242	PUMA	M	189	PUHACZ	25.03	ZOO Warsaw										9 (4, 5) [6]
	M	243	PUPUIL															
	F		Maja															
1935	F	246	PUSTA	M	186	**Björnson**	21.03	ZOO	F	211	PUSTÓŁKA	11.02	ZOO	M	186	**Björnson**	18.07	10 (4, 6) [1]
	F	247	PUSTÓŁKA	F	89	**BILMA**	2.04	Stockholm					Stockholm					
									F	226	PUSTELNICA	11.03	ZOO Stockholm					
									F		Faworyta							
									F		Stolce							
									F		Popala							
									F		Puszcza							
									F		Maja							

(continued)

Appendix 1 (continued)

Years	Birth			Bison transported to the breeding centre					Bison transferred outside the breeding centre					Mortality				Number on 31 Dec.
	Sex	Ped No.	Name	Sex	Ped. No.	Name	Date	From	Sex	Ped. No.	Name	Date	Where	Sex	Ped. No.	Name	Date	
1	2	3	4	5	6	7	8	9	10	11	12	13	14	15	16	17	18	19
1936	F	257	PUSTOTA	M	229	**PLISCH**	17.04.	Pszczyna	M	189	PUHACZ	22.01	Smardzewice					11 (4, 7)
	M	258	PURYTANIIN						M	243	PUPIL	18.03	Schorfheide					
									M	–	Huron	0.03	Zoo Warsaw					
1937	M	478	PULCHNY															13 (5, 8)
	F	479	**Polka**															
1938	F	500	PURYTANKA						M	228	PUK	6.12	Niepołomice					13 (3, 10)
	F	501	PUŻANKA						F	246	PUSTA	6.12	Niepołomice					
	F	502	**Pokorna**						M	258	PURYTANIN	26.06	Spała					
1939	F	577	PULA															
	M	578	PUSZCZYK											F	89	**BILMA**	4.09	16 (4, 12)
	F	519	**POGANKA**															
	F	520	**POLANA**															
1940	F	541	PURGA						M	478	PULCHNY	8.09	Złota Wieś			**Poldi**	19.06	19 (6, 13)
	M	542	PUZON															
	M	543	PUT															
	M	544	**Połdień**															
	F	545	**Poldi**											F	545			
1941	F	568	**Poba**											F	568	**Poba**	23.05	21 (7, 14)
	M	569	PUB											M	569	PUB	31.05	
	F	570	PUCA											F	570	PUCA	12.12	
	F	571	PUBIA											F	571	PUBIA	30.07	
	M	572	**Połamaniec**															
	F	573	PUNARIA															

(continued)

Appendix 1 (continued)

Years	Birth			Bison transported to the breeding centre					Bison transferred outside the breeding centre					Mortality				Number on 31 Dec.
	Sex	Ped No.	Name	Sex	Ped. No.	Name	Date	From	Sex	Ped. No.	Name	Date	Where	Sex	Ped. No.	Name	Date	
1	2	3	4	5	6	7	8	9	10	11	12	13	14	15	16	17	18	19
1942	F	596	PUTKA											F	479	**Polka**	7.09	23 (10, 13)
	M	597	PUNKT											F	573	PUNARIA	24.09	
	M	598	**Pob**											M	598	**Pob**	25.09	
	M	599	PUG											?	601	Pox	6.09	
	M	600	PUGINAŁ											?	602	Poy	6.12	
	?	601	**Pox**															
	?	602	**Poy**															
1943	F	624	PURPURA											F	93	**BISERTA**	16.08	27 (12, 15)
	M	625	PURKUL											M	599	PUG	9.11	
	M	626	**Pok**											F	630	PUFA	6.11	
	M	627	PURA															
	M	628	PUD															
	7	629	PUDA															
	7	630	PUFA															
1944	M	652	PUCUŁ											F	161	**BISCAYA**	23.07	17 (8, 9)
														M	163	BORUSSE	07	
														F	257	PUSTOTA	14.05	
														F	500	PURYTANKA	17.07	
														F	502	**Pokorna**	14.05	
														M	518	PUSZCZYK	27.07	
														F	541	PURGA	7.05	
														M	544	**Połdień**	.07	
														M	626	**Pok**	.07	
														M	628	PUD	.07	
														F	629	PUDA	.07	

(continued)

Appendix 1 (continued)

Years	Birth			Bison transported to the breeding centre					Bison transferred outside the breeding centre					Mortality				Number on 31 Dec.
	Sex	Ped No.	Name	Sex	Ped. No.	Name	Date	From	Sex	Ped. No.	Name	Date	Where	Sex	Ped. No.	Name	Date	
1	2	3	4	5	6	7	8	9	10	11	12	13	14	15	16	17	18	19
1945	F	673	PUSZCZA															22 (10, 12)
	F	674	**POZIOMKA**															
	M	675	PUSTOSZ															
	F	676	PUSŁAW															
		677	PURTA															
1946	M	694	PUSTELNIK						F	242	PUMA	8.04	Niepołomice	M	625	PURKUŁ	4.01	15 (7, 8)
	M	695	PUSZCZANIN						M	542	PUZON	8.04	Niepołomice					
	F	696	PUSTUŁECZKA						F	596	PUTKA	8.04	Niepołomice					
	F	697	PUSZCZYSŁAWA						F	673	PUSZCZA	8.04	Niepołomice					
									F	677	PURTA	8.04	Niepołomice					
									F	517	PULA	18.07	Soviet Union					
									F	543	PUF	18.07	Soviet Union					
									M	600	PUGINAŁ	18.07	Soviet Union					
									F	624	PURPURA	18.07	Soviet Union					
									M	625	PUSTOSZ	18.07	Soviet Union					
1947	F	714	PURCHAWKA	F	219	**Planarie**	18.09	Pszczyna	F	247	PUPILKA	20.11	Niepołomice					17 (9, 8)
	F	715	**POŻOGA**	F	524	**BESTE**	13.05	ZOO Łódż	F	501	PUŻANKA	29.05	Niepołomice					
	M	716	**POGANIN**	F	547	**PLESSE**	18.11	Pszczyna	F	627	PURA	29.05	Niepołomice					
				M	575	**PLATO**	2.06	Pszczyna	F	696	PUSTUŁECZKA	29.05	Niepołomice					
				F	631	**PLARKA**	2.06	Pszczyna	F	697	PUSZCZYSŁAWA	29.05	Niepołomice					
									F	714	PURCHAWKA	20.11	Niepołomice					
1948	F	735	**POJATA**	M	657	**Planetnik**	1.06	Pszczyna	M	679	**Pleciuch**	29.11	Soviet Union					20 (9, 11)
	F	736	**PODWIKA**	M	679	**Pleciuch II**	2.06	Pszczyna	M	676	PUSŁAW	21.11	Soviet Union					
	F	737	**PLAMKA II**						M	696	PUSZCZANIN	21.11	Soviet Union					
	M	738	**PODBIPIĘTA**						F	547	PLESSE	29.11	Soviet Union					
	M	739	**Plaskacz**						M	652	PUCUŁ	25.01	Sweden					
	F	740	**POŚWIATA**															

(continued)

Appendix 1 (continued)

Years	Birth			Bison transported to the breeding centre					Bison transferred outside the breeding centre					Mortality				Number on 31 Dec.
	Sex	Ped No.	Name	Sex	Ped. No.	Name	Date	From	Sex	Ped. No.	Name	Date	Where	Sex	Ped. No.	Name	Date	
1	2	3	4	5	6	7	8	9	10	11	12	13	14	15	16	17	18	19
1949	M	762	**POLEL**						M	597	PUNKT	4.11	ZOO Wrocław	M	657	**Planetnik**	22.01	21 (7, 12)
	M	764	**Pobug**											M	765	**Kap**	14.06	
	M	765	**Kap**											M	771	**Pot**	7.10	
	F	770	**POPIELICA**															
	M	771	**Pot**															
1950	F	788	**Pomsta**						F	219	**Planarie**	25.09	ZOO Warsaw	M	794	**Pogrom**	13.09	19 (7, 12)
	M	792	**KARPACZ**						M	572	**Połamaniec**	25.09	ZOO Warsaw					
	M	794	**Pogrom**						M	694	PUSTELNIK	25.09	ZOO Warsaw					
									M	739	**Plaskacz**	6.11	Smardzewice					
1951	F	812	**PORĘBA**						F	715	**POŻOGA**	15.10	Soviet Union	M	229	**PLISCH**	6.11	21 (7, 14)
	M	816	**POMRUK**						M	764	**Pobug**	15.10	Soviet Uni	M	818	**Polonus**	2.06	
	M	817	**POPAS**						F	788	**Pomsta**	15.10	Soviet Uni					
	M	818	**Polonus**															
	F	827	**POGOŃ**															
	F	831	**POZŁOTA**															
	F	832	**POLATUCHA**															
1952	F	859	**POBUDKA**	M	546	**PLUVIUS II**	22.11	Pszczyna	F	519	POGANKA	20.05	ZOO Rotterdam	F	865	**Kaba**	31.07	21 (8, 13)
														F	866	**Poda**	17.07	
	F	865	**Kaba**						F	674	POZIOMKA	15.06	ZOO Warsaw					
	F	866	**Poda**															

Key names of bison written in capitals and bold (e.g., **BISERTA**) denote lowland (Białowieża) bison that played an active part in the reintroduction, while small bolded names (e.g., **Polka**) denote lowland bison that did not participate. Capital letters (e.g. BORUSSE) denote Białowieża-Caucasus bison, small lower-case letters (Kobold)—the European bison/American bison ("buffalo") crosses whose numbers are given in square brackets [2]. Ped. No—Pedigree number, Dec.—December. *Source* Krasińska ans Krasiński (2007)

Appendix 2 Pedigree data for lowland bison released into the wild from reserves in the years 1952–1968, with a view to a free-living population being reinstated in the Polish part of the Białowieża Forest

Year of introduction	Bison released in the Białowieża Forest		
	Sex	Pedigree number	Name
1952	M	816	*POMRUK*
	M	817	*POPAS*
	F	812	**PORĘBA**
1953	F	735	**POJATA**
	M	897	*PODAREK*
1955	F	910	**POGODNA**
1956	0	0	0
1957	F	812	*PORĘBA*
	M	762	**POLEL**
	F	940	**POŁUDNICA**
1958	M	980	**PORANEK**
	M	985	**POLAS**
	M	1019	**POCZET**
	M	1022	**POŚCIG**
	F	1015	*POCHODNIA*
	F	1025	**POWABNA**
	F	1028	**POHULANKA**
	F	1032	**POGARDA**
	F	1036	**PONOWA**
	F	1072	**POSAŻNA**
1959	F	827	**POGOŃ**
	F	1137	**PODOLANKA**
	F	1144	**POGRANICZNA**
	F	1145	**POMYŁKA**
	F	1146	**PODOBNA**
1960	F	1140	**POCIESZNA**
1962	M	1202	**POTOP**
	M	1231	**POLONEZ**
	M	1225	**PORUBANEK**
	F	832	*POLATUCHA*
	M	1301	**POMNIK**
1964	M	902	*PONURY*
	M	762	*POLEL*
	M	912	*PODBIPIĘTA*
	F	975	*POKUSA*
	F	912	**KAMIONKA**
1965	F	737	*PLAMKA II*
	F	740	*POŚWIATA*
	F	1387	**POROHA**
	F	1085	**PODKOMORZANKA**
1966	F	831	**POZŁOTA**

The bison whose names are written in italics did not participate in the reintroduction (being recaptured or dying). *Source* Krasińska and Krasiński (2007)

Appendix 3 Numbers and structure of the free-ranging population in the Polish part of the Białowieża Forest, 1952–2011

Years	Number on 31 December			Age and sex structure of the population							
	n	Males	Females	Males ≥4 years	%	Females ≥4 years	%	Juveniles 203 years	%	Calves <1 year	%
1952	2	2	0	0	0	0	0	0	0	0	0
1953	5	3	2	0	0	0	0	0	0	0	0
1954	5	3	2	0	0	0	0	0	0	0	0
1955	4	2	2	0	0	0	0	0	0	0	0
1956	3	1	2	0	0	0	0	0	0	0	0
1957	6	1	5	0	0	0	0	0	0	1	0
1958	18	7	11	0	0	0	0	0	0	2	0
1959	28	12	16	0	0	0	0	0	0	6	0
1960	34	14	20	5	14.7	11	32.4	13	38.2	5	14.7
1961	44	19	25	7	15.9	16	36.4	11	25.0	10	22.7
1962	56	26	30	14	25.0	18	32.1	16	28.6	8	14.3
1963	68	33	35	17	25.0	20	29.4	18	26.5	13	19.1
1964	87	44	43	23	26.4	27	31.0	21	24.2	16	18.4
1965	102	52	50	27	26.5	32	31.4	28	27.4	15	14.7
1966	119	55	64	31	26.1	38	31.9	31	26.1	19	15.9
1967	141	63	78	36	25.5	44	31.2	34	24.1	27	19.2
1968	157	74	83	41	26.1	48	30.6	43	27.4	25	15.9
1969	180	86	94	43	23.9	60	33.3	48	26.7	29	16.1
1970	197	89	108	47	23.9	72	36.5	50	25.4	28	14.2
1971	211	94	117	58	27.5	73	34.6	51	24.2	29	13.7
1972	236	104	132	64	27.1	82	34.7	53	22.5	37	15.7
1973	253	112	141	69	27.3	91	36.0	55	21.7	38	15.0
1974	245	113	132	67	27.3	95	38.8	56	22.9	27	11.0
1975	234	103	131	53	22.6	89	38.0	46	19.7	46	19.7
1976	210	97	113	50	23.8	77	36.7	51	24.3	32	15.2
1977	218	102	116	44	20.2	72	33.0	64	29.4	38	17.4
1978	219	98	121	55	25.1	79	36.1	57	26.0	28	12.8
1979	230	106	124	62	27.0	84	36.5	52	22.6	32	13.9
1980	242	115	127	69	28.5	88	36.4	47	19.4	38	15.7
1981	250	117	133	64	25.6	95	38.0	57	22.8	34[a]	13.6
1982	247	119	128	59	23.9	98	39.7	54	21.8	36	14.6
1983	261	123	138	71	27.2	92	35.2	57	21.9	41	15.7
1984	256	118	138	72	28.1	91	35.5	59	23.1	34	13.3
1985	243	110	133	74	30.5	87	35.8	48	19.7	34	14.0
1986	224	99	125	67	29.9	94	42.0	41	18.3	22	9.8
1987	235	101	134	63	26.8	94	40.0	31	13.2	47	20.0
1988	228	98	130	64	28.1	90	39.5	42	18.4	32	14.0
1989	254	107	147	60	23.6	98	38.6	61	24.0	35	13.8
1990	272	121	151	67	24.6	100	36.8	62	22.8	43	15.8
1991	271	119	152	72	26.6	105	38.7	57	21.0	37	13.7
1992	259	114	145	69	26.6	102	39.4	43	16.6	45	17.4

(continued)

Appendix 3 (continued)

Years	Number on 31 December			Age and sex structure of the population							
	n	Males	Females	Males ≥4 years	%	Females ≥4 years	%	Juveniles 203 years	%	Calves <1 year	%
1993	251	105	146	70	28.0	96	38.2	49	19.5	36	14.3
1994	265	114	151	64	24.1	98	37.0	48	18.1	55	20.8
1995	289	111	178	63	21.8	120	41.5	64	22.2	42	14.5
1996	266	101	165	59	22.2	110	41.3	63	23.7	34	12.8
1997	290	117	173	70	24.1	123	42.4	46	15.9	51	17.6
1998	298	117	181	68	22.8	124	41.6	66	22.2	40	13.4
1999	311	123	188	64	20.6	126	40.5	71	22.8	50	16.1
2000	306	130	176	73	23.9	135	44.1	63	20.6	35	11.4
2001	348	141	207	83	23.8	137	39.4	57	16.4	71	20.4
2002	349	146	203	90	25.8	139	39.8	70	20.1	50	14.3
2003	351	151	200	90	25.6	142	40.5	72	20.5	47	13.4
2004	379	158	221	101	26.6	153	40.4	63	16.6	62	16.4
2005	397	167	230	107	27.0	152	38.3	64	16.1	74	18.6
2006	402	167	235	114	28.4	154	38.3	80	19.9	54	13.4
2007	439	187	252	113	25.7	158	36.0	94	21.4	74	16.9
2008	456	201	255	127	27.8	168	36.9	108	23.7	53	11.6
2009	451	197	254	126	27.9	181	40.1	92	20.4	52	11.6
2010	473	199	274	132	27.9	196	41.4	74	15.7	71	15.0
2011	481	211	270	137	28.5	192	39.9	94	19.5	58	12.1
Mean 1960–2011				**25.4 ± 3.0**		**37.2 ± 3.5**		**22.1 ± 4.2**		**15.3 ± 2.7**	

[a]Including two calves born in breeding centre. *Source* Krasińska and Krasiński (2007), updated

Appendix 4 Characteristics of breeding in the free-ranging population of European bison in the Polish part of Białowieża Forest

Years	Number of calves born			Coefficient of fecundity (%)	Coefficient of births (%)
	n	Males	Females		
1957	1	0	1	0	0
1958	2	2	0	0	0
1959	6	5	1	0	0
1960	6	3	3	75.0	21.4
1961	11	6	5	100.0	32.3
1962	8	4	4	80.0	18.2
1963	13	7	6	72.2	23.2
1964	16	10	6	80.0	23.5
1965	15	9	6	55.5	17.2
1966	20	6	14	62.5	19.6
1967	28	13	15	73.6	23.5
1968	25	17	8	59.5	17.7
1969	29	15	14	63.0	18.5
1970	28	12	16	47.4	15.5
1971	29	13	16	40.3	14.7
1972	37	17	20	50.7	17.5
1973	42	22	20	51.2	17.8
1974	27	18	9	29.7	10.7
1975	46	24	22	48.4	18.8
1976	32	14	18	35.9	13.7
1977	39	23	16	50.6	18.6
1978	28	12	16	38.9	12.8
1979	32	16	16	40.5	14.6
1980	40	24	16	47.6	17.4
1981	32	18	14	36.4	13.2
1982	36	24	12	37.9	14.4
1983	54	21	33	55.1	21.9
1984	35	17	18	38.0	13.4
1985	34	18	16	37.4	13.3
1986	24	9	15	27.6	9.9
1987	48	23	25	51.1	21.4
1988	34	17	17	36.2	14.5
1989	36	18	18	40.0	15.8
1990	48	27	21	49.0	18.9
1991	37	14	23	37.0	13.6
1992	45	21	24	42.9	16.6
1993	39	17	22	38.2	15.1
1994	57	24	33	59.4	22.7
1995	45	19	26	45.9	17.0
1996	34	17	17	28.3	11.8
1997	60	31	29	54.5	22.6
1998	41	20	21	33.3	14.1

(continued)

Appendix 4 (continued)

Years	Number of calves born			Coefficient of fecundity (%)	Coefficient of births (%)
	n	Males	Females		
1999	52	24	28	41.6	16.7
2000	35	17	18	25.9	11.4
2001	71	29	42	51.8	20.4
2002	54	24	30	39.0	15.5
2003	48	27	21	37.2	13.8
2004	63	22	41	41.2	17.9
2005	79	28	51	51.6	20.8
2006	57	23	34	37.0	14.2
2007	76	30	46	48.1	17.3
2008	53	25	28	31.5	12.1
2009	52	25	27	28.7	11.6
2010	73	31	42	37.2	16.2
2011	58	30	28	30.2	12.9
Altogether 1957–2011	**2070**	**982**	**1088**		
Mean 1960–2011				**47.1 ± 15.5**	**16.9 ± 4.2**

Coefficient of fecundity—number of calves born as a percentage of the number of females capable of reproduction (≥4 years old) in the population; Coefficient of births-number of calves born as a percentage of the population size at the beginning of the year.
Source Krasińska and Krasiński (2007), updated

Appendix 5 Natural mortality among European bison in the Polish part of the Białowieża Forest, 1959–2011

Years	Number on 31 December	Number of bison dead			Coefficient of mortality (%)
		n	Males	Females	
1959	28(12, 16)	1	0	1	3.6
1960	34(14, 20)	1	1	0	2.9
1961	44(19, 25)	1	1	0	2.3
1962	56(26, 30)	0	0	0	0
1963	68(33, 35)	1	0	1	1.5
1964	87(44, 43)	1	1	0	1.1
1965	102(52, 50)	4	1	3	3.9
1966	119(55, 64)	3	2	1	2.5
1967	141(63, 78)	6	5	1	4.2
1968	157(74, 83)	9	6	3	5.7
1969	180(86, 94)	5	3	2	2.7
1970	197(89, 108)	11	9	2	5.6
1971	211(94, 117)	7	5	2	3.3
1972	236(104, 132)	3	2	1	1.3
1973	253(112, 141)	8	7	1	3.2
1974	245(113, 132)	13	8	5	5.3
1975	234(103, 131)	7	5	2	3.0
1976	210(97, 113)	5	4	1	2.4
1977	218(102, 116)	6	4	2	2.7
1978	219(98, 121)	4	2	2	1.8
1979	230(106, 124)	5	2	3	2.2
1980	242(115, 127)	4	1	3	1.6
1981	250(117, 133)	11	7	4	4.4
1982	247(119, 128)	7	4	3	2.8
1983	261(123, 138)	7	2	5	2.7
1984	256(118, 138)	6	4	2	2.3
1985	243(110, 133)	11	5	6	4.5
1986	224(99, 125)	11	5	6	4.9
1987	235(101, 134)	9	6	3	3.8
1988	228(98, 130)	8	7	1	3.5
1989	254(107, 147)	2	1	1	0.8
1990	272(121, 151)	7	5	2	2.6
1991	271(119, 152)	8	6	2	2.9
1992	259(114, 145)	5	3	2	1.9
1993	251(105, 146)	5	3	2	2.0
1994	265(114, 151)	4	4	0	1.5
1995	289(111, 178)	7	4	3	2.4
1996	266(101, 165)	15	9	6	5.6
1997	290(117, 173)	7	5	2	2.4
1998	298(117, 181)	7	4	3	2.3
1999	311(123, 188)	7	5	2	2.2
2000	306(130, 176)	9	4	5	3.0
2001	348(141, 207)	6	5	1	1.7

(continued)

Appendix 5 (continued)

Years	Number on 31 December	Number of bison dead			Coefficient of mortality (%)
		n	Males	Females	
2002	349(146, 203)	7	5	2	2.0
2003	351(151, 200)	16	8	8	4.6
2004	379(158, 221)	3	0	3	0.8
2005	397(167, 230)	10	4	6	2.5
2006	402(167, 235)	9	6	3	2.2
2007	439(187, 252)	11	8	3	2.5
2008	456(201, 255)	13	4	9	2.9
2009	451(197, 254)	17	8	9	3.8
2010	473(199, 272)	27	16	11	5.7
2011	481(211, 270)	21	7	14	4.4
Mean 1959–2011		**398**	**233 (58.5 %)**	**165 (41.5 %)**	**2,9 ± 1,3**

Coefficient of mortality—number of bison dying as percentage of the population size at the end of the year. In brackets are the numbers of males (first) and the number of females (second).
Source Krasińska and Krasiński (2007), updated

Appendix 6 Mortality among European bison in the Polish part of the Białowieża Forest 195902002 in relation to sex and age ($n = 271$)

Age classes	Number of dead bison				Mean percent of dead for one year of life in each age class
	n	Males	Females	%	
		Period 1959–1973			
Calves up to one years	14	10	4	22.6	22.6
Juveniles 203 yrs	13	10	3	21.0	10.5
Adults 4010 yrs	19	16	3	30.6	4.4
Adults 11015 yrs	8	5	3	12.9	2.6
Old $\geq$ 16 yrs	8	2	6	12.9	2.1
Altogether	62	43 (70 %)	19 (30 %)	100	
		Period 1974–2002			
Calves up to one years	32	13	19	15.3	15.2
Juveniles 203 yrs	39	25	14	18.7	9.3
Adults 4010 yrs	73	57	16	35.0	5.0
Adults 11015 yrs	36	26	10	17.1	3.4
Old $\geq$ 16 yrs	29	7	22	13.9	3.4
Altogether	209	128	81		
		(61 %)	(39 %)	100	

Data concerning the period 1959–1973 from Krasiński (1976,1978a), yrs—years.
Source Krasińska and Krasiński (2007)

Appendix 7 Losses of European bison in the free-ranging population in the Polish part of Białowieża Forest 1952–2011

Years	Captures (transportation or introduction)		Culling		Altogether captured and culled		Natural mortality		Missing		Altogether losses	
	n	%	n	%	n	%	n	%	n	%	n	%
1952	0	0	0	0	0	0	0	0	0	0, 8x	0	0
1953	0	0	0	0	0	0	0	0	0	0, 8x	0	0
1954	0	0	0	0	0	0	0	0	0	0, 8x	0	0
1955	2	0	0	0	2	0	0	0	0	0, 8x	2	0
1956	1	0	0	0	1	0	0	0	0	0, 8x	1	0
1957	1	0	0	0	1	0	0	0	0	0, 8x	1	0
1958	0	0	0	0	0	0	0	0	0	0, 8x	0	0
1959	0	0	0	0	0	0	1	0	0	0, 8x	1	0
1960	0	0	0	0	0	0	1(1, 0)	2.9	0	0	1	2.9
1961	0	0	0	0	0	0	1(1, 0)	2.3	0	0	1	2.3
1962	1	1.8	0	0	1	1.8	0	0	0	0	1	1.8
1963	0	0	0	0	0	0	1(0, 1)	1.5	0	0	1	1.5
1964	1	1.1	0	0	1	1.1	1(1, 0)	1.1	0	0	2	2.3
1965	0	0	0	0	0	0	4(1, 3)	3.9	0	0	4	3.9
1966	1	0.8	0	0	1	0.8	3(2, 1)	2.5	0	0	4	3.4
1967	0	0	0	0	0	0	6(5, 1)	4.2	0	0	6	4.2
1968	0	0	0	0	0	0	9(6, 3)	5.7	0	0	9	5.7
1969	0	0	1	0.6	1	0.6	5(3, 2)	2.7	0	0	6	3.3
1970	0	0	0	0	0	0	11(9, 2)	5.6	0	0	11	5.6
1971	7	3.3	1	0.5	8	3.8	7(5, 2)	3.3	0	0	15	7.1
1972	8	3.4	2	0.8	10	6.4	3(2, 1)	1.3	0	0	13	5.5
1973	15	5.9	2	0.8	17	6.7	8(7, 1)	3.2	0	0	25	9.9
1974	18	7.3	4	1.6	22	9.0	13(8, 5)	5.3	0	0	35	14.3
1975	42	17.9	2	0.8	44	18.7	7(5, 2)	3.0	10	4.3	61	26.0
1976	57	27.1	0	0	57	27.1	5(4, 1)	2.4	0	0	62	29.5
1977	25	11.5	0	0	25	11.5	6(4, 2)	2.7	0	0	1	14.2
1978	20	9.1	3	1.4	23	10.5	4(2, 2)	1.8	0	0	27	2.3
1979	12	5.2	4	1.7	16	6.9	5(2, 3)	2.2	0	0	21	9.1
1980	12	5.0	8	3.3	20	8.3	4(1, 3)	1.6	4	1.7	28	11.6
1981	4	1.6	11	4.4	15	6.0	11(7, 4)	4.4	6	2.4	32	12.8
1982	7	2.8	25	10.1	32	12.9	7(4, 3)	2.8	0	0	39	15.7
1983	8	3.0	25	9.6	33	12.6	7(2, 5)	2.7	0	0	40	15.3
1984	4	1.6	23	9.0	27	10.6	6(4, 2)	2.3	9	3.5	42	16.4
1985	8	3.3	27	11.1	35	14.4	11(5, 6)	4.5	1	0.4	47	19.3
1986	0	0	27	12.1	27	12.1	11(5, 6)	4.9	5	2.2	43	19.2
1987	0	0	28	11.9	28	11.9	9(6, 3)	3.8	2	0.8	39	16.5
1988	0	0	33	14.5	33	14.5	8(7, 1)	3.5	0	0	41	18.0
1989	1	0.4	7	2.7	8	3.1	2(1, 1)	0.8	0	0	10	3.9
1990	0	0	15	5.5	15	5.5	7(5, 2)	2.6	8	2.9	30	11.0
1991	0	0	30	11.1	30	11.1	8(6, 2)	2.9	0	0	38	14.0

(continued)

Appendix 7 (continued)

Years	Captures (transportation or introduction)		Culling		Altogether captured and culled		Natural mortality		Missing		Altogether losses	
	n	%	n	%	n	%	n	%	n	%	n	%
1992	17	6.6	35	13.5	52	20.1	5(3, 2)	1.9	0	0	57	22.0
1993	0	0	36	14.3	36	14.3	5(3, 2)	2.0	6	2.4	47	18.7
1994	0	0	39	14.7	39	14.7	4(4, 0)	1.5	0	0	42	16.2
1995	0	0	23	8.0	23	8.0	7(4, 3)	2.4	0	0	30	10.4
1996	0	0	46	17.3	46	17.3	15(9, 6)	5.6	0	0	61	22.9
1997	0	0	29	10.0	29	10.0	7(5, 2)	2.4	0	0	36	12.4
1998	0	0	23	7.7	23	7.7	7(4,3)	2.3	0	0	30	10.0
1999	0	0	32	10.3	32	10.3	7(5, 2)	2.2	0	0	39	12.5
2000	0	0	31	10.1	31	10.1	9(4, 5)	3.0	0	0	40	13.1
2001	0	0	23	6.6	23	6.6	6(5, 1)	1.7	0	0	29	8.3
2002	0	0	46	13.2	46	13.2	7(5, 2)	2.0	0	0	53	15.2
2003	0	0	30	8.5	30	8.5	16(8, 8)	4.6	0	0	46	13.1
2004	0	0	36	9.5	36	9.5	3(0, 3)	0.8	0	0	39	10.3
2005	13	3.3	38	9.6	51	12.9	10(4, 6)	2.5	0	0	61	15.4
2006	13	3.2	30	7.5	43	10.7	9(6, 3)	2.2	0	0	52	12.9
2007	9	2.5	28	6.4	37	8.4	11(8, 3)	2.5	0	0	48	10.9
2008	5	1.1	23	5.0	28	6.1	13(4, 9)	2.8	0	0	41	9.0
2009	10	2.2	30	6.6	40	8.8	17(8, 9)	3.8	0	0	57	12.6
2010	6	1.3	18	3.8	24	5.1	27(14,11)	5.7	0	0	51	10.8
2011	11	2.3	18	3.7	29	6.0	21(7,14)	4.4	0	0	50	10.4
Total 1960–2011	**335**		**892**		**1227**		**397(233, 164)**		**51**		**1675**	
1955–2011	**339**		**892**		**1231**		**398(233, 165)**		**51**		**1680**	

In brackets are the numbe the numbers of males (first) and the number of females (second).
Source Krasińska and Krasiński (2007), updated

Appendix 8 Age and sex structure of groups captured and culled European bison in the free-ranging population of the Polish part of the Białowieża Forest, 1971–2011

Years	Altogether captured and culled		Age and sex classes							
			Males ≥4 years		Females ≥4 years		Juveniles 203 years		Calves <1 year	
	n	%	n	%	n	%	n	%	n	%
1971	8(3, 5)	3.8	2	25.0	4	50.0	2(1, 1)	25.0	0	0
1972	10(6, 4)	4.2	2	20.0	1	10.0	6(3, 3)	60.0	1(1, 0)	10.0
1973	17(7, 10)	6.7	3	17.6	4	23.5	6(2, 4)	35.3	4(2, 2)	23.6
1974	22(9, 13)	9.0	4	18.2	3	17.6	13(3, 10)	59.1	2(2, 0)	9. 1
1975	44(26, 18)	18.7	17	38.6	8	18.2	17(9, 8)	38.6	2(0, 2)	4.6
1976	57(20, 37)	27.1	9	15.8	17	29.8	28(11, 17)	49.1	3(0, 3)	5.3
1977	25(14, 11)	11.5	11	44.0	4	16.0	10(3, 7)	40.0	0	0
1978	23(14, 9)	10.5	8	34.8	4	17.4	10(5, 5)	43.5	1(1, 0)	4.3
1979	16(6, 10)	6.9	2	12.5	4	25.0	7(4, 3)	43.8	3(0, 3)	18.7
1980	20(12, 8)	8.3	5	25.0	3	15.0	8(6, 2)	40.0	4(1, 3)	20.0
1981	15(11, 4)	6.0	7	46.7	1	6.7	1(0, 1)	6.6	6(4, 2)	40.0
1982	32(18, 14)	12.9	13	40.6	7	21.9	5(3, 2)	15.6	7(2, 5)	21.9
1983	33(15, 18)	12.6	7	21.2	8	24.2	6(1, 5)	18.2	12(7, 5)	36.4
1984	27(14, 13)	10.6	7	25.9	5	18.5	5(3, 2)	18.5	10(4, 6)	37.1
1985	35(21, 14)	14.4	12	34.3	10	28.6	8(7, 1)	22.9	5(2, 3)	14.2
1986	27(15, 12)	12.1	7	25.9	7	25.9	10(7, 3)	37.1	3(1, 2)	11.1
1987	28(15, 13)	11.9	10	35.7	4	14.3	8(3, 5)	28.6	6(2, 4)	21.4
1988	33(13, 20)	14.5	5	15.1	8	24.2	6(2, 4)	18.2	14(6, 8)	42.5
1989	8(8, 0)	3.1	5	62.5	0	0	2(2, 0)	25.0	1(1, 0)	12.5
1990	15(8, 7)	5.5	4	26.7	3	20.0	4(2, 2)	26.7	4(2, 2)	26.6
1991	30(10, 20)	11.1	7	23.3	8	26.7	5(0, 5)	16.7	10(3, 7)	33.3
1992	52(23, 29)	20.1	14	26.9	14	26.9	23(8, 15)	44.2	1(1, 0)	2.0
1993	36(20, 16)	14.3	7	19.4	7	19.4	9(8, 1)	25.0	13(5, 8)	36.2
1994	39(11, 28)	14.7	7	17.9	11	28.2	6(2, 4)	15.4	15(2, 13)	38.5
1995	23(16, 7)	8.0	8	34.8	0	0	4(3, 1)	17.4	11(5, 6)	47.8
1996	46(20, 26)	17.3	8	17.4	13	28.3	6(3, 3)	13.0	19(9, 10)	41.3
1997	29(10, 19)	10.3	1	3.5	7	24.1	7(3, 4)	24.1	14(6, 8)	48.3
1998	23(15, 8)	7.7	5	21.7	6	26.1	5(4, 1)	21.7	7(6, 1)	30.5
1999	32(13, 19)	10.3	9	28.1	9	28.1	7(4, 3)	21.9	7(0, 7)	21.9
2000	31(6, 25)	10.1	1	3.2	6	19.3	11(2, 9)	35.5	13(3, 10)	42.0
2001	23(13, 10)	6.6	3	13.1	4	17.4	9(7, 2)	39.1	7(3, 4)	30.4
2002	46(14, 32)	13.2	7	15.2	10	21.7	4(2, 2)	8.7	25(5, 20)	54.4
2003	30(14, 16)	8.5	7	23.3	4	13.3	3(2, 1)	10.0	16(5, 11)	53.4
2004	36(19, 17)	9.5	8	22.2	9	25.0	5(4, 1)	13.9	14(7, 7)	38.9
2005	51(15, 36)	12.8	7	13.7	11	21.7	4(1, 3)	7.8	29(7, 22)	56.8
2006	43(17, 26)	10.7	4	9.3	7	16.3	7(6, 1)	16.3	25(7, 18)	58.1
2007	37(11, 26)	8.4	4	10.8	12	32.4	5(4, 1)	13.5	16(3, 13)	43.3
2008	28(12, 16)	6.1	7	25.0	12	43.0	9(5, 4)	32.0	0	0
2009	40(21,19)	8.8	12	30.0	3	7.5	12(5, 7)	30.0	13(4, 9)	32.5
2010	24(13, 11)	5.1	7	29.2	6	25.0	4(3, 1)	16.6	7(3, 4)	29.2
2011	29(11, 18)	6.0	9	31.0	8	27.6	6(0,6)	20.7	6(2, 4)	20.7
Total	**1223(559, 664)**		**282**		**272**		**313(153, 160)**		**356(124, 232)**	
Mean	**10.5 ± 4.7**		**24.5 ± 11.8**		**21.5 ± 9.6**		**26.7 ± 13.6**		**27.3 ± 17.1**	

In brackets are the numbers of males (first) and the number of females (second).

Source Krasińska an Krasińska and Krasiński (2007), updated

Appendix 9 Founders of the free-ranging population of Lowland European bison in the Belarussian part of the Białowieża Forest

No.	Name	P.N.	Year of birth	Breeder	Year of release	Parents
				MALES		
1.	**POPIEL**	**945**	1954	Białowieża	1961	**M 738 PODBIPIĘTA, F 770 POPIELICA**
2.	**PODCHORĄŻY**	**1031**	1956	Białowieża	1961	**M 902 PONURY, F 736 PODWIKA**
3.	**MOŁOCH**	**1157**	1958	PTZ	1961	**M 764 POBUG, F 838 MOŁNIJA**
4.	**MOCHOWIK**	**1237**	1959	PTZ	1961	**M 764 POBUG, F 877 MOSZKA**
5.	**POSZUM**	**1201**	1959	Białowieża	1962	**M 738 PODBIPIĘTA, F 740 POŚWIATA**
6.	**BASZTAN**	**1506**	1962	BP free herd	1962	**M 1074 POBÓR, F 718 PLISZKA**
7.	**POBÓR**	**1074**	1957	Białowieża	1962	**M 738 PODBIPIĘTA, F 737 PLAMKA II**
8.	**POMYŚLNY**	**1229**	1959	Białowieża	1962	**M 902 PONURY, F 736 PODWIKA**
9.	**MOŻAJSK**	**1320**	1960	PTZ	1963	**M1092 MOTYLOK, F 1094 MOŻAJKA**
10.	**POWÓD**	**1366**	1961	Białowieża	1964	**M 738 PODBIPIĘTA, F 740 POŚWIATA**
11.	**POZŁOTY**	**1578**	1963	Białowieża	1964	**M 738 PODBIPIĘTA, F 831 POZŁOTA**
12.	**MORSZAŃSK**	**1402**	1961	PTZ	1965	**M 894 POKÓJ, F 1039 MOROSZKA**
13.	**MOSTOK**	**1508**	1962	PTZ	1965	**M 1092 MOTYLOK, F 1046 MOSKOWKA**
14.	**POLNY**	**1685**	1964	Białowieża	1965	**M 762 POLEL, F 737 PLAMKA II**
15.	**POTĘŻNY**	**1687**	1964	Białowieża	1965	**M 792 KARPACZ, F 1206 POTĘGA**
16.	**POCHMURNY**	**1690**	1964	Białowieża	1965	**M 762 POLEL, F 1136 POCHMURNA**
17.	**POLIP**	**1694**	1964	Białowieża	1965	**M 792 KARPACZ, F 939 POLANKA**
18.	**POŁÓW**	**1696**	1964	Białowieża	1965	**M 762 POLEL, F 1143 PORĘCZNA**
19.	**PORON**	**1697**	1964	Białowieża	1965	**M 738 PODBIPIĘTA, F 831 POZŁOTA**
20.	**MORIAK**	**1725**	1964	PTZ	1965	**M 1092 MOTYLOK, F 1307 MORIANKA**
21.	**MOTEL'**	**1914**	1965	PTZ	1967	**M 1092 MOTYLOK, F 1322 MOTOVKA**
22.	**MORGUN**	**1918**	1965	PTZ	1967	**M 1092 MOTYLOK, F 1325 MOSTA**
23.	**MONUMENT**	**2034**	1966	PTZ	1967	**M 1092 MOTYLOK, F 1046 MOSKOWKA**
24.	**MONAH**	**2037**	1966	PTZ	1967	**M 1092 MOTYLOK, F 1325 MOSTA**
25.	**MONOMAH**	**6700**	1988	PTZ	1991	**M 5221 MONTAG, F 2867 PLUMPA**
26.	**MONSEN'OR**	**6969**	1989	PTZ	1991	**M 5221 MONTAG, F 4966 MONŠERI**
				FEMALES		
1.	**PLISZKA**	**718**	1947	Pszczyna	1961	**M 546 PLUVIUS II, F 268 PLOTKARKA**
2.	**PLENNICA**	**775**	1949	Pszczyna	1961	**M 540 PLUVIUS II, F 268 PLOTKARKA**
3.	**BARCHATKA**	**948**	1954	BP reserve	1961	**M 720 PLUSZ, F 718 PLISZKA**
4.	**POSTAWNA**	**1224**	1959	Białowieża	1962	**M 902 PONURY, F 977 POWAGA**
5.	**POLNA**	**1283**	1960	Białowieża	1962	**M 738 PODBIPIĘTA, F 770 POPIELICA**
6.	**MOZAIKA**	**1250**	1959	PTZ	1963	**M 894 POKÓJ, F 715 POŻOGA**
7.	**BANZAJKA**	**1598 (1445)**	1963	BP free herd	1963	**M 1092 MOTYLOK, F 1250 MOZAIKA**
8.	**POKUTA**	**1361**	1961	Białowieża	1964	**M 902 PONURY, F 977 POWAGA**
9.	**POSADA**	**1362**	1961	Białowieża	1964	**M 902 PONURY, F 939 POLANKA**
10.	**POLFA**	**1464**	1962	Białowieża	1964	**M 738 PODBIPIĘTA, F 740 POŚWIATA**
11.	**PORĘKA**	**1571**	1963	Białowieża	1964	**M 792 KARPACZ, F 1143 PORĘCZNA**
12.	**POŚWIĘTNA**	**1572**	1963	Białowieża	1964	**M 738 PODBIPIĘTA, F 740 POŚWIATA**
13.	**MOROŠZKA**	**1039**	1956	PTZ	1965	**M 679 PLECIUCH II, F 715 POŻOGA**
14.	**MORAWA**	**1515**	1962	PTZ	1965	**M 1092 MOTYLOK, F 1039 MOROSZKA**
15.	**MORZJANKA**	**1633**	1963	PTZ	1965	**M 1092 MOTYLOK, F 1039 MOROSZKA**
16.	**POSUCHA**	**1681**	1964	Białowieża	1965	**M 792 KARPACZ, F 736 PODWIKA**
17.	**POŚREDNIA**	**1686**	1964	Białowieża	1965	**M 738 PODBIPIĘTA, F 740 POŚWIATA**
18.	**PLAZMA**	**1700**	1964	Białowieża	1965	**M 789 PLAMIEC, F 1131 KANIA**
19.	**MODNICA**	**2040**	1966	PTZ	1967	**M 913 BARHAT, F1512 MOLDOVA**
20.	**MONAŠKA**	**2042**	1966	PTZ	1967	**M 1092 MOTYLOK, F 877 MOSZKA**

Data after: annual reports of L. N. Korochkina from archives of the "Belovezhskaya Pushcha" N P, Bunevich (2003b), Bunevich et al. (2006), European Bison Pedigree Book (1955–1991). Breeders: European Bison Breeding Centres in Białowieża, Pszczyna (Poland), and *PTZ*—Prioksko-Terrasnyjj zapovednik (Russia), *BP*—BELOVEZHSKAYA PUSHCHA reserve and free herd —bison born in the free-ranging population in the Belarussian part of the Forest, but parents known, P.N.—pedigree number. *Source* Krasińska and Krasiński (2007)

Appendix 10 Sizes of the free-ranging population of Lowland European bison in the Belarussian part of the Białowieża Forest

Years	Numbers on 31 December data after EBPB[a]	Numbers on 31 December data after NPBP[b]	Births	Deaths	Culling	Catching/ transport	Introduction
1961	0	9(6, 3)	0	0	0	0	9(6, 3)
1962	0	16(10, 6)	2(1, 1)	0	0	0	5(3, 2)
1963	0	17(10, 7)	2(1, 1)	2(1, 1)	0	1(1, 0)	2(1, 1)
1964	24(11, 13)	28(15, 13)	2(1, 1)	0	0	0	8(3, 5)
1965	45(23, 22)	49(29, 20)	6(5, 1)	0		0	15(9, 6)
1966	48(24, 24)	51(29, 22)	4(1, 3)	2(1, 1)	1(1, 0)	0	
1967	60(31, 29)	58(31, 27)	7(2, 5)	5(4, 1)	0	1(1, 0)	6(4, 2)
1968	62(33, 29)	60(30, 30)	7(2, 5)	3(1, 2)	1(1, 0)	0	0
1969	63(33, 30)	63(33, 30)	7(4, 3)	5(3, 2)	1(1, 0)	0	0
1970	63(34, 29	63(34, 29)	6(3, 3)	6(4, 2)	0	0	0
1971	66(34, 32)	66(34, 32)	9(4, 5)	4(2, 2)	2(2, 0)	0	0
1972	77(40, 37)	77(40, 37)	9(5, 4)	2(1, 1)	0	0	0
1973	82(43, 39)	82(43, 39)	11(8, 3)	3(2, 1)	3(3, 0)	0	0
1974	87(43, 44)	87(43, 44)	9(4, 5)	1(1, 0)	3(2, 1)	0	0
1975	90(43, 47)	90(43, 47)	8(5, 3)	3(3, 0)	3(3, 0)	0	0
1976	101(46, 55)	102(46, 56)	17(7, 10)	4(2, 2)	2(1, 1)	0	0
1977	114(51, 63)	114(51, 63)	18(9, 9)	4(2, 2)	0	0	0
1978	132(57, 75)	132(56, 76)	22(9, 13)	3(3, 0)	1(1, 0)	0	0
1979	154(74, 80)	144(60, 84)	15(7, 8)	1(1, 0)	3(2, 1)	0	0
1980	169(80, 89)	159(?)	24(9, 15)	4(2, 2)	0	0	0
1981	171	156(?)	9(5, 4)	11(6, 5)	1(0, 1)	0	0
1862	169	169(78, 91)	25(12,13)	12(9, 3)	1(1, 0)	0	0
1983	199	192(78, 114)	33(16,17)	2(2,0)	6(4, 2)	0	0
1984	211	202(84, 118)	32(18, 14)	13(7, 6)	7(3, 4)	3(2, 1)	0
1985	204	200(81, 119)	27(17,10)	3(2, 0)	15(12, 3)	1(1, 0)	0
1986	227	221(85, 136)	30(13,17)	3(3, 0)	9(5, 4)	3(1, 2)	0
1987	242	237(88, 149)	43(18, 25)	4(3, 1)	18(11, 7)	5(2, 3)	0
1988	255	252(?)	37	5(3, 2)	20(13, 7)	0	0
1989	263	260(?)	30	2(1, 1)	15(8, 7)	0	0
1990	315	313(116, 197)	72(36, 36)	2(2, 0)	0	0	0
1991	315	313	48(25, 23)	23(12, 11)	25(16, 9)	0	2(2, 0)
1992	295	291(101, 190)	26(12, 14)	19(6, 13)	22(13, 9)	0	0
1993	308	303(106, 197)	52(24, 28)	13(8, 5)	20(15, 5)	0	0
1994	290	286(98, 188)	40(18, 22)	7(3, 4)	15(7, 8)	15(5, 10)	0
1995	280	275(92, 183)	46(18, 28)	7(4, 3)	36(14, 22)		0
1996	251	246(?)	34(15,19)	13(9, 4)	12(7, 5)	16(3, 13)	0
1997	232	228(82, 146)	32(9, 23)	12(6, 6)	10(3, 7)	17(4, 13)	0
1998	238(82, 156)	232(79, 153)	30(13, 17)	5(4, 1)	3(2, 1)	10(2, 8)	0
1999	248(87, 161)	243(84,159)	31(11, 20)	2(1, 1)	5(1, 4)	10(3, 7)	0
2000	260	257(?)	42(?, ?)	3(1, 2)	7(5, 2)	13(3, 10)	0
2001	265	262(110,152)	39(19, 20)	8(4, 4)	10(3, 7)	0	
2002	275	270(112,158)	38(18, 20)	8(1, 7)	9(6, 3)	1(0, 0)	0
2003	277	273(?,?)	38(18, 20)	8(4, 4)	15(10,5)	1(1, 0)	0
2004	299	294(109,185)	46(22, 24)	4(4, 0)	10(5, 5)	0	0
2005	0	312	47(19, 28)	9(5, 4)	8(5, 3)	21(5,16)	0
1961–2005		**Total**	**1112**	**251**	**319**	**124**	**47**
			1003*(463, 540)	**(145, 106)**	**(186, 133)**	**(39, 85)**	**(28,19)**

In brackets are the numbers of males (first) and females (second). Data after EBPB[a]—European bison Pedigree Book, NPBP** NPBP[b]—"Belovezhskaya Pushcha" National Park (Belarus)—annual reports of L.N. Korochkina from 1961–1967, Koroch Korochkina 1969, 1975, Korochkina and Kochko (1983), Bunevich and Kochko (1988), Bunevich (2003b), Bunevich et al. (2006), Kozlo and Bunevich (2009); [a]—except for the years 1988, 1989, 2000. Source: Krasińska and Krasiński (2007)

Appendix 11 Characteristics of breeding of the free-ranging population of Lowland European bison in the Belarussian part of the Białowieża Forest 1962–2005

Years	Number of calves	Males	Females	Coefficient of fecundity (%)	Coefficient of birth (%)
1962	2	1	1	0	22.2
1963	2	1	1	0	12.5
1964	2	1	1	0	11.8
1965	6	5	1	0	21.4
1966	4	1	3	0	8.2
1967	7	2	5	0	14.6
1968	7	2	5	0	12.1
1969	7	4	3	0	11.1
1970	6	3	3	33.3	9.5
1971	9	4	5	43.6	14.3
1972	9	5	4	45.3	13.6
1973	11	8	3	46.1	14.3
1974	9	4	5	40.9	11.0
1975	8	5	3	38.6	9.2
1976	17	7	10	51.5	18.9
1977	18	9	9	48.7	17.6
1978	22	9	13	55.0	19.3
1979	15	7	8	34.9	11.4
1980	24	9	15	47.6	16.8
1981	9	5	4	15.7	5.7
1982	25	12	13	44.6	16.0
1983	33	16	17	41.7	19.5
1984	32	18	14	38.3	16.3
1985	27	17	10	31.8	13.1
1986	30	13	17	34.9	13.3
1987	43	18	25	41.0	17.8
1988	37	?	?	41.6	15.3
1989	30	?	?	28.9	11.8
1990	72	36	36	51.4	27.4
1991	48	25	23	38.4	15.2
1992	26	12	14	20.8	8.3
1993	52	24	28	39.7	17.6
1994	40	18	22	34.2	13.8
1995	46	18	28	39.0	16.1
1996	34	15	19	30.1	12.4
1997	32	9	23	29.4	13.0
1998	30	13	17	31.2	13.2
1999	31	11	20	31.3	13.4
2000	42 (?)	?	?	47.2	17.3
2001	39	19	20	37.9	14.4
2002	38	18	20	31.0	14.9
2003	38	18	20	33.9	14.1
2004	46	22	24	35.7	16.9
2005	47	19	28	34.1	16.0
Total	**1112**	**0**	**0**	**Mean 1970–2005**	**Mean 1962–2005**
	1003[a]	**463**	**540**	**37.8**	**14.7**

Sources annual reports of L. N. Korochkina from 1961–1967, Korochkina (1975), Korochkina and Kochko (1983), Bunevich and Kochko (1988), Krasochko et al. (2004), Buneviech et al. (2006), Kozlo and Bunevich (2009), [a]—except for the years 1988, 1989, 2000. *Source* Krasińska and Krasiński (2007)

MIX
Papier aus verantwortungsvollen Quellen
Paper from responsible sources
FSC® C105338

If you have any concerns about our products,
you can contact us on
ProductSafety@springernature.com

In case Publisher is established outside the EU,
the EU authorized representative is:
Springer Nature Customer Service Center GmbH
Europaplatz 3, 69115 Heidelberg, Germany

Printed by Libri Plureos GmbH
in Hamburg, Germany